Genetic Algorithms + Data Structures
= Evolution Programs

Springer
Berlin
Heidelberg
New York
Barcelona
Budapest
Hong Kong
London
Milan
Paris
Santa Clara
Singapore
Tokyo

Zbigniew Michalewicz

Genetic Algorithms
+ Data Structures
= Evolution Programs

Third, Revised and Extended Edition

With 68 Figures and 36 Tables

 Springer

Zbigniew Michalewicz
Department of Computer Science
University of North Carolina
Charlotte, NC 28223, USA
zbyszek@uncc.edu

The first edition appeared 1992
in the Springer series *Artificial Intelligence*

Library of Congress Cataloging-in-Publication Data

Michalewicz, Zbigniew.
    Genetic algorithms + data structures = evolution programs /
Zbigniew Michalewicz. -- 3rd rev. and extended ed.
      p.   cm.
    Includes bibliographical references and index.
    ISBN 3-540-60676-9 (hardcover)
    1. Evolutionary programming (Computer science) 2. Genetic
algorithms. 3. Data structures (Computer science)   I. Title.
QA76.618.M53   1996
005.1--dc20   95-48027
                                                               CIP

ISBN 3-540-60676-9 Springer-Verlag Berlin Heidelberg New York
ISBN 3-540-58090-5 2nd ed. Springer-Verlag Berlin Heidelberg New York

© Springer-Verlag Berlin Heidelberg 1992, 1994, 1996
Printed in the United States of America

The use of general descriptive names, registered names, trademarks, etc. in this publication does not imply, even in the absence of a specific statement, that such names are exempt from the relevant protective laws and regulations and therefore free for general use.

Cover design: Struve & Partner, Heidelberg
Cover photographs: Bilderagentur Tony Stone Bilderwelten
Typesetting: Camera ready by author

SPIN 10519108   45/3140 – 5 4 3 2 1 0 – Printed on acid-free paper

To the next generation:
Matthew, Katherine, Michael,
Thomas, and Irene

Preface to the Third Edition

It is always the latest song
that an audience applauds the most.

Homer, *Odyssey*

During the World Congress on Computational Intelligence (Orlando, 27 June – 2 July 1994), Ken De Jong, one of the plenary speakers and the current editor-in-chief of *Evolutionary Computation*, said that these days a large majority of implementations of evolutionary techniques used non-binary representations. It seems that the *general concept* of an evolutionary method (or evolutionary algorithm, evolution program, etc.) was widely accepted by most practitioners in the field. Consequently, in most applications of evolutionary techniques, a population of individuals is processed, where each individual represents a potential solution to the problem at hand, and a selection process introduce some bias: better individuals have better chances to survive and reproduce. In the same time, a particular representation of the individuals and the set of operators which alter their genetic code are often problem-specific. Hence, there is really little point in arguing any further that the incorporation of problem-specific knowledge, by means of representation and specialized operators, may enhance the performance of an evolutionary system in a significant way. On the other hand, many successful implementations of such a hybrid system [89]:

> "... had pushed the application of simple GAs well beyond our initial theories and understanding, creating a need to revisit and extend them."

I believe this is one of the most challenging tasks for researchers in the field of evolutionary computation. Some recent results support the experimental developments by providing some theoretical foundations (see, for example, the work of Nick Radcliffe [315, 316, 317] on formal analysis and respectful recombinations). However, further studies on various factors affecting the ability of evolutionary techniques to solve optimization problems are necessary.

Despite this change in the perception of evolutionary techniques, the original organization of the book is left unchanged in this edition; for example, the

Introduction is kept almost without any alternation (with an argument for a departure from binary-coded genetic algorithms towards more complex, problem-specific systems). The book still consists of three parts, which discuss genetic algorithms (the best known technique in the area of evolutionary computation), numerical optimization, and various applications of evolution programs, respectively. However, there are several changes between this and the previous edition of the book. Apart from some minor changes, corrections, and modifications present in most chapters (including Appendix A), the main differences can be summarized as follows:

- due to some new developments connected with constrained optimization in numerical domains, Chapter 7 was totally rewritten;

- Chapter 11 was modified in a significant way, several new developments were included;

- there is a new Chapter 13 which discusses the original evolutionary programming techniques and quite recent paradigm of genetic programming;

- Chapter 14 incorporates material from Conclusions of the second edition;

- Chapter 15 provides a general overview on heuristic methods and constraint handling techniques in evolutionary methods;

- Conclusions were rewritten to discuss the current directions of research in evolutionary techniques; because of this change, it was necessary to change also the citation used at the beginning of this chapter;

- Appendices B and C contain a few test functions (unconstrained and constrained, respectively) which might be used in various experiments with evolutionary techniques; and

- Appendix D discusses a few possible projects; this part might be useful if the book is adopted as a text for a project-oriented course.

I do hope that these changes would further enhance the popularity of the text.

As with the first and second editions, I am pleased to acknowledge the assistance of several co-authors who worked with me during the last two years; many results of this collaboration were included in this volume. The list of new co-authors (not listed in the prefaces to the previous editions) include (in alphabetical order): Tom Cassen, Michael Cavaretta, Dipankar Dasgupta, Susan Esquivel, Raul Gallard, Sridhar Isukapalli, Rodolphe Le Riche, Li-Tine Li, Hoi-Shan Lin, Rafic Makki, Maciej Michalewicz, Mohammed Moinuddin, Subbu Muddappa, Girish Nazhiyath, Robert Reynolds, Marc Schoenauer, and Kalpathi Subramanian. Thanks are due to Sita Raghavan, who improved the simple real-coded genetic algorithm included in the Appendix A, and Girish Nazhiyath, who developed a new version of the GENOCOP III system (described in Chapter 7) to handle nonlinear constraints. I thank all the individuals who took their time to share their thoughts on the text with me; they are primarily responsible for most changes incorporated in this edition. In particular, I express my gratitude to Thomas Bäck and David Fogel, which whom I have been working on a volume entitled *Handbook of Evolutionary Computation* [17], to the executive editor at Springer-Verlag, Hans Wössner, for his help throughout the project, and to Gabi Fischer, Frank Holzwarth, and Andy Ross at Springer-Verlag for all their efforts on this project. I would like also to acknowledge a grant (IRI-9322400) from the National Science Foundation, which helped me in preparing this edition. I was able to incorporate many results (revision of Chapter 7, new Chapter 15) obtained with the support of this grant. I greatly appreciate the assistance of Larry Reeker, Program Director at National Science Foundation. Also, I would like to thank all my graduate students from UNC-Charlotte, Universidad Nacional de San Luis, and Linköping University, who took part in my courses offered during 1994/95; as usual, I enjoyed each offering and found them very rewarding.

Charlotte   Zbigniew Michalewicz
October 1995

Preface to the Second Edition

> As natural selection works solely
> by and for the good of each being,
> all corporeal and mental endowments
> will tend to progress toward perfection.
>
> Charles Darwin, *Origin of Species*

The field of evolutionary computation has reached a stage of some maturity. There are several, well established international conferences that attract hundreds of participants (International Conferences on Genetic Algorithms—ICGA [167, 171, 344, 32, 129], Parallel Problem Solving from Nature—PPSN [351, 251], Annual Conferences on Evolutionary Programming—EP [123, 124, 378]); new annual conferences are getting started (IEEE International Conferences on Evolutionary Computation [275, 276]). Also, there are tens of workshops, special sessions, and local conferences every year, all around the world. A new journal, *Evolutionary Computation* (MIT Press) [87], is devoted entirely to evolutionary computation techniques; many other journals organized special issues on evolutionary computation (e.g., [118, 263]). Many excellent tutorial papers [28, 29, 320, 397, 119] and technical reports provide more-or-less complete bibliographies of the field [161, 336, 297]. There is also *The Hitch-Hiker's Guide to Evolutionary Computation* prepared by Jörg Heitkötter [177] from University of Dortmund, available on comp.ai.genetic interest group (Internet).

This trend prodded me to prepare the second, extended edition of the book. As it was the case with the first edition, the volume consists mostly of articles I published over the last few years—because of that, the book represents a personal overview of the evolutionary computation area rather than a balanced survey of all activities in this field. Consequently, the book is not really a textbook; however, many universities used the first edition as a text for an "evolutionary computation" course. To help potential future students, I have incorporated a few additional items into this volume (an appendix with a simple genetic code, brief references to other developments in the field, an index, etc.). At the same time, I did not provide any exercises at the end of chapters. The reason is that the field of evolutionary computation is still very young and there are many areas worthy of further study—these should be easy to identify in the text, which still triggers more questions than provides answers. The aim of this volume is to talk about the field of evolutionary computation in simple terms, and discuss their simplicity and elegance on many interesting test cases. Writing an evolution program for a given problem should be an enjoyable experience; the book may serve as a guide in this task.

I have also used this opportunity to correct many typos present in the first edition. However, I left the organization of the book unchanged: there are still twelve chapters, apart from the introduction and conclusions, but some chapters were extended in a significant way. Three new subsections were added to Chapter 4 (on contractive mapping genetic algorithms, on genetic algorithms with varying population size, and on constraint handling techniques for the knapsack problem). Some information on Gray coding was included in Chapter 5. Chapter 7 was extended by sections on implementation of the GENOCOP system. These include a description of first experiments and results, a discussion on further modifications of the system, and a description of first experiments on optimization problems with nonlinear constraints (GENOCOP II). A brief section on multimodal and multiobjective optimization was added to Chapter 8. In Chapter 11, a section on path planning in a mobile robot environment was added. The Conclusions were extended by the results of recent experiments, confirming a hypothesis that the problem-specific knowledge enhances an algorithm in terms of performance (time and precision) and, at the same time, narrows its applicability. Also, some information on cultural algorithms was included. This edition has an Index (missing in the first edition) and an Appendix, where a code for a simple real-coded genetic algorithm is given. This C code might be useful for beginners in the field; the efficiency has been sacrificed for clarity. There are also some other minor changes throughout the whole text: several paragraphs were deleted, inserted, or modified. Also, this volume has over one hundred added references.

As with the first edition, it is a pleasure to acknowledge the assistance of several co-authors who worked with me during the last two years; many results of this collaboration were included in this volume. The list of recent co-authors include: Jarosław Arabas, Naguib Attia, Hoi-Shan Lin, Thomas Logan, Jan Mulawka, Swarnalatha Swaminathan, Andrzej Szałas, and Jing Xiao. Thanks are due to Denis Cormier, who wrote the simple real-coded genetic algorithm included in the Appendix. I would like to thank all the individuals who took their time to share their thoughts on the text with me; they are primarily responsible for most changes incorporated in this edition. Also, I would like to thank all my graduate students from UNC-Charlotte and North Carolina State University who took part in the teleclass offered in Fall 1992 (and consequently had to use the first edition of the book as the text); it was a very enjoyable and rewarding experience for me.

Charlotte   Zbigniew Michalewicz
March 1994

Preface to the First Edition

'What does your Master teach?'
asked a visitor.
'Nothing,' said the disciple.
'Then why does he give discourses?'
'He only points the way — he teaches
nothing.'

Anthony de Mello, *One Minute Wisdom*

During the last three decades there has been a growing interest in algorithms which rely on analogies to natural processes. The emergence of massively parallel computers made these algorithms of practical interest. The best known algorithms in this class include evolutionary programming, genetic algorithms, evolution strategies, simulated annealing, classifier systems, and neural networks. Recently (1–3 October 1990) the University of Dortmund, Germany, hosted the First Workshop on Parallel Problem Solving from Nature [351].

This book discusses a subclass of these algorithms — those which are based on the principle of evolution (survival of the fittest). In such algorithms a population of individuals (potential solutions) undergoes a sequence of unary (mutation type) and higher order (crossover type) transformations. These individuals strive for survival: a selection scheme, biased towards fitter individuals, selects the next generation. After some number of generations, the program converges — the best individual hopefully represents the optimum solution.

There are many different algorithms in this category. To underline the similarities between them we use the common term "evolution programs".

Evolution programs can be perceived as a generalization of genetic algorithms. Classical genetic algorithms operate on fixed-length binary strings, which need not be the case for evolution programs. Also, evolution programs usually incorporate a variety of "genetic" operators, whereas classical genetic algorithms use binary crossover and mutation.

The beginnings of genetic algorithms can be traced back to the early 1950s when several biologists used computers for simulations of biological systems [154]. However, the work done in late 1960s and early 1970s at the University of Michigan under the direction of John Holland led to genetic algorithms as they are known today. The interest in genetic algorithms is growing rapidly — the recent Fourth International Conference on Genetic Algorithms (San Diego, 13–16 July 1991) attracted around 300 participants.

The book describes the results of three years' research from early 1989 at Victoria University of Wellington, New Zealand, through late 1991 (since July 1989 I have been at the University of North Carolina at Charlotte). During this time I built and experimented with various modifications of genetic algorithms using different data structures for chromosome representation of individuals, and various "genetic" operators which operated on them. Because of my background in databases [256], [260], where the constraints play a central role, most evolution programs were developed for constrained problems.

The idea of evolution programs (in the sense presented in this book), was conceived quite early [204], [277] and was supported later by a series of experiments. Despite the fact that evolution programs, in general, lack a strong theoretical background, the experimental results were more than encouraging: very often they performed much better than classical genetic algorithms, than commercial systems, and than other, best-known algorithms for a particular class of problems.

Some other researchers, at different stages of their research, performed experiments which were perfect examples of the "evolution programming" technique — some of them are discussed in this volume. Chapter 8 presents a survey of evolution strategies — a technique developed in Germany by I. Rechenberg and H.-P. Schwefel [319], [348] for parameter optimization problems. Many researchers investigated the properties of evolution systems for ordering problems, including the widely known, "traveling salesman problem" (Chapter 10). In Chapter 11 we present systems for a variety of problems including problems on graphs, scheduling, and partitioning. Chapter 12 describes the construction of an evolution program for inductive learning in attribute based spaces, developed by C. Janikow [200]. In the Conclusions, we briefly discuss evolution programs for generating LISP code to solve problems, developed by J. Koza [228], and present an idea for a new programming environment.

The book is organized as follows. The introduction provides a general discussion on the motivation and presents the main idea of the book. Since evolution programs are based on the principles of genetic algorithms, Part I of this book serves as survey on this topic. We explain what genetic algorithms are, how they work, and why (Chapters 1–3). The last chapter of Part I (Chapter 4) presents some selected issues (selection routines, scaling, etc.) for genetic algorithms.

In Part II we explore a single data structure: a vector in a floating point representation, only recently widely accepted in the GA community [78]. We talk only about numerical optimization. We present some experimental comparison of binary and floating point representations (Chapter 5) and discuss new "genetic" operators responsible for fine local tuning (Chapter 6). Chapter 7 presents two evolution programs to handle constrained problems: the GENO-COP system, for optimizing functions in the presence of linear constraints, and the GAFOC[1] system, for optimal control problems. Various tests cases are considered; the results of the evolution programs are compared with a commercial system. The last chapter of this part (Chapter 8) presents a survey of evolution strategies and describes some other methods.

Part III discusses a collection of evolution programs built over recent years to explore their applicability to a variety of hard problems. We present further experiments with order–based evolution programs, evolution programs with matrices and graphs as a chromosome structure. Also, we discuss an application of evolution program to machine learning, comparing it with other approaches.

The title of this book rephrases the famous expression used by N. Wirth fifteen years ago for his book, *Algorithms + Data Structures = Programs* [406]. Both books share a common idea. To build a successful program (in particular, an evolution program), appropriate data structures should be used (the data structures, in the case of the evolution program, correspond to the chromosome representation) together with appropriate algorithms (these correspond to "genetic" operators used for transforming one or more individual chromosomes).

The book is aimed at a large audience: graduate students, programmers, researchers, engineers, designers — everyone who faces challenging optimization problems. In particular, the book will be of interest to the Operations Research community, since many of the problems considered (traveling salesman, scheduling, transportation problems) are from their area of interest. An understanding of introductory college level mathematics and of basic concepts of programming is sufficient to follow all presented material.

Acknowledgements

It is a pleasure to acknowledge the assistance of several people and institutions in this effort.

Thanks are due to many co-authors who worked with me at different stages of this project; these are (in alphabetical order): Paul Elia, Lindsay Groves, Matthew Hobbs, Cezary Janikow, Andrzej Jankowski, Mohammed Kazemi, Jacek Krawczyk, Zbigniew Ras, Joseph Schell, David Seniw, Don Shoff, Jim Stevens, Tony Vignaux, and George Windholz.

I would like to thank all my graduate students who took part in a course *Genetic Algorithms* I offered at UNC–Charlotte in Fall 1990, and my Master students, who wrote their thesis on various modifications of genetic algorithms: Jason Foodman, Jay Livingston, Jeffrey B. Rayfield, David Seniw, Jim Stevens, Charles Strickland, Swarnalatha Swaminathan, and Keith Wharton. Jim Stevens provided a short story on rabbits and foxes (Chapter 1).

Thanks are due to Abdollah Homaifar from North Carolina State University, Kenneth Messa from Loyola University (New Orleans), and to several colleagues at UNC–Charlotte: Mike Allen, Rick Lejk, Zbigniew Ras, Harold Reiter, Joe Schell, and Barry Wilkinson, for reviewing selected parts of the book.

I would like to acknowledge the assistance of Doug Gullett, Jerry Holt, and Dwayne McNeil (Computing Services, UNC–Charlotte) for recovering all

---

[1]This system is not discussed in this edition of the book.

chapters of the book (accidentally deleted). I would like also to thank Jan Thomas (Academic Computing Services, UNC–Charlotte) for overall help.

I would like to extend my thanks to the editors of the Artificial Intelligence Series, Springer-Verlag: Leonard Bolc, Donald Loveland, and Hans Wössner for initiating the idea of the book and their help throughout the project. Special thanks are due to J. Andrew Ross, the English Copy Editor, Springer-Verlag, for his precious assistance with writing style.

I would like to acknowledge a series of grants from North Carolina Supercomputing Center (1990–1991), which allowed me to run the hundreds of experiments described in this text.

In the book I have included many citations from other published work, as well as parts of my own published articles. Therefore, I would like to acknowledge all permissions to use such material in this publication I received from the following publishers: Pitman Publishing Company; Birkhäuser Verlag AG; John Wiley and Sons, Ltd; Complex Systems Publications, Inc.; Addison–Wesley Publishing Company; Kluwer Academic Publishers; Chapman and Hall Ltd, Scientific, Technical and Medical Publishers; Prentice Hall; IEEE; Association for Computing Machinery; Morgan Kaufmann Publishers, Inc.; and individuals: David Goldberg, Fred Glover, John Grefenstette, Abdollah Homaifar, and John Koza.

Finally, I would like to thank my family for their patience and support during the (long) summer of 1991.

Charlotte   Zbigniew Michalewicz
January 1992

Table of Contents

Introduction

Again I saw that under the sun
the race is not to the swift,
nor the battle to the strong,
nor bread to the wise,
nor riches to the intelligent,
nor favor to the man of skill;
but time and chance
happen to them all.

The Bible, Ecclesiastes, 9

During the last thirty years there has been a growing interest in problem solving systems based on principles of evolution and hereditary: such systems maintain a population of potential solutions, they have some selection process based on fitness of individuals, and some "genetic" operators. One type of such systems is a class of Evolution Strategies i.e., algorithms which imitate the principles of natural evolution for parameter optimization problems [319, 348] (Rechenberg, Schwefel). Fogel's Evolutionary Programming [126] is a technique for searching through a space of small finite-state machines. Glover's Scatter Search techniques [142] maintain a population of reference points and generate offspring by weighted linear combinations. Another type of evolution based systems are Holland's Genetic Algorithms (GAs) [188]. In 1990, Koza [231] proposed an evolution based systems, Genetic Programming, to search for the most fit computer program to solve a particular problem.

We use a common term, **Evolution Programs (EP)**, for all evolution-based systems (including systems described above). The structure of an evolution program is shown in Figure 0.1.

The evolution program is a probabilistic algorithm which maintains a population of individuals, $P(t) = \{x_1^t, \ldots, x_n^t\}$ for iteration t. Each individual represents a potential solution to the problem at hand, and, in any evolution program, is implemented as some (possibly complex) data structure S. Each solution x_i^t is evaluated to give some measure of its "fitness". Then, a new population (iteration $t + 1$) is formed by selecting the more fit individuals (select step). Some members of the new population undergo transformations (alter step) by means of "genetic" operators to form new solutions. There are unary transformations

procedure evolution program
begin
  $t \leftarrow 0$
  initialize $P(t)$
  evaluate $P(t)$
  while (**not** termination-condition) **do**
  begin
    $t \leftarrow t + 1$
    select $P(t)$ from $P(t-1)$
    alter $P(t)$
    evaluate $P(t)$
  end
end

Fig. 0.1. The structure of an evolution program

m_i (mutation type), which create new individuals by a small change in a single individual ($m_i : S \rightarrow S$), and higher order transformations c_j (crossover type), which create new individuals by combining parts from several (two or more) individuals ($c_j : S \times \ldots \times S \rightarrow S$). After some number of generations the program converges — it is hoped that the best individual represents a near-optimum (reasonable) solution.

Let us consider the following example. Assume we search for a graph which should satisfy some requirements (say, we search for the optimal topology of a communication network accordingly to some criteria: cost of sending messages, reliability, etc.). Each individual in the evolution program represents a potential solution to the problem, i.e., each individual represents a graph. The initial population of graphs $P(0)$ (either generated randomly or created as a result of some heuristic process) is a starting point ($t = 0$) for the evolution program. The evaluation function usually is given — it incorporates the problem requirements. The evaluation function returns the fitness of each graph, distinguishing between better and worse individuals. Several mutation operators can be designed which would transform a single graph. A few crossover operators can be considered which combine the structure of two (or more) graphs into one. Very often such operators incorporate the problem-specific knowledge. For example, if the graph we search for is connected and acyclic (i.e., it is a tree), a possible mutation operator may delete an edge from the graph and add a new edge to connect two disjoint subgraphs. The other possibility would be to design a problem-independent mutation and incorporate this requirement into the evaluation function, penalizing graphs which are not trees.

Clearly, many evolution programs can be formulated for a given problem. Such programs may differ in many ways; they can use different data structures for implementing a single individual, "genetic" operators for transforming individuals, methods for creating an initial population, methods for handling constraints of the problem, and parameters (population size, probabilities of applying different operators, etc.). However, they share a common principle: a population of individuals undergoes some transformations, and during this evolution process the individuals strive for survival.

As mentioned earlier, the idea of evolution programming is not new and has been around for at least thirty years [126, 188, 348]. Many different evolutionary systems have emerged since then; however, in this text we discuss these various paradigms of evolutionary programs from the perspective of their similarities. After all, the main differences between them are hidden on a lower level. We will not discuss any philosophical differences between various evolutionary techniques (e.g., whether they operate on the genotype or phenotype level), but rather we discuss them from the perspective of building an evolutionary program for a particular class of problems. Thus we advocate the use of proper (possibly complex) data structures (for chromosome representation) together with an expanded set of genetic operators, whereas, for example, classical genetic algorithms use fixed-length binary strings (as a chromosome, data structure S) for its individuals and two operators: binary mutation and binary crossover. In other words, the structure of a genetic algorithm is the same as the structure of an evolution program (Figure 0.1) and the differences are hidden on the lower level. In EPs chromosomes need not be represented by bit-strings and the alteration process includes other "genetic" operators appropriate for the given structure and the given problem.

This is not entirely a new direction. In 1985 De Jong wrote [84]:

> "What should one do when elements in the space to be searched are most naturally represented by more complex data structures such as arrays, trees, digraphs, etc. Should one attempt to 'linearize' them into a string representation or are there ways to creatively redefine crossover and mutation to work directly on such structures. I am unaware of any progress in this area."

As mentioned earlier, genetic algorithms use fixed-length binary strings and only two basic genetic operators. Two major (early) publications on genetic algorithms [188, 82] describe the theory and implementations of such GAs. As stated in [155]:

> "The contribution of this work [82] was in its ruthless abstraction and simplification; De Jong got somewhere not in spite of his simplification but because of it. [...] Holland's book [188] laid the theoretical foundation for De Jong's and all subsequent GA work by mathematically identifying the combined role in genetic search of similarity subsets (schemata), minimal operator disruption, and reproductive selection. [...] Subsequent researchers have tended to take the theoretical suggestions in [188] quite literally, thereby reinforcing the implementation success of De Jong's neat codings and operators."

However, in the next paragraph Goldberg [155] says:

"It is interesting, if not ironic, that neither man intended for his
work to be taken so literally. Although De Jong's implementations
established usable technique in accordance with Holland's theoretical simplifications, subsequent researchers have tended to treat both
accomplishments as inviolate gospel."

It seems that a "natural" representation of a potential solution for a given
problem plus a family of applicable "genetic" operators might be quite useful
in the approximation of solutions of many problems, and this nature-modeled
approach (evolution programming) is a promising direction for problem solving in general. Apart from other paradigms of evolutionary computation (e.g.,
evolution strategies, evolutionary programming, genetic programming), some
researchers in genetic algorithms community have explored the use of other
representations as ordered lists (for bin-packing), embedded lists (for factory
scheduling problems), variable-element lists (for semiconductor layout). During
the last ten years, various application–specific variations on the genetic algorithm were reported [73, 167, 171, 173, 278, 363, 364, 392]. These variations
include variable length strings (including strings whose elements were *if–then–
else* rules [363]), richer structures than binary strings (for example, matrices
[392]), and experiments with modified genetic operators to meet the needs of
particular applications [270]. In [285] there is a description of a genetic algorithm which uses backpropagation (a neural network training technique) as an
operator, together with mutation and crossover that were tailored to the neural network domain. Davis and Coombs [65, 76] described a genetic algorithm
that carried out one stage in the process of designing packet-switching communication network; the representation used was not binary and five "genetic"
operators (knowledge based, statistical, numerical) were used. These operators
were quite different to binary mutation and crossover. Other researchers, in
their study on solving a job shop scheduling problem [21], wrote:

"To enhance the performance of the algorithm and to expand the
search space, a chromosome representation which stores problem
specific information is devised. Problem specific recombination operators which take advantage of the additional information are also
developed."

There are numerous similar citations available. It seems that most researches
modified their implementations of genetic algorithms either by using non-string
chromosome representation or by designing problem specific genetic operators
to accommodate the problem to be solved. In [228] Koza observed:

"Representation is a key issue in genetic algorithm work because the
representation scheme can severely limit the window by which the
system observes its world. However, as Davis and Steenstrup [74]
point out, 'In all of Holland's work, and in the work of many of his
students, chromosomes are bit strings.' String-based representations
schemes are difficult and unnatural for many problems and the need

for more powerful representations has been recognized for some time [84, 85, 86]."

Various nonstandard implementations were created for particular problems — simply, the classical GAs were difficult to apply directly to a problem and some modifications in chromosome structures were required. In this book we have consciously departed from classical genetic algorithms which operate on strings of bits: we searched for richer data structures and applicable "genetic" operators for these structures for variety of problems. By experimenting with such structures and operators, we obtained systems which were not genetic algorithms any more, or, at least, not classical GAs. The titles of several reports started with: "A Modified Genetic Algorithm ..." [271], "Specialized Genetic Algorithms..." [201], "A Non-Standard Genetic Algorithm..." [278]. Also, there is a feeling that the name "genetic algorithms" might be quite misleading with respect to the developed systems. Davis developed several non-standard systems with many problem-specific operators. He observed in [77]:

"I have seen some head-shaking about that system from other researchers in the genetic algorithm field [...] a frank disbelief that the system we built was a genetic algorithm (since we didn't use binary representation, binary crossover, and binary mutation)."

Additionally, we can ask, for example, whether an evolution strategy is a genetic algorithm? Is the opposite true? To avoid all issues connected with classification of evolutionary systems, we call them simply "evolution programs" (EPs).

Why do we depart from genetic algorithms towards more flexible evolution programs? Even though nicely theorized, GA failed to provide for successful applications in many areas. It seems that the major factor behind this failure is the same one responsible for their success: domain independence.

One of the consequences of the neatness of GAs (in the sense of their domain independence) is their inability to deal with nontrivial constraints. As mentioned earlier, in most work in genetic algorithms, chromosomes are bit strings — lists of 0s and 1s. An important question to be considered in designing a chromosome representation of solutions to a problem is the implementation of constraints on solutions (problem-specific knowledge). As stated in [74]:

"Constraints that cannot be violated can be implemented by imposing great penalties on individuals that violate them, by imposing moderate penalties, or by creating decoders of the representation that avoid creating individuals violating the constraint. Each of these solutions has its advantages and disadvantages. If one incorporates a high penalty into the evaluation routine and the domain is one in which production of an individual violating the constraint is likely, one runs the risk of creating a genetic algorithm that spends most of its time evaluating illegal individuals. Further, it can happen that when a legal individual is found, it drives the others out and the population converges on it without finding better individuals, since the likely paths to other legal individuals require the production of illegal individuals as intermediate structures, and the penalties for violating the constraint make it unlikely that such intermediate structures will reproduce. If one imposes moderate penalties, the system may evolve individuals that violate the constraint but are rated better than those that do not because the rest of the evaluation function can be satisfied better by accepting the moderate constraint penalty than by avoiding it. If one builds a "decoder" into the evaluation procedure that intelligently avoids building an illegal individual from the chromosome, the result is frequently computation-intensive to run. Further, not all constraints can be easily implemented in this way."

(An example of decoders and repair algorithms, together with several penalty functions is given in section 4.5, where a knapsack problem is considered).

In evolution programming, the problem of constraint satisfaction has a different flavor. It is not the issue of selecting an evaluation function with some penalties, but rather selecting "the best" chromosomal representation of solutions together with meaningful genetic operators to satisfy all constraints imposed by the problem. Any genetic operator should pass some characteristic structure from parent to offspring, so the representation structure plays an important role in defining genetic operators. Moreover, different representation structures have different characteristics of suitability for constraint representation, which complicates the problem even more. These two components (representation and operators) influence each other; it seems that any problem would require careful analysis which would result in appropriate representation for which there are meaningful genetic operators.

Glover in his study on solving a complex keyboard configuration problem [141] wrote:

"Although the robust character of the GA search paradigm is well suited to the demands of the keyboard configuration problem, the bit string representation and idealized operators are not properly matched to the [...] required constraints. For instance, if three bits are used to represent each component of a simple keyboard of only 40 components, it is easy to show that only one out of every 10^{16} arbitrarily selected 120-bit structures represents a legal configuration map structure."

Another citation is from the work of De Jong [88], where the traveling salesman problem is briefly discussed:

"Using the standard crossover and mutation operators, a GA will explore the space of all *combinations* of city names when, in fact, it is the space of all *permutations* which is of interest. The obvious problem is that as N [the number of cities in the tour] increases, the space of permutations is a vanishingly small subset of the space of combinations, and the powerful GA sampling heuristic has been rendered impotent by a poor choice of representation."

At early stages of AI, the general problem solvers (GPSs) were designed as generic tools for approaching complex problems. However, as it turned out, it was necessary to incorporate problem-specific knowledge due to unmanageable complexity of these systems. Now the history repeated itself: until recently genetic algorithms were perceived as generic tools useful for optimization of many hard problems. However, the need for the incorporation of the problem-specific knowledge in genetic algorithms has been recognized in some research articles for some time [10, 128, 131, 170, 370]. It seems that GAs (as GPS) are too domain independent to be useful in many applications. So it is not surprising that evolution programs, incorporating problem-specific knowledge in the chromosomes' data structures and specific "genetic" operators, perform much better.

The basic conceptual difference between classical genetic algorithms and evolution programs is presented in Figures 0.2 and 0.3. Classical genetic algorithms, which operate on binary strings, require a modification of an original problem into appropriate (suitable for GA) form; this would include mapping between potential solutions and binary representation, taking care of decoders or repair algorithms, etc. This is not usually an easy task.

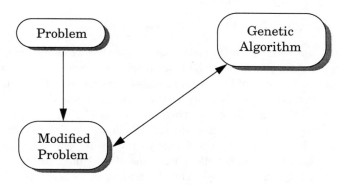

Fig. 0.2. Genetic algorithm approach

On the other hand, evolution programs would leave the problem unchanged, modifying a chromosome representation of a potential solution (using "natural" data structures), and applying appropriate "genetic" operators.

In other words, to solve a nontrivial problem using an evolution program, we can either transform the problem into a form appropriate for the genetic algorithm (Figure 0.2), or we can transform the genetic algorithm to suit the problem (Figure 0.3). Clearly, classical GAs take the former approach, EPs the

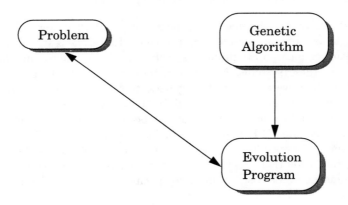

Fig. 0.3. Evolution program approach

latter. So the idea behind evolution programs is quite simple and is based on the following motto:

> "If the mountain will not come to Mohammed, then Mohammed will go to the mountain."

This is not a very new idea. In [77] Davis wrote:

> "It has seemed true to me for some time that we cannot handle most real-world problems with binary representations and an operator set consisting only of binary crossover and binary mutation. One reason for this is that nearly every real-world domain has associated domain knowledge that is of use when one is considering a transformation of a solution in the domain [...] I believe that genetic algorithms are the appropriate algorithms to use in a great many real-world applications. I also believe that one should incorporate real-world knowledge in one's algorithm by adding it to one's decoder or by expanding one's operator set."

Here, we call such modified genetic algorithms "evolution programs".

It is quite hard to draw a line between genetic algorithms and evolution programs. What is required for an evolution program to be a genetic algorithm? Maintaining population of potential solutions? Binary representation of potential solutions? Selection process based on fitness of individuals? Recombination operators? The existence of a Schema Theorem? Building-block hypothesis? All of the above? Is an evolution program for the traveling salesman problem with integer vector representation and PMX operator (Chapter 10) a genetic algorithm? Is an evolution program for the transportation problem with matrix representation and arithmetical crossover operator (Chapter 9) a genetic algorithm? In this book we will not provide answers for the above question, instead we present some interesting results of using evolution programming techniques on variety of problems.

As mentioned earlier, several researchers recognized potential behind various modifications. In [78] Davis wrote:

> "When I talk to the user, I explain that my plan is to hybridize the genetic algorithm technique and the current algorithm by employing the following three principles:
>
> - *Use the Current Encoding.* Use the current algorithm's encoding technique in the hybrid algorithm.
>
> - *Hybridize Where Possible.* Incorporate the positive features of the current algorithm in the hybrid algorithm.
>
> - *Adapt the Genetic Operators.* Create crossover and mutation operators for the new type of encoding by analogy with bit string crossover and mutation operators. Incorporate domain-based heuristics as operators as well.
>
> [...] I use the term *hybrid genetic algorithm* for algorithms created by applying these three principles."

It seems that hybrid genetic algorithms and evolution programs share a common idea: departure from classical, bit-string genetic algorithms towards more complex systems, involving the appropriate data structures (Use the Current Encoding) and suitable genetic operators (Adapt the Genetic Operators). On the other hand, Davis assumed the existence of one or more current (traditional) algorithms available on the problem domain — on the basis of such algorithms a construction of a hybrid genetic algorithm is discussed. In our approach of evolution programming, we do not make any assumption of this sort: all evolution systems discussed later in the book were built from scratch.

What are the strengths and weaknesses of evolution programming? It seems that the major strength of EP technique is its wide applicability. In this book, we try to describe a variety of different problems and discuss a construction of an evolution program for each of them. Very often, the results are outstanding: the systems perform much better than commercially available software. Another strong point connected with evolution programs is that they are parallel in nature. As stated in [154]:

> "In a world where serial algorithms are usually made parallel through countless tricks and contortions, it is no small irony that genetic algorithms (highly parallel algorithms) are made serial through equally unnatural tricks and turns."

Of course, this is also true for any (population based) evolution program. On the other hand, we have to admit the poor theoretical basis of evolution programs. Experimenting with different data structures and modifying crossover and mutation requires a careful analysis, which would guarantee reasonable performance. This has not been done yet.

In general, AI problem solving strategies are categorized into "strong" and "weak" methods. A weak method makes few assumptions about the problem domain; hence it usually enjoys wide applicability. On the other hand, it can suffer from combinatorially explosive solution costs when scaling up to larger problems [90]. This can be avoided by making strong assumptions about the problem domain, and consequently exploiting these assumptions in the problem solving method. But a disadvantage of such strong methods is their limited applicability: very often they require significant redesign when applied even to related problems.

Evolution programs fit somewhere between weak and strong methods. Some evolution programs (as genetic algorithms) are quite weak without making any assumption of a problem domain. Some other programs (e.g., GENOCOP or GENETIC-2) are more problem specific with a varying degree of problem dependence. For example, GENOCOP (Chapter 7), like all evolution strategies (Chapter 8), was build to solve parameter optimization problems. The system can handle any objective function with any set of linear constraints. The GENETIC-2 (Chapter 9) work for transportation problems. Other systems (see Chapters 10 and 11) are suitable for combinatorial optimization problems (like scheduling problems, traveling salesman problems, graph problems). An interesting application of an evolution program for inductive learning of decision rules is discussed in Chapter 12.

It is little bit ironic: genetic algorithms are perceived as weak methods; however, in the presence of nontrivial constraints, they change rapidly into strong methods. Whether we consider a penalty function, decoder, or a repair algorithm, these must be tailored for a specific application. On the other hand, the evolution programs (perceived as much stronger, problem-dependent methods) suddenly seem much weaker (we discuss further this issue in Chapter 14). This demonstrates a huge potential behind the evolution programming approach.

All these observations triggered my interest in investigating the properties of different genetic operators defined on richer structures than bit strings — further, this research would lead to the creation of a new programming methodology (in [277] such a proposed programming methodology was called EVA for "EVolution progrAmming").[1] Roughly speaking, a programmer in such an environment would select data structures with appropriate genetic operators for a given problem as well as selecting an evaluation function and initializing the population (the other parameters are tuned by another genetic process).

However, a lot of research should be done before we can propose the basic constructs of such a programming environment. This book provides just the first step towards this goal by investigating different structures and genetic operators building evolution programs for many problems.

---

[1] We shall return to the idea of a new programming environment towards the end of the book (Chapter 14).

Part I
Genetic Algorithms

1. GAs: What Are They?

Paradoxical as it seemed, the Master
always insisted that the true reformer
was one who was able to see that everything
is perfect as it is — and able to
leave it alone.

Anthony de Mello, *One Minute Wisdom*

There is a large class of interesting problems for which no reasonably fast algorithms have been developed. Many of these problems are optimization problems that arise frequently in applications. Given such a hard optimization problem it is often possible to find an efficient algorithm whose solution is approximately optimal. For some hard optimization problems we can use probabilistic algorithms as well — these algorithms do not guarantee the optimum value, but by randomly choosing sufficiently many "witnesses" the probability of error may be made as small as we like.

There are a lot of important practical optimization problems for which such algorithms of high quality have become available [73]. For instance we can apply simulated annealing for wire routing and component placement problems in VLSI design or for the traveling salesman problem. Moreover, many other large-scale combinatorial optimization problems (many of which have been proved NP-hard) can be solved approximately on present-day computers by this kind of Monte Carlo technique.

In general, any abstract task to be accomplished can be thought of as solving a problem, which, in turn, can be perceived as a search through a space of potential solutions. Since we are after "the best" solution, we can view this task as an optimization process. For small spaces, classical exhaustive methods usually suffice; for larger spaces special artificial intelligence techniques must be employed. Genetic Algorithms (GAs) are among such techniques; they are stochastic algorithms whose search methods model some natural phenomena: genetic inheritance and Darwinian strife for survival. As stated in [74]:

> "... the metaphor underlying genetic algorithms is that of natural evolution. In evolution, the problem each species faces is one of searching for beneficial adaptations to a complicated and changing environment. The 'knowledge' that each species has gained is embodied in the makeup of the chromosomes of its members."

The idea behind genetic algorithms is to do what nature does. Let us take rabbits as an example: at any given time there is a population of rabbits. Some of them are faster and smarter than other rabbits. These faster, smarter rabbits are less likely to be eaten by foxes, and therefore more of them survive to do what rabbits do best: make more rabbits. Of course, some of the slower, dumber rabbits will survive just because they are lucky. This surviving population of rabbits starts breeding. The breeding results in a good mixture of rabbit genetic material: some slow rabbits breed with fast rabbits, some fast with fast, some smart rabbits with dumb rabbits, and so on. And on the top of that, nature throws in a 'wild hare' every once in a while by mutating some of the rabbit genetic material. The resulting baby rabbits will (on average) be faster and smarter than these in the original population because more faster, smarter parents survived the foxes. (It is a good thing that the foxes are undergoing similar process — otherwise the rabbits might become too fast and smart for the foxes to catch any of them).

A genetic algorithm follows a step-by-step procedure that closely matches the story of the rabbits. Before we take a closer look at the structure of a genetic algorithm, let us have a quick look at the history of genetics (from [380]):

"The fundamental principle of natural selection as the main evolutionary principle has been formulated by C. Darwin long before the discovery of genetic mechanisms. Ignorant of the basic heredity principles, Darwin hypothesized fusion or blending inheritance, supposing that parental qualities mix together like fluids in the offspring organism. His selection theory arose serious objections, first stated by F. Jenkins: crossing quickly levels off any hereditary distinctions, and there is no selection in homogeneous populations (the so-called 'Jenkins nightmare').

It was not until 1865, when G. Mendel discovered the basic principles of transference of hereditary factors from parent to offspring, which showed the discrete nature of these factors, that the 'Jenkins nightmare' could be explained, since because of this discreteness there is no 'dissolution' of hereditary distinctions.

Mendelian laws became known to the scientific community after they had been independently rediscovered in 1900 by H. de Vries, K. Correns and K. von Tschermak. Genetics was fully developed by T. Morgan and his collaborators, who proved experimentally that chromosomes are the main carriers of hereditary information and that genes, which present hereditary factors, are lined up on chromosomes. Later on, accumulated experimental facts showed Mendelian laws to be valid for all sexually reproducing organisms.

However, Mendel's laws, even after they had been rediscovered, and Darwin's theory of natural selection remained independent, unlinked concepts. And moreover, they were opposed to each other. Not until the 1920s (see, for instance the classical work by Četverikov [61]) was it proved that Mendel's genetics and Darwin's theory of natural selection are in no way conflicting and that their happy marriage yields modern evolutionary theory."

Genetic algorithms use a vocabulary borrowed from natural genetics. We would talk about *individuals* (or *genotypes, structures*) in a population; quite often these individuals are called also *strings* or *chromosomes*. This might be a little bit misleading: each cell of every organism of a given species carries a certain number of chromosomes (man, for example, has 46 of them); however, in this book we talk about one-chromosome individuals only, i.e., *haploid* chromosomes (for additional information on *diploidy* — pairs of chromosomes — dominance, and other related issues, in connection with genetic algorithms, the reader is referred to [154]; see also a very recent work by Greene [165] and Ng with Wong [298]). Chromosomes are made of units — *genes* (also *features, characters, or decoders*) — arranged in linear succession; every gene controls the inheritance of one or several characters. Genes of certain characters are located at certain places of the chromosome, which are called *loci* (string positions). Any character of individuals (such as hair color) can manifest itself differently; the gene is said to be in several states, called *alleles* (feature values).

Each genotype (in this book a single chromosome) would represent a potential solution to a problem (the meaning of a particular chromosome, i.e., its *phenotype*, is defined externally by the user); an evolution process run on a population of chromosomes corresponds to a search through a space of potential solutions. Such a search requires balancing two (apparently conflicting) objectives: exploiting the best solutions and exploring the search space [46]. Hillclimbing is an example of a strategy which exploits the best solution for possible improvement; on the other hand, it neglects exploration of the search space. Random search is a typical example of a strategy which explores the search space ignoring the exploitations of the promising regions of the space. Genetic algorithms are a class of general purpose (domain independent) search methods which strike a remarkable balance between exploration and exploitation of the search space.

GAs have been quite successfully applied to optimization problems like wire routing, scheduling, adaptive control, game playing, cognitive modeling, transportation problems, traveling salesman problems, optimal control problems, database query optimization, etc. (see [15, 34, 45, 84, 121, 154, 167, 170, 171, 273, 344, 129, 103, 391, 392]). However, De Jong [84] warned against perceiving GAs as optimization tools:

"...because of this historical focus and emphasis on function optimization applications, it is easy to fall into the trap of perceiving GAs *themselves* as optimization algorithms and then being surprised and/or disappointed when they fail to find an 'obvious' optimum in a particular search space. My suggestion for avoiding this perceptual trap is to think of GAs as a (highly idealized) simulation of a natural process and as such they embody the goals and purposes (if any) of that natural process. I am not sure if anyone is up to the task of defining the goals and purpose of evolutionary systems; however, I think it's fair to say that such systems are *not* generally perceived as functions optimizers".

On the other hand, optimization is a major field of GA's applicability. In [348] (1981) Schwefel said:

"There is scarcely a modern journal, whether of engineering, economics, management, mathematics, physics, or the social sciences, in which the concept 'optimization' is missing from the subject index. If one abstracts from all specialist points of view, the recurring problem is to select a better or best (according to Leibniz, optimal) alternative from among a number of possible states of affairs."

During the last decade, the significance of optimization has grown even further — many important large-scale combinatorial optimization problems and highly constrained engineering problems can only be solved approximately on present day computers.

Genetic algorithms aim at such complex problems. They belong to the class of probabilistic algorithms, yet they are very different from random algorithms as they combine elements of directed and stochastic search. Because of this, GA are also more robust than existing directed search methods. Another important property of such genetic based search methods is that they maintain a population of potential solutions — all other methods process a single point of the search space.

Hillclimbing methods use the iterative improvement technique; the technique is applied to a single point (the current point) in the search space. During a single iteration, a new point is selected from the neighborhood of the current point (this is why this technique is known also as neighborhood search or local search [233]). If the new point provides a better[1] value of the objective function, the new point becomes the current point. Otherwise, some other neighbor is selected and tested against the current point. The method terminates if no further improvement is possible.

It is clear that the hillclimbing methods provide local optimum values only and these values depend on the selection of the starting point. Moreover, there is no information available on the relative error (with respect to the global optimum) of the solution found.

To increase the chances to succeed, hillclimbing methods usually are executed for a (large) number of different starting points (these points need not be selected randomly — a selection of a starting point for a single execution may depend on the result of the previous runs).

The simulated annealing technique [1] eliminates most disadvantages of the hillclimbing methods: solutions do not depend on the starting point any longer and are (usually) close to the optimum point. This is achieved by introducing

---

[1]smaller, for minimization, and larger, for maximization problems.

a probability p of acceptance (i.e., replacement of the current point by a new point): $p = 1$, if the new point provides a better value of the objective function; however, $p > 0$, otherwise. In the latter case, the probability of acceptance p is a function of the values of objective function for the current point and the new point, and an additional control parameter, "temperature", T. In general, the lower temperature T is, the smaller the chances for the acceptance of a new point are. During execution of the algorithm, the temperature of the system, T, is lowered in steps. The algorithm terminates for some small value of T, for which virtually no changes are accepted anymore.

As mentioned earlier, a GA performs a multi-directional search by maintaining a population of potential solutions and encourages information formation and exchange between these directions. The population undergoes a simulated evolution: at each generation the relatively "good" solutions reproduce, while the relatively "bad" solutions die. To distinguish between different solutions we use an objective (evaluation) function which plays the role of an environment.

An example of hillclimbing, simulated annealing, and genetic algorithm techniques is given later in this chapter (section 1.4).

The structure of a simple genetic algorithm is the same as the structure of any evolution program (see Figure 0.1, Introduction). During iteration t, a genetic algorithm maintains a population of potential solutions (chromosomes, vectors), $P(t) = \{x_1^t, \ldots, x_n^t\}$. Each solution x_i^t is evaluated to give some measure of its "fitness". Then, a new population (iteration $t + 1$) is formed by selecting the more fit individuals. Some members of this new population undergo alterations by means of crossover and mutation, to form new solutions. Crossover combines the features of two parent chromosomes to form two similar offspring by swapping corresponding segments of the parents. For example, if the parents are represented by five-dimensional vectors $(a_1, b_1, c_1, d_1, e_1)$ and $(a_2, b_2, c_2, d_2, e_2)$, then crossing the chromosomes after the second gene would produce the offspring $(a_1, b_1, c_2, d_2, e_2)$ and $(a_2, b_2, c_1, d_1, e_1)$. The intuition behind the applicability of the crossover operator is information exchange between different potential solutions.

Mutation arbitrarily alters one or more genes of a selected chromosome, by a random change with a probability equal to the mutation rate. The intuition behind the mutation operator is the introduction of some extra variability into the population.

A genetic algorithm (as any evolution program) for a particular problem must have the following five components:

- a genetic representation for potential solutions to the problem,

- a way to create an initial population of potential solutions,

- an evaluation function that plays the role of the environment, rating solutions in terms of their "fitness",

- genetic operators that alter the composition of children,

- values for various parameters that the genetic algorithm uses (population size, probabilities of applying genetic operators, etc.).

We discuss the main features of genetic algorithms by presenting three examples. In the first one we apply a genetic algorithm for optimization of a simple function of one real variable. The second example illustrates the use of a genetic algorithm to learn a strategy for a simple game (the prisoner's dilemma). The third example discusses one possible application of a genetic algorithm to approach a combinatorial NP-hard problem, the traveling salesman problem.

1.1 Optimization of a simple function

In this section we discuss the basic features of a genetic algorithm for optimization of a simple function of one variable. The function is defined as

$$f(x) = x \cdot \sin(10\pi \cdot x) + 1.0$$

and is drawn in Figure 1.1. The problem is to find x from the range $[-1..2]$ which maximizes the function f, i.e., to find x_0 such that

$$f(x_0) \geq f(x), \text{ for all } x \in [-1..2].$$

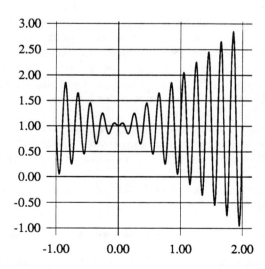

Fig. 1.1. Graph of the function $f(x) = x \cdot \sin(10\pi \cdot x) + 1.0$

It is relatively easy to analyse the function f. The zeros of the first derivative f' should be determined:

$$f'(x) = \sin(10\pi \cdot x) + 10\pi x \cdot \cos(10\pi \cdot x) = 0;$$

the formula is equivalent to

$$\tan(10\pi \cdot x) = -10\pi x.$$

It is clear that the above equation has an infinite number of solutions,

$x_i = \frac{2i-1}{20} + \epsilon_i$, for $i = 1, 2, \ldots$
$x_0 = 0$
$x_i = \frac{2i+1}{20} - \epsilon_i$, for $i = -1, -2, \ldots,$

where terms ϵ_is represent decreasing sequences of real numbers (for $i = 1, 2, \ldots,$ and $i = -1, -2, \ldots$) approaching zero.

Note also that the function f reaches its local maxima for x_i if i is an odd integer, and its local minima for x_i if i is an even integer (see Figure 1.1).

Since the domain of the problem is $x \in [-1..2]$, the function reaches its maximum for for $x_{19} = \frac{37}{20} + \epsilon_{19} = 1.85 + \epsilon_{19}$, where $f(x_{19})$ is slightly larger than $f(1.85) = 1.85 \cdot \sin(18\pi + \frac{\pi}{2}) + 1.0 = 2.85$.

Assume that we wish to construct a genetic algorithm to solve the above problem, i.e., to maximize the function f. Let us discuss the major components of such a genetic algorithm in turn.

1.1.1 Representation

We use a binary vector as a chromosome to represent real values of the variable x. The length of the vector depends on the required precision, which, in this example, is six places after the decimal point.

The domain of the variable x has length 3; the precision requirement implies that the range $[-1..2]$ should be divided into at least $3 \cdot 1000000$ equal size ranges. This means that 22 bits are required as a binary vector (chromosome):

$$2097152 = 2^{21} < 3000000 \le 2^{22} = 4194304.$$

The mapping from a binary string $\langle b_{21}b_{20} \ldots b_0 \rangle$ into a real number x from the range $[-1..2]$ is straightforward and is completed in two steps:

- convert the binary string $\langle b_{21}b_{20} \ldots b_0 \rangle$ from the base 2 to base 10:

$$((\langle b_{21}b_{20} \ldots b_0 \rangle)_2 = (\textstyle\sum_{i=0}^{21} b_i \cdot 2^i)_{10} = x',$$

- find a corresponding real number x:

$$x = -1.0 + x' \cdot \frac{3}{2^{22}-1},$$

where -1.0 is the left boundary of the domain and 3 is the length of the domain.

For example, a chromosome

$$(1000101110110101000111)$$

represents the number 0.637197, since

$$x' = (1000101110110101000111)_2 = 2288967$$

and

$$x = -1.0 + 2288967 \cdot \tfrac{3}{4194303} = 0.637197.$$

Of course, the chromosomes

$$(0000000000000000000000) \text{ and } (1111111111111111111111)$$

represent boundaries of the domain, -1.0 and 2.0, respectively.

1.1.2 Initial population

The initialization process is very simple: we create a population of chromosomes, where each chromosome is a binary vector of 22 bits. All 22 bits for each chromosome are initialized randomly.

1.1.3 Evaluation function

Evaluation function *eval* for binary vectors v is equivalent to the function f:

$$eval(v) = f(x),$$

where the chromosome v represents the real value x.

As noted earlier, the evaluation function plays the role of the environment, rating potential solutions in terms of their fitness. For example, three chromosomes:

$$v_1 = (1000101110110101000111),$$
$$v_2 = (0000001110000000010000),$$
$$v_3 = (1110000000111111000101),$$

        •

correspond to values $x_1 = 0.637197$, $x_2 = -0.958973$, and $x_3 = 1.627888$, respectively. Consequently, the evaluation function would rate them as follows:

$$eval(v_1) = f(x_1) = 1.586345,$$
$$eval(v_2) = f(x_2) = 0.078878,$$
$$eval(v_3) = f(x_3) = 2.250650.$$

Clearly, the chromosome v_3 is the best of the three chromosomes, since its evaluation returns the highest value.

1.1.4 Genetic operators

During the alteration phase of the genetic algorithm we would use two classical genetic operators: mutation and crossover.

As mentioned earlier, mutation alters one or more genes (positions in a chromosome) with a probability equal to the mutation rate. Assume that the fifth gene from the v_3 chromosome was selected for a mutation. Since the fifth gene in this chromosome is 0, it would be flipped into 1. So the chromosome v_3 after this mutation would be

$$v_3' = (1110100000111111000101).$$

This chromosome represents the value $x_3' = 1.721638$ and $f(x_3') = -0.082257$. This means that this particular mutation resulted in a significant decrease of the value of the chromosome v_3. On the other hand, if the 10th gene was selected for mutation in the chromosome v_3, then

$$v_3'' = (1110000001111111000101).$$

The corresponding value $x_3'' = 1.630818$ and $f(x_3'') = 2.343555$, an improvement over the original value of $f(x_3) = 2.250650$.

Let us illustrate the crossover operator on chromosomes v_2 and v_3. Assume that the crossover point was (randomly) selected after the 5th gene:

$$v_2 = (00000|01110000000010000),$$
$$v_3 = (11100|00000111111000101).$$

The two resulting offspring are

$$v_2' = (00000|00000111111000101),$$
$$v_3' = (11100|01110000000010000).$$

These offspring evaluate to

$$f(v_2') = f(-0.998113) = 0.940865,$$
$$f(v_3') = f(1.666028) = 2.459245.$$

Note that the second offspring has a better evaluation than both of its parents.

1.1.5 Parameters

For this particular problem we have used the following parameters: population size $pop\_size = 50$, probability of crossover $p_c = 0.25$, probability of mutation $p_m = 0.01$. The following section presents some experimental results for such a genetic system.

1.1.6 Experimental results

In Table 1.1 we provide the generation number for which we noted an improvement in the evaluation function, together with the value of the function. The best chromosome after 150 generations was

$$v_{max} = (1111001101000100000101),$$

which corresponds to a value $x_{max} = 1.850773$.

As expected, $x_{max} = 1.85 + \epsilon$, and $f(x_{max})$ is slightly larger than 2.85.

Generation number	Evaluation function
1	1.441942
6	2.250003
8	2.250283
9	2.250284
10	2.250363
12	2.328077
39	2.344251
40	2.345087
51	2.738930
99	2.849246
137	2.850217
145	2.850227

Table 1.1. Results of 150 generations

1.2 The prisoner's dilemma

In this section, we explain how a genetic algorithm can be used to learn a strategy for a simple game, known as the prisoner's dilemma. We present the results obtained by Axelrod [14].

Two prisoners are held in separate cells, unable to communicate with each other. Each prisoner is asked, independently, to defect and betray the other prisoner. If only one prisoner defects, he is rewarded and the other is punished. If both defect, both remain imprisoned and are tortured. If neither defects, both receive moderate rewards. Thus, the selfish choice of defection always yields a higher payoff than cooperation — no matter what the other prisoner does — but if both defect, both do worse than if both had cooperated. The prisoner's dilemma is to decide whether to defect or cooperate with the other prisoner.

The prisoner's dilemma can be played as a game between two players, where at each turn, each player either defects or cooperates with the other prisoner. The players then score according to the payoffs listed in the Table 1.2.

Player 1	Player 2	P_1	P_2	Comment
Defect	Defect	1	1	Punishment for mutual defection
Defect	Cooperate	5	0	Temptation to defect and sucker's payoff
Cooperate	Defect	0	5	Sucker's payoff, and temptation to defect
Cooperate	Cooperate	3	3	Reward for mutual cooperation

Table 1.2. Payoff table for prisoner's dilemma game: P_i is the payoff for Player i

We will now consider how a genetic algorithm might be used to learn a strategy for the prisoner's dilemma. A GA approach is to maintain a population of "players", each of which has a particular strategy. Initially, each player's strategy is chosen at random. Thereafter, at each step, players play games and their scores are noted. Some of the players are then selected for the next generation, and some of those are chosen to mate. When two players mate, the new player created has a strategy constructed from the strategies of its parents (crossover). A mutation, as usual, introduces some variability into players' strategies by random changes on representations of these strategies.

1.2.1 Representing a strategy

First of all, we need some way to represent a strategy (i.e., a possible solution). For simplicity, we will consider strategies that are deterministic and use the outcomes of the three previous moves to make a choice in the current move. Since there are four possible outcomes for each move, there are $4 \times 4 \times 4 = 64$ different histories of the three previous moves.

A strategy of this type can be specified by indicating what move is to be made for each of these possible histories. Thus, a strategy can be represented by a string of 64 bits (or Ds and Cs), indicating what move is to be made for each of the 64 possible histories. To get the strategy started at the beginning of the game, we also need to specify its initial premises about the three hypothetical moves which preceded the start of the game. This requires six more genes, making a total of seventy loci on the chromosome.

This string of seventy bits specifies what the player would do in every possible circumstance and thus completely defines a particular strategy. The string of 70 genes also serves as the player's chromosome for use in the evolution process.

1.2.2 Outline of the genetic algorithm

Axelrod's genetic algorithm to learn a strategy for the prisoner's dilemma works in four stages, as follows:

1. Choose an initial population. Each player is assigned a random string of seventy bits, representing a strategy as discussed above.

2. Test each player to determine its effectiveness. Each player uses the strategy defined by its chromosome to play the game with other players. The player's score is its average over all the games it plays.

3. Select players to breed. A player with an average score is given one mating; a player scoring one standard deviation above the average is given two matings; and a player scoring one standard deviation below the average is given no matings.

4. The successful players are randomly paired off to produce two offspring per mating. The strategy of each offspring is determined from the strategies of its parents. This is done by using two genetics operators: crossover and mutation.

After these four stages we get a new population. The new population will display patterns of behavior that are more like those of the successful individuals of the previous generation, and less like those of the unsuccessful ones. With each new generation, the individuals with relatively high scores will be more likely to pass on parts of their strategies, while the relatively unsuccessful individuals will be less likely to have any parts of their strategies passed on.

1.2.3 Experimental results

Running this program, Axelrod obtained quite remarkable results. From a strictly random start, the genetic algorithm evolved populations whose median member was just as successful as the best known heuristic algorithm. Some behavioral patterns evolved in the vast majority of the individuals; these are:

1. Don't rock the boat: continue to cooperate after three mutual cooperations (i.e., C after $(CC)(CC)(CC)^2$).

2. Be provokable: defect when the other player defects out of the blue (i.e., D after receiving $(CC)(CC)(CD)$).

3. Accept an apology: continue to cooperate after cooperation has been restored
(i.e., C after $(CD)(DC)(CC)$).

4. Forget: cooperate when mutual cooperation has been restored after an exploitation (i.e., C after $(DC)(CC)(CC)$).

5. Accept a rut: defect after three mutual defections (i.e., D after $(DD)(DD)$ (DD)).

---

[2]The last three moves are described by three pairs $(a_1b_1)(a_2b_2)(a_3b_3)$, where the a's are this player's moves (C for cooperate, D for defect) and the b's are the other player's moves.

For more details, see [14]. The prisoner's dilemma problem can be generalized to more than two players; for details and interesting experimental results see [413]. [3]

1.3 Traveling salesman problem

In this section, we explain how a genetic algorithm can be used to approach the Traveling Salesman Problem (TSP). Note that we shall discuss only one possible approach. In Chapter 10 we discuss other approaches to the TSP as well.

Simply stated, the traveling salesman must visit every city in his territory exactly once and then return to the starting point; given the cost of travel between all cities, how should he plan his itinerary for minimum total cost of the entire tour?

The TSP is a problem in combinatorial optimization and arises in numerous applications. There are several branch-and-bound algorithms, approximate algorithms, and heuristic search algorithms which approach this problem. During the last few years there have been several attempts to approximate the TSP by genetic algorithms [154, pages 166–179]; here we present one of them.

First, we should address an important question connected with the chromosome representation: should we leave a chromosome to be an integer vector, or rather we should transform it into a binary string? In the previous two examples (optimization of a function and the prisoner's dilemma) we represented a chromosome (in a more or less natural way) as a binary vector. This allowed us to use binary mutation and crossover; applying these operators we got legal offspring, i.e., offspring within the search space. This is not the case for the traveling salesman problem. In a binary representation of a n cities TSP problem, each city should be coded as a string of $\lceil \log_2 n \rceil$ bits; a chromosome is a string of $n \cdot \lceil \log_2 n \rceil$ bits. A mutation can result in a sequence of cities, which is not a tour: we can get the same city twice in a sequence. Moreover, for a TSP with 20 cities (where we need 5 bits to represent a city), some 5-bit sequences (for example, 10101) do not correspond to any city. Similar problems are present when applying crossover operator. Clearly, if we use mutation and crossover operators as defined earlier, we would need some sort of a "repair algorithm"; such an algorithm would "repair" a chromosome, moving it back into the search space.

It seems that the integer vector representation is better: instead of using repair algorithms, we can incorporate the knowledge of the problem into operators: in that way they would "intelligently" avoid building an illegal individual. In this particular approach we accept integer representation: a vector $\boldsymbol{v} = \langle i_1 i_2 \ldots i_n \rangle$ represents a tour: from i_1 to i_2, etc., from i_{n-1} to i_n and back to i_1 ($\boldsymbol{v}$ is a permutation of $\langle\, 1\ 2 \ldots n\, \rangle$).

---

[3]See Chapter 13 and [120] for a discussion on evolutionary programming technique for this problem.

For the initialization process we can either use some heuristics (for example, we can accept a few outputs from a greedy algorithm for the TSP, starting from different cities), or we can initialize the population by a random sample of permutations of $\langle$ 1 2 ... n $\rangle$.

The evaluation of a chromosome is straightforward: given the cost of travel between all cities, we can easily calculate the total cost of the entire tour.

In the TSP we search for the best ordering of cities in a tour. It is relatively easy to come up with some unary operators (unary type operators) which would search for better string orderings. However, using only unary operators, there is a little hope of finding even good orderings (not to mention the best one) [160]. Moreover, the strength of genetic algorithms arises from the structured information exchange of crossover combinations of highly fit individuals. So what we need is a crossover-like operator that would exploit important similarities between chromosomes. For that purpose we use a variant of a OX operator [71], which, given two parents, builds offspring by choosing a subsequence of a tour from one parent and preserving the relative order of cities from the other parent. For example, if the parents are

$$\langle \text{ 1 2 3 4 5 6 7 8 9 10 11 12 } \rangle \text{ and}$$
$$\langle \text{ 7 3 1 11 4 12 5 2 10 9 6 8 } \rangle$$

and the chosen part is

$$(4 \ 5 \ 6 \ 7),$$

the resulting offspring is

$$\langle \text{ 1 11 12 4 5 6 7 2 10 9 8 3 } \rangle.$$

As required, the offspring bears a structural relationship to both parents. The roles of the parents can then be reversed in constructing a second offspring.

A genetic algorithm based on the above operator outperforms random search, but leaves much room for improvements. Typical (average over 20 random runs) results from the algorithm, as applied to 100 randomly generated cities, gave (after 20000 generations) a value of the whole tour 9.4% above optimum.

For full discussion on the TSP, the representation issues and genetic operators used, the reader is referred to Chapter 10.

1.4 Hillclimbing, simulated annealing, and genetic algorithms

In this section we discuss three algorithms, i.e., hillclimbing, simulated annealing, and the genetic algorithm, applied to a simple optimization problem. This example underlines the uniqueness of the GA approach.

The search space is a set of binary strings v of the length 30. The objective function f to be maximized is given as

$$f(\boldsymbol{v}) = |11 \cdot one(\boldsymbol{v}) - 150|,$$

where the function $one(\boldsymbol{v})$ returns the number of 1s in the string $\boldsymbol{v}$.
For example, the following three strings

$$\boldsymbol{v}_1 = (1101101011101011111111011011011),$$
$$\boldsymbol{v}_2 = (1110001001001101110010101100011),$$
$$\boldsymbol{v}_3 = (0000100000110010000000010001000),$$

would evaluate to

$$f(\boldsymbol{v}_1) = |11 \cdot 22 - 150| = 92,$$
$$f(\boldsymbol{v}_2) = |11 \cdot 15 - 150| = 15,$$
$$f(\boldsymbol{v}_3) = |11 \cdot \ \ 6 - 150| = 84,$$

$(one(\boldsymbol{v}_1) = 22$, $one(\boldsymbol{v}_2) = 15$, and $one(\boldsymbol{v}_3) = 6)$.

The function f is linear and does not provide any challenge as an optimization task. We use it only to illustrate the ideas behind these three algorithms. However, the interesting characteristic of the function f is that it has one global maximum for

$$\boldsymbol{v}_g = (1111111111111111111111111111111),$$

$f(\boldsymbol{v}_g) = |11 \cdot 30 - 150| = 180$, and one local maximum for

$$\boldsymbol{v}_l = (0000000000000000000000000000000),$$

$f(\boldsymbol{v}_l) = |11 \cdot 0 - 150| = 150$.

There are a few versions of hillclimbing algorithms. They differ in the way a new string is selected for comparison with the current string. One version of a simple (iterated) hillclimbing algorithm (MAX iterations) is given in Figure 1.2 (steepest ascent hillclimbing). Initially, all 30 neighbors are considered, and the one $\boldsymbol{v}_n$ which returns the largest value $f(\boldsymbol{v}_n)$ is selected to compete with the current string $\boldsymbol{v}_c$. If $f(\boldsymbol{v}_c) < f(\boldsymbol{v}_n)$, then the new string becomes the current string. Otherwise, no local improvement is possible: the algorithm has reached (local or global) optimum ($local$ = TRUE). In a such case, the next iteration ($t \leftarrow t + 1$) of the algorithm is executed with a new current string selected at random.

It is interesting to note that the success or failure of the single iteration of the above hillclimber algorithm (i.e., return of the global or local optimum) is determined by the starting string (randomly selected). It is clear that if the starting string has thirteen 1s or less, the algorithm will always terminate in the local optimum (failure). The reason is that a string with thirteen 1s returns a value 7 of the objective function, and any single-step improvement towards the global optimum, i.e., increase the number of 1s to fourteen, decreases the value of the objective function to 4. On the other hand, any decrease of the number of 1s would increase the value of the function: a string with twelve 1s yields a

procedure iterated hillclimber
begin
  $t \leftarrow 0$
  repeat
    $local \leftarrow$ FALSE
    select a current string $\boldsymbol{v}_c$ at random
    evaluate $\boldsymbol{v}_c$
    repeat
      select 30 new strings in the neighborhood of $\boldsymbol{v}_c$
        by flipping single bits of $\boldsymbol{v}_c$
      select the string $\boldsymbol{v}_n$ from the set of new strings
        with the largest value of objective function f
      if $f(\boldsymbol{v}_c) < f(\boldsymbol{v}_n)$
        then $v_c \leftarrow v_n$
        else $local \leftarrow$ TRUE
    until $local$
    $t \leftarrow t + 1$
  until $t = MAX$
end

Fig. 1.2. A simple (iterated) hillclimber

value of 18, a string with eleven 1s yields a value of 29, etc. This would push the search in the "wrong" direction, towards the local maximum.

For problems with many local optima, the chances of hitting the global optimum (in a single iteration) are slim.

The structure of the simulated annealing procedure is given in Figure 1.3.

The function $random[0, 1)$ returns a random number from the range $[0, 1)$. The (termination-condition) checks whether 'thermal equilibrium' is reached, i.e., whether the probability distribution of the selected new strings approaches the Boltzmann distribution [1]. However, in some implementations [4], this repeat loop is executed just k times (k is an additional parameter of the method).

The temperature T is lowered in steps ($g(T, t) < T$ for all t). The algorithm terminates for some small value of T: the (stop-criterion) checks whether the system is 'frozen', i.e., virtually no changes are accepted anymore.

As mentioned earlier, the simulated annealing algorithm can escape local optima. Let us consider a string

$$\boldsymbol{v}_4 = (1110000001001101011001010100000),$$

with twelve 1s, which evaluates to $f(\boldsymbol{v}_4) = |11 \cdot 12 - 150| = 18$. For $\boldsymbol{v}_4$ as the starting string, the hillclimbing algorithm (as discussed earlier) would approach the local maximum

$$\boldsymbol{v}_l = (0000000000000000000000000000000),$$

procedure simulated annealing
begin
    $t \leftarrow 0$
    initialize temperature T
    select a current string $\boldsymbol{v}_c$ at random
    evaluate $\boldsymbol{v}_c$
    repeat
        repeat
            select a new string $\boldsymbol{v}_n$
                in the neighborhood of $\boldsymbol{v}_c$
                by flipping a single bit of $\boldsymbol{v}_c$
            if $f(\boldsymbol{v}_c) < f(\boldsymbol{v}_n)$
                then $\boldsymbol{v}_c \leftarrow \boldsymbol{v}_n$
                else if $random[0,1) < \exp\{(f(\boldsymbol{v}_n) - f(\boldsymbol{v}_c))/T\}$
                    then $\boldsymbol{v}_c \leftarrow \boldsymbol{v}_n$
        until (termination-condition)
        $T \leftarrow g(T, t)$
        $t \leftarrow t + 1$
    until (stop-criterion)
end

Fig. 1.3. Simulated annealing

since any string with thirteen 1s (i.e., a step 'towards' the global optimum) evaluates to 7 (less than 18). On the other hand, the simulated annealing algorithm would accept a string with thirteen 1s as a new current string with probability

$$p = \exp\{(f(\boldsymbol{v}_n) - f(\boldsymbol{v}_c))/T\} = \exp\{(7 - 18)/T\},$$

which, for some temperature, say, $T = 20$, gives

$$p = e^{-\frac{11}{20}} = 0.57695,$$

i.e., the chances for acceptance are better than 50%.

Genetic algorithms, as discussed in section 1.1, maintain a population of strings. Two relatively poor strings

$$\boldsymbol{v}_5 = (111110000000110111001110100000) \text{ and}$$
$$\boldsymbol{v}_6 = (000000000001101110010101111111)$$

each of which evaluate to 16, can produce much better offspring (if the crossover point falls anywhere between the 5th and the 12th position):

$$\boldsymbol{v}_7 = (111110000001101110010101111111).$$

The new offspring $\boldsymbol{v}_7$ evaluates to

$$f(\boldsymbol{v}_7) = |11 \cdot 19 - 150| = 59.$$

For a detailed discussion on these and other algorithms (various variants of hillclimbers, genetic search, and simulated annealing) tested on several functions with different characteristics, the reader is referred to [4]. Also, it is possible to construct hybrids which combine several techniques (including genetic algorithms) — see, for example, dynamic hill climbing technique [92]. We conclude this section by citing a funny message which was presented recently on the *Internet* (comp.ai.neural-nets [337]): it provides a nice comparison between hill-climbing, simulated annealing, and genetic algorithm techniques:

"Notice that in all [hill-climbing] methods discussed so far, the kangaroo can hope at best to find the top of a mountain close to where he starts. There's no guarantee that this mountain will be Everest, or even a very high mountain. Various methods are used to try to find the actual global optimum.

In simulated annealing, the kangaroo is drunk and hops around randomly for a long time. However, he gradually sobers up and tends to hop up hill.

In genetic algorithms, there are lots of kangaroos that are parachuted into the Himalayas (if the pilot didn't get lost) at random places. These kangaroos do not know that they are supposed to be looking for the top of Mt. Everest. However, every few years, you shoot the kangaroos at low altitudes and hope the ones that are left will be fruitful and multiply".

1.5 Conclusions

The three examples of genetic algorithms for function optimization, the prisoner's dilemma, and the traveling salesman problem, show a wide applicability of genetic algorithms. However, at the same time we should observe first signs of potential difficulties. The representation issues for the traveling salesman problem were not obvious. The new operator used (OX crossover) was far from trivial. What kind of further difficulties may we have for some other (hard) problems? In the first and third examples (optimization of a function and the traveling salesman problem) the evaluation function was clearly defined; in the second example (the prisoner's dilemma) a simple simulation process would give us an evaluation of a chromosome (we test each player to determine its effectiveness: each player uses the strategy defined by its chromosome to play the game with other players and the player's score is its average over all the games it plays). How should we proceed in a case where the evaluation function is not clearly defined? For example, the Boolean Satisfiability Problem (SAT) seems to have a natural string representation (the i-th bit represents the truth value of the i-th Boolean variable), however, the process of choosing an evaluation function is far from obvious [90].

The first example of optimization of an unconstrained function allows us to use a convenient representation, where any binary string would correspond to a value from the domain of the problem (i.e., $[-1..2]$). This means that any mutation and any crossover would produce a legal offspring. The same was true in the second example: any combination of bits represents a legal strategy. The third problem has a single constraint: each city should appear precisely once in a legal tour. This caused some problems: we used vectors of integers (instead of binary representation) and we modified the crossover operator. But how should we approach a constrained problem in general? What possibilities do we have?

The answers are not easy; we explore these issues later in the book.

2. GAs: How Do They Work?

> To every thing there is a season,
> and a time to every purpose under the heaven:
>
> A time to be born and a time to die;
> a time to plant, and a time to pluck up
> that which is planted;
>
> A time to kill, and a time to heal;
> a time to break down, and a time to build up.
>
> *The Bible, Ecclesiastes, 3*

In this chapter we discuss the actions of a genetic algorithm for a simple parameter optimization problem. We start with a few general comments; a detailed example follows.

Let us note first that, without any loss of generality, we can assume maximization problems only. If the optimization problem is to minimize a function f, this is equivalent to maximizing a function g, where $g = -f$, i.e.,

$$\min f(x) = \max g(x) = \max\{-f(x)\}.$$

Moreover, we may assume that the objective function f takes positive values on its domain; otherwise we can add some positive constant C, i.e.,

$$\max g(x) = \max\{g(x) + C\}.$$

Now suppose we wish to maximize a function of k variables, $f(x_1, \ldots, x_k)$: $R^k \rightarrow R$. Suppose further that each variable x_i can take values from a domain $D_i = [a_i, b_i] \subseteq R$ and $f(x_1, \ldots, x_k) > 0$ for all $x_i \in D_i$. We wish to optimize the function f with some required precision: suppose six decimal places for the variables' values is desirable.

It is clear that to achieve such precision each domain D_i should be cut into $(b_i - a_i) \cdot 10^6$ equal size ranges. Let us denote by m_i the smallest integer such that $(b_i - a_i) \cdot 10^6 \leq 2^{m_i} - 1$. Then, a representation having each variable x_i coded as a binary string of length m_i clearly satisfies the precision requirement. Additionally, the following formula interprets each such string:

$$x_i = a_i + decimal(1001...001_2) \cdot \frac{b_i - a_i}{2^{m_i} - 1},$$

where $decimal(string_2)$ represents the decimal value of that binary string.

Now, each chromosome (as a potential solution) is represented by a binary string of length $m = \sum_{i=1}^{k} m_i$; the first m_1 bits map into a value from the range $[a_1, b_1]$, the next group of m_2 bits map into a value from the range $[a_2, b_2]$, and so on; the last group of m_k bits map into a value from the range $[a_k, b_k]$.

To initialize a population, we can simply set some pop_size number of chromosomes randomly in a bitwise fashion. However, if we do have some knowledge about the distribution of potential optima, we may use such information in arranging the set of initial (potential) solutions.

The rest of the algorithm is straightforward: in each generation we evaluate each chromosome (using the function f on the decoded sequences of variables), select new population with respect to the probability distribution based on fitness values, and alter the chromosomes in the new population by mutation and crossover operators. After some number of generations, when no further improvement is observed, the best chromosome represents an (possibly the global) optimal solution. Often we stop the algorithm after a fixed number of iterations depending on speed and resource criteria.

For the selection process (selection of a new population with respect to the probability distribution based on fitness values), a roulette wheel with slots sized according to fitness is used. We construct such a roulette wheel as follows (we assume here that the fitness values are positive, otherwise, we can use some scaling mechanism — this is discussed in Chapter 4):

- Calculate the fitness value $eval(\boldsymbol{v}_i)$ for each chromosome $\boldsymbol{v}_i$ ($i = 1, \ldots,$ pop_size).

- Find the total fitness of the population

  $$F = \sum_{i=1}^{pop\_size} eval(\boldsymbol{v}_i).$$

- Calculate the probability of a selection p_i for each chromosome $\boldsymbol{v}_i$ ($i = 1, \ldots, pop\_size$):

  $$p_i = eval(\boldsymbol{v}_i)/F.$$

- Calculate a cumulative probability q_i for each chromosome $\boldsymbol{v}_i$ ($i = 1, \ldots, pop\_size$):

  $$q_i = \sum_{j=1}^{i} p_j.$$

The selection process is based on spinning the roulette wheel pop_size times; each time we select a single chromosome for a new population in the following way:

- Generate a random (float) number r from the range $[0..1]$.

- If $r < q_1$ then select the first chromosome ($\boldsymbol{v}_1$); otherwise select the i-th chromosome $\boldsymbol{v}_i$ ($2 \leq i \leq pop\_size$) such that $q_{i-1} < r \leq q_i$.

Obviously, some chromosomes would be selected more than once. This is in accordance with the Schema Theorem (see next chapter): the best chromosomes get more copies, the average stay even, and the worst die off.

Now we are ready to apply the recombination operator, crossover, to the individuals in the new population. As mentioned earlier, one of the parameters of a genetic system is probability of crossover p_c. This probability gives us the expected number $p_c \cdot pop\_size$ of chromosomes which undergo the crossover operation. We proceed in the following way:

For each chromosome in the (new) population:

- Generate a random (float) number r from the range $[0..1]$;

- If $r < p_c$, select given chromosome for crossover.

Now we mate selected chromosomes randomly: for each pair of coupled chromosomes we generate a random integer number pos from the range $[1..m-1]$ (m is the total length — number of bits — in a chromosome). The number pos indicates the position of the crossing point. Two chromosomes

$(b_1 b_2 \ldots b_{pos} b_{pos+1} \ldots b_m)$ and
$(c_1 c_2 \ldots c_{pos} c_{pos+1} \ldots c_m)$

are replaced by a pair of their offspring:

$(b_1 b_2 \ldots b_{pos} c_{pos+1} \ldots c_m)$ and
$(c_1 c_2 \ldots c_{pos} b_{pos+1} \ldots b_m)$.

The next operator, mutation, is performed on a bit-by-bit basis. Another parameter of the genetic system, probability of mutation p_m, gives us the expected number of mutated bits $p_m \cdot m \cdot pop\_size$. Every bit (in all chromosomes in the whole population) has an equal chance to undergo mutation, i.e., change from 0 to 1 or vice versa. So we proceed in the following way.

For each chromosome in the current (i.e., after crossover) population and for each bit within the chromosome:

- Generate a random (float) number r from the range $[0..1]$;

- If $r < p_m$, mutate the bit.

Following selection, crossover, and mutation, the new population is ready for its next evaluation. This evaluation is used to build the probability distribution (for the next selection process), i.e., for a construction of a roulette wheel with slots sized according to current fitness values. The rest of the evolution is just cyclic repetition of the above steps (see Figure 0.1 in the Introduction).

The whole process is illustrated by an example. We run a simulation of a genetic algorithm for function optimization. We assume that the population size $pop\_size = 20$, and the probabilities of genetic operators are $p_c = 0.25$ and $p_m = 0.01$.

Let us assume also that we maximize the following function:

$$f(x_1, x_2) = 21.5 + x_1 \cdot \sin(4\pi x_1) + x_2 \cdot \sin(20\pi x_2),$$

where $-3.0 \leq x_1 \leq 12.1$ and $4.1 \leq x_2 \leq 5.8$. The graph of the function f is given in Figure 2.1.

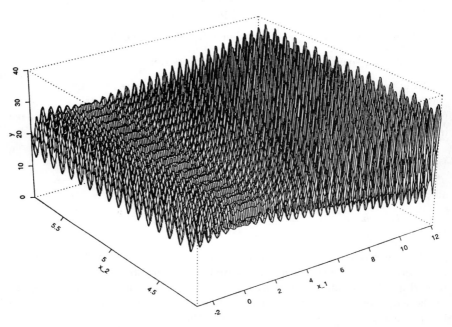

Fig. 2.1. Graph of the function $f(x_1, x_2) = 21.5 + x_1 \cdot \sin(4\pi x_1) + x_2 \cdot \sin(20\pi x_2)$

Let assume further that the required precision is four decimal places for each variable. The domain of variable x_1 has length 15.1; the precision requirement implies that the range $[-3.0, 12.1]$ should be divided into at least $15.1 \cdot 10000$ equal size ranges. This means that 18 bits are required as the first part of the chromosome:

$$2^{17} < 151000 \leq 2^{18}.$$

The domain of variable x_2 has length 1.7; the precision requirement implies that the range $[4.1, 5.8]$ should be divided into at least $1.7 \cdot 10000$ equal size ranges. This means that 15 bits are required as the second part of the chromosome:

$$2^{14} < 17000 \leq 2^{15}.$$

The total length of a chromosome (solution vector) is then $m = 18+15 = 33$ bits; the first 18 bits code x_1 and remaining 15 bits (19–33) code x_2. Let us consider an example chromosome:

(010001001011010000111110010100010).

The first 18 bits,

010001001011010000,

represent $x_1 = -3.0 + decimal(010001001011010000_2) \cdot \frac{12.1-(-3.0)}{2^{18}-1} = -3.0 +$ $70352 \cdot \frac{15.1}{262143} = -3.0 + 4.052426 = 1.052426$.
The next 15 bits,

111110010100010,

represent $x_2 = 4.1 + decimal(111110010100010_2) \cdot \frac{5.8-4.1}{2^{15}-1} = 4.1 + 31906 \cdot \frac{1.7}{32767} =$ $4.1 + 1.655330 = 5.755330$.
So the chromosome

(010001001011010000111110010100010)

corresponds to $\langle x_1, x_2 \rangle = \langle 1.052426, 5.755330 \rangle$. The fitness value for this chromosome is

$$f(1.052426, 5.755330) = 20.252640.$$

To optimize the function f using a genetic algorithm, we create a population of $pop\_size = 20$ chromosomes. All 33 bits in all chromosomes are initialized randomly.
Assume that after the initialization process we get the following population:

$$\boldsymbol{v}_1 = (100110100000001111111010011011111)$$
$$\boldsymbol{v}_2 = (111000100100110111001010100011010)$$
$$\boldsymbol{v}_3 = (000010000011001000001010111011101)$$
$$\boldsymbol{v}_4 = (100011000101101001111000001110010)$$
$$\boldsymbol{v}_5 = (000111011001010011010111111000101)$$
$$\boldsymbol{v}_6 = (000101000010010101001010111111011)$$
$$\boldsymbol{v}_7 = (001000100000110101111011011111011)$$
$$\boldsymbol{v}_8 = (100001100001110100010110101100111)$$
$$\boldsymbol{v}_9 = (010000000101100010110000001111100)$$
$$\boldsymbol{v}_{10} = (000001111000110000011010000111011)$$
$$\boldsymbol{v}_{11} = (011001111110110101100001101111000)$$
$$\boldsymbol{v}_{12} = (110100010111101101000101010000000)$$
$$\boldsymbol{v}_{13} = (111011111010001000110000001000110)$$
$$\boldsymbol{v}_{14} = (010010011000001010100111100101001)$$
$$\boldsymbol{v}_{15} = (111011101101110000100011111011110)$$
$$\boldsymbol{v}_{16} = (110011110000011111100001101001011)$$
$$\boldsymbol{v}_{17} = (011010111111001111010001101111101)$$
$$\boldsymbol{v}_{18} = (011101000000001110100111110101101)$$
$$\boldsymbol{v}_{19} = (000101010011111111110000110001100)$$
$$\boldsymbol{v}_{20} = (101110010110011110011000101111110)$$

During the evaluation phase we decode each chromosome and calculate the fitness function values from (x_1, x_2) values just decoded. We get:

$$
\begin{aligned}
eval(v_1) &= f(6.084492, 5.652242) = 26.019600 \\
eval(v_2) &= f(10.348434, 4.380264) = 7.580015 \\
eval(v_3) &= f(-2.516603, 4.390381) = 19.526329 \\
eval(v_4) &= f(5.278638, 5.593460) = 17.406725 \\
eval(v_5) &= f(-1.255173, 4.734458) = 25.341160 \\
eval(v_6) &= f(-1.811725, 4.391937) = 18.100417 \\
eval(v_7) &= f(-0.991471, 5.680258) = 16.020812 \\
eval(v_8) &= f(4.910618, 4.703018) = 17.959701 \\
eval(v_9) &= f(0.795406, 5.381472) = 16.127799 \\
eval(v_{10}) &= f(-2.554851, 4.793707) = 21.278435 \\
eval(v_{11}) &= f(3.130078, 4.996097) = 23.410669 \\
eval(v_{12}) &= f(9.356179, 4.239457) = 15.011619 \\
eval(v_{13}) &= f(11.134646, 5.378671) = 27.316702 \\
eval(v_{14}) &= f(1.335944, 5.151378) = 19.876294 \\
eval(v_{15}) &= f(11.089025, 5.054515) = 30.060205 \\
eval(v_{16}) &= f(9.211598, 4.993762) = 23.867227 \\
eval(v_{17}) &= f(3.367514, 4.571343) = 13.696165 \\
eval(v_{18}) &= f(3.843020, 5.158226) = 15.414128 \\
eval(v_{19}) &= f(-1.746635, 5.395584) = 20.095903 \\
eval(v_{20}) &= f(7.935998, 4.757338) = 13.666916
\end{aligned}
$$

It is clear, that the chromosome v_{15} is the strongest one, and the chromosome v_2 the weakest.

Now the system constructs a roulette wheel for the selection process. The total fitness of the population is

$$F = \sum_{i=1}^{20} eval(v_i) = 387.776822.$$

The probability of a selection p_i for each chromosome v_i $(i = 1, \ldots, 20)$ is:

$$
\begin{aligned}
p_1 &= eval(v_1)/F = 0.067099 & p_2 &= eval(v_2)/F = 0.019547 \\
p_3 &= eval(v_3)/F = 0.050355 & p_4 &= eval(v_4)/F = 0.044889 \\
p_5 &= eval(v_5)/F = 0.065350 & p_6 &= eval(v_6)/F = 0.046677 \\
p_7 &= eval(v_7)/F = 0.041315 & p_8 &= eval(v_8)/F = 0.046315 \\
p_9 &= eval(v_9)/F = 0.041590 & p_{10} &= eval(v_{10})/F = 0.054873 \\
p_{11} &= eval(v_{11})/F = 0.060372 & p_{12} &= eval(v_{12})/F = 0.038712 \\
p_{13} &= eval(v_{13})/F = 0.070444 & p_{14} &= eval(v_{14})/F = 0.051257 \\
p_{15} &= eval(v_{15})/F = 0.077519 & p_{16} &= eval(v_{16})/F = 0.061549 \\
p_{17} &= eval(v_{17})/F = 0.035320 & p_{18} &= eval(v_{18})/F = 0.039750 \\
p_{19} &= eval(v_{19})/F = 0.051823 & p_{20} &= eval(v_{20})/F = 0.035244
\end{aligned}
$$

The cumulative probabilities q_i for each chromosome v_i $(i = 1, \ldots, 20)$ are:

$$q_1 = 0.067099 \quad q_2 = 0.086647 \quad q_3 = 0.137001 \quad q_4 = 0.181890$$
$$q_5 = 0.247240 \quad q_6 = 0.293917 \quad q_7 = 0.335232 \quad q_8 = 0.381546$$
$$q_9 = 0.423137 \quad q_{10} = 0.478009 \quad q_{11} = 0.538381 \quad q_{12} = 0.577093$$
$$q_{13} = 0.647537 \quad q_{14} = 0.698794 \quad q_{15} = 0.776314 \quad q_{16} = 0.837863$$
$$q_{17} = 0.873182 \quad q_{18} = 0.912932 \quad q_{19} = 0.964756 \quad q_{20} = 1.000000$$

Now we are ready to spin the roulette wheel 20 times; each time we select a single chromosome for a new population. Let us assume that a (random) sequence of 20 numbers from the range [0..1] is:

$$0.513870 \quad 0.175741 \quad 0.308652 \quad 0.534534 \quad 0.947628$$
$$0.171736 \quad 0.702231 \quad 0.226431 \quad 0.494773 \quad 0.424720$$
$$0.703899 \quad 0.389647 \quad 0.277226 \quad 0.368071 \quad 0.983437$$
$$0.005398 \quad 0.765682 \quad 0.646473 \quad 0.767139 \quad 0.780237$$

The first number $r = 0.513870$ is greater than q_{10} and smaller than q_{11}, meaning the chromosome v_{11} is selected for the new population; the second number $r = 0.175741$ is greater than q_3 and smaller than q_4, meaning the chromosome v_4 is selected for the new population, etc.

Finally, the new population consists of the following chromosomes:

$$v'_1 = (0110011111101101011000011011111000) \ (v_{11})$$
$$v'_2 = (1000110001011010011110000011110010) \ (v_4)$$
$$v'_3 = (0010001000001101011110110111111011) \ (v_7)$$
$$v'_4 = (0110011111101101011000011011111000) \ (v_{11})$$
$$v'_5 = (0001010100111111111100001100011000) \ (v_{19})$$
$$v'_6 = (1000110001011010011110000011110010) \ (v_4)$$
$$v'_7 = (1110111011011100001000111111011110) \ (v_{15})$$
$$v'_8 = (0001110110010100110101111110000101) \ (v_5)$$
$$v'_9 = (0110011111101101011000011011111000) \ (v_{11})$$
$$v'_{10} = (0000100000110010000010101110111010) \ (v_3)$$
$$v'_{11} = (1110111011011100001000111111011110) \ (v_{15})$$
$$v'_{12} = (0100000001011000101100000011111100) \ (v_9)$$
$$v'_{13} = (0001010000100101010010101111111011) \ (v_6)$$
$$v'_{14} = (1000011000011010001011010110100111) \ (v_8)$$
$$v'_{15} = (1011100101100111100110001011111110) \ (v_{20})$$
$$v'_{16} = (1001101000000011111110100110111111) \ (v_1)$$
$$v'_{17} = (0000011110001100000110100001110111) \ (v_{10})$$
$$v'_{18} = (1110111110100010001100000010001101) \ (v_{13})$$
$$v'_{19} = (1110111011011100001000111111011110) \ (v_{15})$$
$$v'_{20} = (1100111100000111111000011010010111) \ (v_{16})$$

Now we are ready to apply the recombination operator, crossover, to the individuals in the new population (vectors v'_i). The probability of crossover $p_c = 0.25$, so we expect that (on average) 25% of chromosomes (i.e., 5 out of 20) undergo crossover. We proceed in the following way: for each chromosome in the (new) population we generate a random number r from the range $[0..1]$; if $r < 0.25$, we select a given chromosome for crossover.

Let us assume that the sequence of random numbers is:

$$\begin{array}{ccccc}
0.822951 & 0.151932 & 0.625477 & 0.314685 & 0.346901 \\
0.917204 & 0.519760 & 0.401154 & 0.606758 & 0.785402 \\
0.031523 & 0.869921 & 0.166525 & 0.674520 & 0.758400 \\
0.581893 & 0.389248 & 0.200232 & 0.355635 & 0.826927
\end{array}$$

This means that the chromosomes v'_2, v'_{11}, v'_{13}, and v'_{18} were selected for crossover. (We were lucky: the number of selected chromosomes is even, so we can pair them easily. If the number of selected chromosomes were odd, we would either add one extra chromosome or remove one selected chromosome — this choice is made randomly as well.) Now we mate selected chromosomes randomly: say, the first two (i.e., v'_2 and v'_{11}) and the next two (i.e., v'_{13} and v'_{18}) are coupled together. For each of these two pairs, we generate a random integer number pos from the range $[1..32]$ (33 is the total length — number of bits — in a chromosome). The number pos indicates the position of the crossing point. The first pair of chromosomes is

$$v'_2 = (100011000|10110100111000001110010)$$
$$v'_{11} = (111011101|10111000010001111011110)$$

and the generated number $pos = 9$. These chromosomes are cut after the 9th bit and replaced by a pair of their offspring:

$$v''_2 = (100011000|10111000010001111011110)$$
$$v''_{11} = (111011101|10110100111000001110010).$$

The second pair of chromosomes is

$$v'_{13} = (00010100001001010100|1010111111011)$$
$$v'_{18} = (11101111101000100011|0000001000110)$$

and the generated number $pos = 20$. These chromosomes are replaced by a pair of their offspring:

$$v''_{13} = (00010100001001010100|0000001000110)$$
$$v''_{18} = (11101111101000100011|1010111111011).$$

The current version of the population is:

$$v'_1 = (011001111110110101100001101111000)$$
$$v''_2 = (100011000101110000100011111011110)$$
$$v'_3 = (001000100000110101111011011111011)$$
$$v'_4 = (011001111110110101100001101111000)$$
$$v'_5 = (000101010011111111110000110001100)$$
$$v'_6 = (100011000101101001111000001110010)$$
$$v'_7 = (111011101101110000100011111011110)$$

$$v'_8 = (000111011001010011010111111000101)$$
$$v'_9 = (011001111110110101100001101111000)$$
$$v'_{10} = (000010000011001000001010111011101)$$
$$v''_{11} = (111011101101101001111000001110010)$$
$$v'_{12} = (010000000101100010110000001111100)$$
$$v''_{13} = (000101000010010101000000001000110)$$
$$v'_{14} = (100001100001110100010110101100111)$$
$$v'_{15} = (101110010110011110011000101111110)$$
$$v'_{16} = (100110100000001111111010011011111)$$
$$v'_{17} = (000001111000110000011010000111011)$$
$$v''_{18} = (111011111010001000111010111111011)$$
$$v'_{19} = (111011101101110000100011111011110)$$
$$v'_{20} = (110011110000011111100001101001011)$$

The next operator, mutation, is performed on a bit-by-bit basis. The probability of mutation $p_m = 0.01$, so we expect that (on average) 1% of bits would undergo mutation. There are $m \times pop\_size = 33 \times 20 = 660$ bits in the whole population; we expect (on average) 6.6 mutations per generation. Every bit has an equal chance to be mutated, so, for every bit in the population, we generate a random number r from the range $[0..1]$; if $r < 0.01$, we mutate the bit.

This means that we have to generate 660 random numbers. In a sample run, 5 of these numbers were smaller than 0.01; the bit number and the random number are listed below:

Bit position	Random number
112	0.000213
349	0.009945
418	0.008809
429	0.005425
602	0.002836

The following table translates the bit position into chromosome number and the bit number within the chromosome:

Bit position	Chromosome number	Bit number within chromosome
112	4	13
349	11	19
418	13	22
429	13	33
602	19	8

This means that four chromosomes are affected by the mutation operator; one of the chromosomes (the 13th) has two bits changed.

The final population is listed below; the mutated bits are typed in boldface. We drop *primes* for modified chromosomes: the population is listed as new vectors v_i:

$$v_1 = (0110011111101101011000011011111000)$$
$$v_2 = (1000110001011100001000111111011110)$$
$$v_3 = (0010001000001101011110110111111011)$$
$$v_4 = (0110011111100101011000011011111000)$$
$$v_5 = (0001010100111111111110000110001100)$$
$$v_6 = (1000110001011010011110000001110010)$$
$$v_7 = (1110111011011100001000111111011110)$$
$$v_8 = (0001110110010100110101111111000101)$$
$$v_9 = (0110011111101101011000011011111000)$$
$$v_{10} = (0000100000110010000010101110111010)$$
$$v_{11} = (1110111011011010010110000011110010)$$
$$v_{12} = (0100000001011000101100000011111100)$$
$$v_{13} = (0001010000100101010001000010000111)$$
$$v_{14} = (1000011000011101000101101011000111)$$
$$v_{15} = (1011100101100111100110001011111110)$$
$$v_{16} = (1001101000000011111110100110111111)$$
$$v_{17} = (0000011110001100001101000011101100)$$
$$v_{18} = (1110111110100010001110101111111011)$$
$$v_{19} = (1110111001011100001000111111011110)$$
$$v_{20} = (1100111100000111111000011010010111)$$

We have just completed one iteration (i.e., one generation) of the **while** loop in the genetic procedure (Figure 0.1 from the Introduction). It is interesting to examine the results of the evaluation process of the new population. During the evaluation phase we decode each chromosome and calculate the fitness function values from (x_1, x_2) values just decoded. We get:

$$eval(v_1) = f(3.130078, 4.996097) = 23.410669$$
$$eval(v_2) = f(5.279042, 5.054515) = 18.201083$$
$$eval(v_3) = f(-0.991471, 5.680258) = 16.020812$$
$$eval(v_4) = f(3.128235, 4.996097) = 23.412613$$
$$eval(v_5) = f(-1.746635, 5.395584) = 20.095903$$
$$eval(v_6) = f(5.278638, 5.593460) = 17.406725$$
$$eval(v_7) = f(11.089025, 5.054515) = 30.060205$$
$$eval(v_8) = f(-1.255173, 4.734458) = 25.341160$$
$$eval(v_9) = f(3.130078, 4.996097) = 23.410669$$
$$eval(v_{10}) = f(-2.516603, 4.390381) = 19.526329$$
$$eval(v_{11}) = f(11.088621, 4.743434) = 33.351874$$
$$eval(v_{12}) = f(0.795406, 5.381472) = 16.127799$$
$$eval(v_{13}) = f(-1.811725, 4.209937) = 22.692462$$
$$eval(v_{14}) = f(4.910618, 4.703018) = 17.959701$$
$$eval(v_{15}) = f(7.935998, 4.757338) = 13.666916$$
$$eval(v_{16}) = f(6.084492, 5.652242) = 26.019600$$
$$eval(v_{17}) = f(-2.554851, 4.793707) = 21.278435$$
$$eval(v_{18}) = f(11.134646, 5.666976) = 27.591064$$
$$eval(v_{19}) = f(11.059532, 5.054515) = 27.608441$$
$$eval(v_{20}) = f(9.211598, 4.993762) = 23.867227$$

Note that the total fitness of the new population F is 447.049688, much higher than total fitness of the previous population, 387.776822. Also, the best chromosome now (v_{11}) has a better evaluation (33.351874) than the best chromosome (v_{15}) from the previous population (30.060205).

Now we are ready to run the selection process again and apply the genetic operators, evaluate the next generation, etc. After 1000 generations the population is:

$$v_1 = (111011111011001101110010101011011)$$
$$v_2 = (111001100110000100010101010111000)$$
$$v_3 = (111011110111011011100101010111011)$$
$$v_4 = (111001100010000110000101010111001)$$
$$v_5 = (111011111011101101110010101010111011)$$
$$v_6 = (111001100110000100001000010100001)$$
$$v_7 = (110101100010010010001100010110000)$$
$$v_8 = (111101100010001010001101010010001)$$
$$v_9 = (111001100010010010001100010110001)$$
$$v_{10} = (111011110111011011100101010111011)$$
$$v_{11} = (110101100000010010001100010110000)$$
$$v_{12} = (110101100010010010001100010110001)$$
$$v_{13} = (111011110111011011100101010111011)$$
$$v_{14} = (111001100110000100000101010111011)$$
$$v_{15} = (111001101010111001010100110110001)$$
$$v_{16} = (111001100110000101000100010100001)$$
$$v_{17} = (111001100110000100000101010111011)$$
$$v_{18} = (111001100110000100000101010111001)$$
$$v_{19} = (111101100010001010001110000010001)$$
$$v_{20} = (111001100110000100000101010111001)$$

The fitness values are:

$$eval(v_1) = f(11.120940, 5.092514) = 30.298543$$
$$eval(v_2) = f(10.588756, 4.667358) = 26.869724$$
$$eval(v_3) = f(11.124627, 5.092514) = 30.316575$$
$$eval(v_4) = f(10.574125, 4.242410) = 31.933120$$
$$eval(v_5) = f(11.124627, 5.092514) = 30.316575$$
$$eval(v_6) = f(10.588756, 4.214603) = 34.356125$$
$$eval(v_7) = f(9.631066, 4.427881) = 35.458636$$
$$eval(v_8) = f(11.518106, 4.452835) = 23.309078$$
$$eval(v_9) = f(10.574816, 4.427933) = 34.393820$$
$$eval(v_{10}) = f(11.124627, 5.092514) = 30.316575$$
$$eval(v_{11}) = f(9.623693, 4.427881) = 35.477938$$
$$eval(v_{12}) = f(9.631066, 4.427933) = 35.456066$$
$$eval(v_{13}) = f(11.124627, 5.092514) = 30.316575$$
$$eval(v_{14}) = f(10.588756, 4.242514) = 32.932098$$
$$eval(v_{15}) = f(10.606555, 4.653714) = 30.746768$$
$$eval(v_{16}) = f(10.588814, 4.214603) = 34.359545$$

$$eval(\boldsymbol{v}_{17}) = f(10.588756, 4.242514) = 32.932098$$
$$eval(\boldsymbol{v}_{18}) = f(10.588756, 4.242410) = 32.956664$$
$$eval(\boldsymbol{v}_{19}) = f(11.518106, 4.472757) = 19.669670$$
$$eval(\boldsymbol{v}_{20}) = f(10.588756, 4.242410) = 32.956664$$

However, if we look carefully at the progress during the run, we may discover that in earlier generations the fitness values of some chromosomes were better than the value 35.477938 of the best chromosome after 1000 generations. For example, the best chromosome in generation 396 had value of 38.827553. This is due to the stochastic errors of sampling — we discuss this issue in Chapter 4.

It is relatively easy to keep track of the best individual in the evolution process. It is customary (in genetic algorithm implementations) to store "the best ever" individual at a separate location; in that way, the algorithm would report the best value found during the whole process (as opposed to the best value in the final population).

3. GAs: Why Do They Work?

> Species do not evolve to perfection, but
> quite the contrary. The weak, in fact,
> always prevail over the strong, not only
> because they are in the majority, but also
> because they are the more crafty.
>
> Friedrich Nietzsche, *The Twilight of the Idols*

The theoretical foundations of genetic algorithms rely on a binary string representation of solutions, and on the notion of a schema (see e.g., [188]) — a template allowing exploration of similarities among chromosomes. A schema is built by introducing a *don't care* symbol ($\star$) into the alphabet of genes. A schema represents all strings (a hyperplane, or subset of the search space), which match it on all positions other than '$\star$'.

For example, let us consider the strings and schemata of the length 10. The schema ($\star$ 1 1 1 1 0 0 1 0 0) matches two strings

$$\{(0111100100), (1111100100)\},$$

and the schema ($\star$ 1 $\star$ 1 1 0 0 1 0 0) matches four strings:

$$\{(0101100100), (0111100100), (1101100100), (1111100100)\}.$$

Of course, the schema (1 0 0 1 1 1 0 0 0 1) represents one string only: (1001110001), and the schema ($\star\star\star\star\star\star\star\star\star\star$) represents all strings of length 10. It is clear that every schema matches exactly 2^r strings, where r is the number of *don't care* symbols '$\star$' in a schema template. On the other hand, each string of the length m is matched by 2^m schemata. For example, let us consider a string (1001110001). This string is matched by the following 2^{10} schemata:

(1 0 0 1 1 1 0 0 0 1)
($\star$ 0 0 1 1 1 0 0 0 1)
(1 $\star$ 0 1 1 1 0 0 0 1)
(1 0 $\star$ 1 1 1 0 0 0 1)
$\vdots$
(1 0 0 1 1 1 0 0 0 $\star$)
($\star\star$ 0 1 1 1 0 0 0 1)

$$(\star\, 0 \star 1\ 1\ 1\ 0\ 0\ 0\ 1)$$

$$\vdots$$

$$(1\ 0\ 0\ 1\ 1\ 1\ 0\ 0 \star \star)$$
$$(\star\star\star 1\ 1\ 1\ 0\ 0\ 0\ 1)$$

$$\vdots$$

$$(\star\star\star\star\star\star\star\star\star\star),$$

Considering strings of the length m, there are in total 3^m possible schemata. In a population of size n, between 2^m and $n \cdot 2^m$ different schemata may be represented.

Different schemata have different characteristics. We have already noticed that the number of *don't care* conditions $\star$ in a schema determines the number of strings matched by the schema. There are two important schema properties, *order* and *defining length*; the Schema Theorem will be formulated on the basis of these properties.

The *order* of the schema S (denoted by $o(S)$) is the number of 0 and 1 positions, i.e., *fixed* positions (non-*don't care* positions), present in the schema. In other words, it is the length of the template minus the number of *don't care* ($\star$) symbols. The order defines the speciality of a schema. For example, the following three schemata, each of length 10,

$$S_1 = (\star\star\star 0\ 0\ 1 \star 1\ 1\ 0),$$
$$S_2 = (\star\star\star\star 0\ 0 \star\star 0 \star),$$
$$S_3 = (1\ 1\ 1\ 0\ 1 \star\star 0\ 0\ 1),$$

have the following orders:

$$o(S_1) = 6,\ o(S_2) = 3,\ \text{and}\ o(S_3) = 8,$$

and the schema S_3 is the most specific one.

The notion of the order of a schema is useful in calculating survival probabilities of the schema for mutations; we discuss it later in the chapter.

The *defining length* of the schema S (denoted by $\delta(S)$) is the distance between the first and the last fixed string positions. It defines the compactness of information contained in a schema. For example,

$$\delta(S_1) = 10 - 4 = 6,\ \delta(S_2) = 9 - 5 = 4,\ \text{and}\ \delta(S_3) = 10 - 1 = 9.$$

Note that the schema with a single fixed position has a defining length of zero.

The notion of the defining length of a schema is useful in calculating survival probabilities of the schema for crossovers; we discuss it later in the chapter.

As discussed earlier, the simulated evolution process of genetic algorithms consists of four consecutively repeated steps:

$$t \leftarrow t + 1$$
select $P(t)$ from $P(t-1)$
recombine $P(t)$
evaluate $P(t)$

The first step $(t \leftarrow t + 1)$ simply moves the evolution clock one unit further; during the last step (evaluate $P(t)$) we just evaluate the current population. The main phenomenon of the evolution process occurs in two remaining steps of the evolution cycle: selection and recombination. We discuss the effect of these two steps on the expected number of schemata represented in the population. We start with the selection step; we illustrate all formulae by a running example.

Let us assume, the population size $pop\_size = 20$, the length of a string (and, consequently, the length of a schema template) is $m = 33$ (as in the running example discussed in the previous chapter). Assume further that (at the time t) the population consists of the following strings:

$$
\begin{aligned}
\boldsymbol{v}_1 &= (100110100000001111111010011011111) \\
\boldsymbol{v}_2 &= (111000100100110111001010100011010) \\
\boldsymbol{v}_3 &= (000010000011001000001010111011101) \\
\boldsymbol{v}_4 &= (100011000101101001111000001110010) \\
\boldsymbol{v}_5 &= (000111011001010011010111111000101) \\
\boldsymbol{v}_6 &= (000101000010010101001010111111011) \\
\boldsymbol{v}_7 &= (001000100000110101111011011111011) \\
\boldsymbol{v}_8 &= (100001100001110100010110101100111) \\
\boldsymbol{v}_9 &= (010000000101100010110000001111100) \\
\boldsymbol{v}_{10} &= (000001111000110000011010000111011) \\
\boldsymbol{v}_{11} &= (011001111110110101100001101111000) \\
\boldsymbol{v}_{12} &= (110100010111101101000101010000000) \\
\boldsymbol{v}_{13} &= (111011111010001000110000001000110) \\
\boldsymbol{v}_{14} &= (010010011000001010100111100101001) \\
\boldsymbol{v}_{15} &= (111011110110111000010001111011110) \\
\boldsymbol{v}_{16} &= (110011110000011111100001101001011) \\
\boldsymbol{v}_{17} &= (011010111111001111010001101111101) \\
\boldsymbol{v}_{18} &= (011101000000001110100111110101101) \\
\boldsymbol{v}_{19} &= (000101010011111111110000110001100) \\
\boldsymbol{v}_{20} &= (101110010110011110011000101111110)
\end{aligned}
$$

Let us denote by $\xi(S, t)$ the number of strings in a population at the time t, matched by schema S. For example, for a given schema

$$
S_0 = (\star\star\star\star 1\ 1\ 1\star\star\star\star\star\star\star\star\star\star\star\star\star\star\star\star\star\star\star\star\star\star\star\star\star),
$$

$\xi(S_0, t) = 3$, since there are 3 strings, namely $\boldsymbol{v}_{13}$, $\boldsymbol{v}_{15}$, and $\boldsymbol{v}_{16}$, matched by the schema S_0. Note that the order of the schema S_0, $o(S_0) = 3$, and its defining length $\delta(S_0) = 7 - 5 = 2$.

Another property of a schema is its *fitness* at time t, $eval(S, t)$. It is defined as the average fitness of all strings in the population matched by the schema S. Assume there are p strings $\{\boldsymbol{v}_{i_1}, \ldots, \boldsymbol{v}_{i_p}\}$ in the population matched by a schema S at the time t. Then

$$
eval(S, t) = \sum_{j=1}^{p} eval(\boldsymbol{v}_{i_j})/p.
$$

During the selection step, an intermediate population is created: *pop_size* = 20 single string selections are made. Each string is copied zero, one, or more times, according to its fitness. As we have seen in the previous chapter, in a single string selection, the string v_i has probability $p_i = eval(v_i)/F(t)$ to be selected ($F(t)$ is the total fitness of the whole population at time t, $F(t) = \sum_{i=1}^{20} eval(v_i)$).

After the selection step, we expect to have $\xi(S, t+1)$ strings matched by schema S. Since (1) for an average string matched by a schema S, the probability of its selection (in a single string selection) is equal to $eval(S, t)/F(t)$, (2) the number of strings matched by a schema S is $\xi(S, t)$, and (3) the number of single string selections is *pop_size*, it is clear that

$$\xi(S, t+1) = \xi(S, t) \cdot pop\_size \cdot eval(S, t)/F(t),$$

We can rewrite the above formula: taking into account that the average fitness of the population $\overline{F(t)} = F(t)/pop\_size$, we can write:

$$\xi(S, t+1) = \xi(S, t) \cdot eval(S, t)/\overline{F(t)}. \tag{3.1}$$

In other words, the number of strings in the population grows as the ratio of the fitness of the schema to the average fitness of the population. This means that an "above average" schema receives an increasing number of strings in the next generation, a "below average" scheme receives decreasing number of strings, and an average schema stays on the same level.

The long-term effect of the above rule is also clear. If we assume that a schema S remains above average by $\epsilon\%$ (i.e., $eval(S, t) = \overline{F(t)} + \epsilon \cdot \overline{F(t)}$), then

$$\xi(S, t) = \xi(S, 0)(1 + \epsilon)^t,$$

and $\epsilon = (eval(S, t) - \overline{F(t)})/\overline{F(t)}$ ($\epsilon > 0$ for above average schemata and $\epsilon < 0$ for below average schemata).

This is a geometric progression equation: now we can say not only that an "above average" schema receives an increasing number of strings in the next generation, but that such a schema receives an *exponentially* increasing number of strings in the next generations.

We call the equation (3.1) the reproductive schema growth equation.

Let us return to the example schema, S_0. Since there are 3 strings, namely v_{13}, v_{15}, and v_{16} (at the time t) matched by the schema S_0, the fitness $eval(S_0)$ of the schema is

$$eval(S_0, t) = (27.316702 + 30.060205 + 23.867227)/3 = 27.081378.$$

At the same time, the average fitness of the whole population is

$$\overline{F(t)} = \sum_{i=1}^{20} eval(v_i)/pop\_size = 387.776822/20 = 19.388841,$$

and the ratio of the fitness of the schema S_0 to the average fitness of the population is

$$eval(S_0, t)/\overline{F(t)} = 1.396751.$$

This means that if the schema S_0 stays above average, then it receives an exponentially increasing number of strings in the next generations. In particular, if the schema S_0 stays above average by the constant factor of 1.396751, then, at time $t + 1$, we expect to have $3 \times 1.396751 = 4.19$ strings matched by S_0 (i.e., most likely 4 or 5), at time $t + 2$: $3 \times 1.396751^2 = 5.85$ such strings (i.e., very likely, 6 strings), etc.

The intuition is that such a schema S_0 defines a promising part of the search space and is being sampled in an exponentially increased manner.

Let us check these predictions on our running example for the schema S_0. In the population at the time t, the schema S_0 matched 3 strings, v_{13}, v_{15}, and v_{16}. In the previous chapter we simulated the selection process using the same population. The new population consists of the following chromosomes:

$$\begin{aligned}
v_1' &= (011001111110110101100001101111000)\ (v_{11})\\
v_2' &= (100011000101101001111000001110010)\ (v_4)\\
v_3' &= (001000100000110101111011011111011)\ (v_7)\\
v_4' &= (011001111110110101100001101111000)\ (v_{11})\\
v_5' &= (000101010011111111110000110001100)\ (v_{19})\\
v_6' &= (100011000101101001111000001110010)\ (v_4)\\
v_7' &= (111011101101110000100011111011110)\ (v_{15})\\
v_8' &= (000111011001010011010111111000101)\ (v_5)\\
v_9' &= (011001111110110101100001101111000)\ (v_{11})\\
v_{10}' &= (000010000011001000001010111011101)\ (v_3)\\
v_{11}' &= (111011101101110000100011111011110)\ (v_{15})\\
v_{12}' &= (010000000101100010110000001111100)\ (v_9)\\
v_{13}' &= (000101000010010101001010111111011)\ (v_6)\\
v_{14}' &= (100001100001110100010110101100111)\ (v_8)\\
v_{15}' &= (101110010110011110011000101111110)\ (v_{20})\\
v_{16}' &= (111001100110000101000100010100001)\ (v_1)\\
v_{17}' &= (111001100110000100000101010111011)\ (v_{10})\\
v_{18}' &= (111011111010001000110000001000110)\ (v_{13})\\
v_{19}' &= (111011101101110000100011111011110)\ (v_{15})\\
v_{20}' &= (110011110000011111100001101001011)\ (v_{16})
\end{aligned}$$

Indeed, the schema S_0 now (time $t + 1$) matches 5 strings: v_7', v_{11}', v_{18}', v_{19}', and v_{20}'.

However, selection alone does not introduce any new points (potential solutions) for consideration from the search space; selection just copies some strings to form an intermediate population. So the second step of the evolution cycle, recombination, takes the responsibility of introducing new individuals in the population. This is done by two genetic operators: crossover and mutation. We discuss the effect of these two operators on the expected number of schemata in the population in turn.

Let us start with crossover and consider the following example. As discussed earlier in the chapter, a single string from the population, say, v_{18}'

(1110111110100010001100000001000110),

is matched by 2^{30} schemata; in particular, the string is matched by these two schemata:

$$S_0 = (\star\star\star\star 1\,1\,1\star\star\star\star\star\star\star\star\star\star\star\star\star\star\star\star\star\star\star\star\star\star\star\star\star\star\star\star) \text{ and}$$
$$S_1 = (1\,1\,1\star\star\star\star\star\star\star\star\star\star\star\star\star\star\star\star\star\star\star\star\star\star\star\star\star\star\star\star 1\,0).$$

Let us assume further that the above string was selected for crossover (as happened in Chapter 2). Assume further (according to experiments from Chapter 2, where v'_{18} was crossed with v'_{13}) that the generated crossing site *pos* = 20. It is clear that the schema S_0 survives such a crossover, i.e., one of the offspring still matches S_0. The reason is that the crossing site preserves the sequence '111' on the fifth, sixth, and seventh positions in the string in one of the offsprings: a pair

$$v'_{18} = (11101111101000100011|0000001000110),$$
$$v'_{13} = (00010100001001010100|1010111111011),$$

would produce

$$v''_{18} = (11101111101000100011|1010111111011),$$
$$v''_{13} = (00010100001001010100|0000001000110).$$

On the other hand, the schema S_1 would be destroyed: none of the offspring would match it. The reason is that the fixed positions '111' at the beginning of the template and the fixed positions '10' at the end are placed in different offspring.

It should be clear that the defining length of a schema plays a significant role in the probability of its destruction and survival. Note, that the defining length of the schema S_0 was $\delta(S_0) = 2$, and the defining length of the schema S_1 was $\delta(S_1) = 32$.

In general, a crossover site is selected uniformly among $m-1$ possible sites. This implies that the probability of destruction of a schema S is

$$p_d(S) = \frac{\delta(S)}{m-1},$$

and consequently, the probability of schema survival is

$$p_s(S) = 1 - \frac{\delta(S)}{m-1}.$$

Indeed, the probabilities of survival and destruction of our example schemata S_0 and S_1 are:

$$p_d(S_0) = 2/32, \; p_s(S_0) = 30/32, \; p_d(S_1) = 32/32 = 1, \; p_d(S_1) = 0,$$

so the outcome was predictable.

It is important to note that only some chromosomes undergo crossover and the selective probability of crossover is p_c. This means that the probability of a schema survival is in fact:

$$p_s(S) = 1 - p_c \cdot \frac{\delta(S)}{m-1}.$$

Again, referring to our example schema S_0 and the running example ($p_c = 0.25$):

$$p_s(S_0) = 1 - 0.25 \cdot \frac{2}{32} = 63/64 = 0.984375.$$

Note also that even if a crossover site is selected between fixed positions in a schema, there is still a chance for the schema to survive. For example, if *both* strings v'_{18} and v'_{13} started with '111' and ended with '10', the schema S_1 would survive crossover (however, the probability of such event is quite small). Because of that, we should modify the formula for the probability of schema survival:

$$p_s(S) \geq 1 - p_c \cdot \frac{\delta(S)}{m-1}.$$

So the combined effect of selection and crossover gives us a new form of the reproductive schema growth equation:

$$\xi(S, t+1) \geq \xi(S, t) \cdot eval(S, t)/\overline{F(t)} \left[1 - p_c \cdot \frac{\delta(S)}{m-1} \right]. \qquad (3.2)$$

The equation (3.2) tells us about the expected number of strings matching a schema S in the next generation as a function of the actual number of strings matching the schema, relative fitness of the schema, and its defining length. It is clear that above-average schemata with short defining length would still be sampled at exponentially increased rates. For the schema S_0:

$$eval(S_0, t)/\overline{F(t)} \left[1 - p_c \cdot \frac{\delta(S)}{m-1} \right] = 1.396751 \cdot 0.984375 = 1.374927.$$

This means that the short, above-average schema S_0 would still receive an exponentially increasing number of strings in the next generations: at time $(t+1)$ we expect to have $3 \times 1.374927 = 4.12$ strings matched by S_0 (only slightly less than 4.19 — a value we got considering selection only), at time $(t+2)$: $3 \times 1.374927^2 = 5.67$ such strings (again, slightly less than 5.85).

The next operator to be considered is mutation. The mutation operator randomly changes a single position within a chromosome with probability p_m. The change is from zero to one or vice versa. It is clear that all of the fixed positions of a schema must remain unchanged if the schema survives mutation. For example, consider again a single string from the population, say, v'_{19}:

$$(11101110110111000010001111011110)$$

and schema S_0:

$$S_0 = (\star\star\star\star 1\,1\,1\star\star\star\star\star\star\star\star\star\star\star\star\star\star\star\star\star\star\star\star\star\star\star\star\star).$$

Assume further that the string v'_{19} undergoes mutation, i.e., at least one bit is flipped, as happened in the previous chapter. (Recall also that four strings underwent mutation there: one of these strings, v'_{13}, was mutated at two positions, three other strings — including v'_{19} — at one.) Since v'_{19} was mutated at the 8th position, its offspring,

$$\boldsymbol{v}''_{19} = (1110111001011100001000111110111110)$$

still is matched by the schema S_0. If the selected mutation positions were from 1 to 4, or from 8 to 33, the resulting offspring would still be matched by S_0. Only 3 bits (fifth, sixth, and seventh — the fixed bit positions in the schema S_0) are "important": mutation of at least one of these bits would destroy the schema S_0. Clearly, the number of such "important" bits is equal to the order of the schema, i.e., the number of fixed positions.

Since the probability of the alteration of a single bit is p_m, the probability of a single bit survival is $1 - p_m$. A single mutation is independent from other mutations, so the probability of a schema S surviving a mutation (i.e., sequence of one-bit mutations) is

$$p_s(S) = (1 - p_m)^{o(S)}.$$

Since $p_m \ll 1$, this probability can be approximated by:

$$p_s(S) \approx 1 - o(S) \cdot p_m.$$

Again, referring to our example schema S_0 and the running example ($p_m = 0.01$):

$$p_s(S_0) \approx 1 - 3 \cdot 0.01 = 0.97.$$

The combined effect of selection, crossover, and mutation gives us a new form of the reproductive schema growth equation:

$$\xi(S, t+1) \geq \xi(S, t) \cdot eval(S, t)/\overline{F(t)} \left[1 - p_c \cdot \frac{\delta(S)}{m-1} - o(S) \cdot p_m \right]. \qquad (3.3)$$

As in the simpler forms (equations (3.1) and (3.2)), equation (3.3) tells us about the expected number of strings matching a schema S in the next generation as a function of the actual number of strings matching the schema, the relative fitness of the schema, and its defining length and order. Again, it is clear that above-average schemata with short defining length and low-order would still be sampled at exponentially increased rates.

For the schema S_0:

$$eval(S_0, t)/\overline{F(t)} \left[1 - p_c \cdot \tfrac{\delta(S)}{m-1} - o(S_0) \cdot p_m \right] = 1.396751 \cdot 0.954375 = 1.333024.$$

This means that the short, low-order, above-average schema S_0 would still receive an exponentially increasing number of strings in the next generations: at time $(t + 1)$ we expect to have $3 \times 1.333024 = 4.00$ strings matched by S_0 (not much less than 4.19 — a value we got considering selection only, or than 4.12 — a value we got considering selections and crossovers), at time $(t + 2)$: $3 \times 1.333024^2 = 5.33$ such strings (again, not much less than 5.85 or 5.67).

Note that equation (3.3) is based on the assumption that the fitness function f returns only positive values; when applying GAs to optimization problems where the optimization function may return negative values, some additional mapping between optimization and fitness functions is required. We discuss these issues in the next chapter.

In summary, the growth equation (3.1) shows that selection increases the sampling rates of the above-average schemata, and that this change is exponential. The sampling itself does not introduce any new schemata (not represented in the initial $t = 0$ sampling). This is exactly why the crossover operator is introduced — to enable structured, yet random information exchange. Additionally, the mutation operator introduces greater variability into the population. The combined (disruptive) effect of these operators on a schema is not significant if the schema is short and low-order. The final result of the growth equation (3.3) can be stated as:

Theorem 1 (Schema Theorem.) *Short, low-order, above-average schemata receive exponentially increasing trials in subsequent generations of a genetic algorithm.*

An immediate result of this theorem is that GAs explore the search space by short, low-order schemata which, subsequently, are used for information exchange during crossover:

Hypothesis 1 (Building Block Hypothesis.) *A genetic algorithm seeks near-optimal performance through the juxtaposition of short, low-order, high-performance schemata, called the building blocks.*

As stated in [154]:

> "Just as a child creates magnificent fortresses through the arrangement of simple blocks of wood, so does a genetic algorithm seek near optimal performance through the juxtaposition of short, low-order, high performance schemata."

We have seen a perfect example of a building block through this chapter:

$$S_0 = (\star \star \star \star 1\ 1\ 1 \star).$$

S_0 is a short, low-order schema, which (at least in early populations) was also above average. This schema contributed towards finding the optimum.

Although some research has been done to prove this hypothesis [38], for most nontrivial applications we rely mostly on empirical results. During the last fifteen years many GAs applications were developed which supported the building block hypothesis in many different problem domains. Nevertheless, this hypothesis suggests that the problem of coding for a genetic algorithm is critical for its performance, and that such a coding should satisfy the idea of short building blocks.

Earlier in the chapter we stated that a population of *pop_size* individuals of length m processes at least 2^m and at most 2^{pop_size} schemata. Some of them are processed in a useful manner: these are sampled at the (desirable) exponentially increasing rate, and are not disrupted by crossover and mutation (which may happen for long defining length and high-order schemata).

Holland [188] showed, that at least pop_size^3 of them are processed usefully — he has called this property an *implicit parallelism*, as it is obtained without any extra memory/processing requirements. It is interesting to note that in a population of pop_size strings there are many more than pop_size schemata represented.[1]

This constitutes possibly the only known example of a combinatorial explosion working to our advantage instead of our disadvantage.

In this chapter we have provided some standard explanations for why genetic algorithms work. Note, however, that the building block hypothesis is just an article of faith, which for some problems is easily violated. For example, assume that the two short, low-order schemata (this time, let us consider schemata of the total length of 11 positions):

$$S_1 = (1\ 1\ 1 \star \star \star \star \star \star \star \star) \text{ and}$$
$$S_2 = (\star \star \star \star \star \star \star \star \star 1\ 1)$$

are above average, but their combination

$$S_3 = (1\ 1\ 1 \star \star \star \star \star \star 1\ 1),$$

is much less fit than

$$S_4 = (0\ 0\ 0 \star \star \star \star \star \star 0\ 0).$$

Assume further that the optimal string is $s_0 = (1111111111111)$ (S_3 matches it). A genetic algorithm may have some difficulties in converging to s_0, since it may tend to converge to points like (00011111100). This phenomenon is called deception [38], [154]: some building blocks (short, low-order schemata) can mislead genetic algorithm and cause its convergence to suboptimal points.

A phenomenon of deception is strongly connected with the concept of *epistasis* , which (in terms of genetic algorithms) means strong interaction among genes in a chromosome.[2] In other words, epistasis measures the extent to which the contribution to fitness of one gene depends on the values of other genes. For a given problem, high degree of epistasis means that building blocks can not form; consequently, the problem is deceptive.

Three approaches were proposed to deal with deception (see [155]). The first one assumes prior knowledge of the objective function to code it in an appropriate way (to get 'tight' building blocks). For example, prior knowledge about the objective function, and consequently about the deception, might result in a different coding, where the five bits required to optimize the function are adjacent, instead of being six positions apart.

---

[1]Recently, Bertoni and Dorigo [36] shown that the pop_size^3 estimate is correct only in the particular case when pop_size is proportional to 2^l and provided a more general analysis.

[2]Geneticists use the term epistasis for masking or switching effect: a gene is epistatic if its presence suppresses the effect of a gene at another locus.

The second approach uses the third genetic operator, *inversion*. Simple inversion is (like mutation) a unary operator: it selects two points within a string and inverts the order of bits between selected points, but remembering the bit's 'meaning'. This means that we have to identify bits in the strings: we do so by keeping bits together with a record of their original positions. For example, a string

$$s = ((1,0)(2,0)(3,0)|(4,1)(5,1)(6,0)(7,1)|(8,0)(9,0)(10,0)(11,1))$$

with two marked points, after inversion becomes

$$s' = ((1,0)(2,0)(3,0)|(7,1)(6,0)(5,1)(4,1)|(8,0)(9,0)(10,0)(11,1)).$$

A genetic algorithm with inversion as one of the operators searches for the best arrangements of bits for forming building blocks. For example, the desirable schema considered earlier

$$S_3 = (1\ 1\ 1\star\star\star\star\star\star 1\ 1),$$

rewritten as

$$S_3 = ((1,1)(2,1)(3,1)(4,\star)(5,\star)(6,\star)(7,\star)(8,\star)(9,\star)(10,1)(11,1)),$$

might be regrouped (after successful inversion) into

$$S_3 = ((1,1)(2,1)(3,1)(11,1)(10,1)(9,\star)(8,\star)(7,\star)(6,\star)(5,\star)(4,\star)),$$

making an important building block. However, as stated in [155]:

> "An earlier study [160] argued that inversion — a unary operator — was incapable of searching efficiently for tight building blocks because it lacked the power of juxtaposition inherent in binary operators. Put another way, inversion is to orderings what mutation is to alleles: both fight the good fight against search-stopping lack of diversity, but neither is sufficiently powerful to search for good structures, allelic or permutational, on its own when good structures require epistatic interaction of the individual parts."

The third approach to fight the deception was proposed recently [155, 159]: a messy genetic algorithm (mGA). Since mGAs have other interesting properties as well, we discuss them briefly in the next chapter (section 4.6).

4. GAs: Selected Topics

A man once saw a butterfly
struggling to emerge from
its cocoon, too slowly
for his taste, so he began
to blow on it gently. The
warmth of his breath speeded
up the process all right. But
what emerged was not a butterfly
but a creature with mangled
wings.

Anthony de Mello, *One Minute Wisdom*

GA theory provides some explanation why, for a given problem formulation, we may obtain convergence to the sought optimal point. Unfortunately, practical applications do not always follow the theory, with the main reasons being:

- the coding of the problem often moves the GA to operate in a different space than that of the problem itself,

- there is a limit on the hypothetically unlimited number of iterations, and

- there is a limit on the hypothetically unlimited population size.

One of the implications of these observations is the inability of GAs, under certain conditions, to find the optimal solutions; such failures are caused by a premature convergence to a local optimum. The premature convergence is a common problem of genetic algorithms and other optimization algorithms. If convergence occurs too rapidly, then the valuable information developed in part of the population is often lost. Implementations of genetic algorithms are prone to converge prematurely before the optimal solution has been found, as stated in [46]:

"...While the performance of most implementations is comparable to or better than the performance of many other search techniques, it [GA] still fails to live up to the high expectations engendered by the theory. The problem is that, while the theory points to sampling rates and search behavior in the limit, any implementation uses a finite population or set of sample points. Estimates based on finite samples inevitably have a sampling error and lead to search trajectories much different from those theoretically predicted. This problem is manifested in practice as a premature loss of diversity in the population with the search converging to a sub-optimal solution."

Eshelman and Schaffer [105] discuss a few strategies for combating premature convergence; these include (1) a mating strategy, called *incest prevention*,[1] (2) a use of uniform crossover (see section 4.6), and (3) detecting duplicate strings in the population (similar to the crowding model; see section 4.1).

However, most of research in this area relates to:

- the magnitude and kind of errors introduced by the sampling mechanism, and

- the characteristics of the function itself.

These two issues are closely related; however, we discuss them in turn (sections 4.1 and 4.2). Additional two sections present a result on the convergence of a class of genetic algorithms (called contractive mapping genetic algorithms), which is based on Banach fixpoint theorem (section 4.3), and the first results of some experiments with genetic algorithms with varying population size (section 4.4). Section 4.5 discusses briefly a few constraint handling methods, and the last section presents some additional ideas for enhancing the genetic search.

4.1 Sampling mechanism

It seems that there are two important issues in the evolution process of the genetic search: population diversity and selective pressure. These factors are strongly related: an increase in the selective pressure decreases the diversity of the population, and vice versa. In other words, strong selective pressure "supports" the premature convergence of the GA search; a weak selective pressure can make the search ineffective. Thus it is important to strike a balance between these two factors; sampling mechanisms are attempt to achieve this goal. As observed by Whitley [395]:

> "It can be argued that there are only two primary factors (and perhaps only two factors) in genetic search: population diversity and selective pressure [...] In some sense this is just another variation on the idea of exploration versus exploitation that has been discussed by Holland and others. Many of the various parameters that are used to 'tune' genetic search are really indirect means of affecting selective pressure and population diversity. As selective pressure is increased, the search focuses on the top individuals in the population, but

---

[1] For additional information on an incest prevention technique applied to the TSP, see the end of Chapter 10.

because of this 'exploitation' genetic diversity is lost. Reducing the selective pressure (or using larger population) increases 'exploration' because more genotypes and thus more schemata are involved in the search."

The first, and possibly the most recognized work, was due to DeJong [82] in 1975. He considered several variations of the simple selection presented in the previous chapter. The first variation, named the *elitist model*, enforces preserving the best chromosome. The second variation, the *expected value model*, reduces the stochastic errors of the selection routine. This is done by introducing a count for each chromosome v, which is set initially to the $f(v)/\bar{f}$ value and decreased by 0.5 or 1 when the chromosome is selected for reproduction with crossover or mutation, respectively. When the chromosome count falls below zero, the chromosome is not available for selection any longer. In the third variation, the *elitist expected value model*, the first two variations are combined together. In the fourth model, the *crowding factor model*, a newly generated chromosome replaces an "old" one and the doomed chromosome is selected from those which resemble the new one.

In 1981 Brindle [49] considered some further modifications: *deterministic sampling, remainder stochastic sampling without replacement, stochastic tournament*, and *remainder stochastic sampling with replacement*. This study confirmed the superiority of some of these modifications over simple selection. In particular, the *remainder stochastic sampling with replacement* method, which allocates samples according to the integer part of the expected value of occurrences of each chromosome in a new population and where the chromosomes compete according to the fractional parts for the remaining places in the population, was the most successful one and adopted by many researchers as standard. In 1987 Baker [23] provided a comprehensive theoretical study of these modifications using some well defined measures, and also presented a new improved version called *stochastic universal sampling*. This method uses a single wheel spin. This wheel, which is constructed in the standard way (Chapter 2), is spun with a number of equally spaced markers equal to the population size as opposed to a single one.

Other methods to sample a population are based on introducing artificial weights: chromosomes are selected proportionally to their rank rather than actual evaluation values (see e.g., [22], [395]). These methods are based on a belief that the common cause of rapid (premature) convergence is the presence of *super individuals*, which are much better than the average fitness of the population. Such super individuals have a large number of offspring and (due to the constant size of the population) prevent other individuals from contributing any offspring in next generations. In a few generations a super individual can eliminate desirable chromosomal material and cause a rapid convergence to (possibly local) optimum.

There are many methods to assign a number of offspring based on ranking. For example, Baker [22] took a user defined value, MAX, as the upper bound for the expected number of offspring, and a linear curve through MAX was taken such that the area under the curve equaled the population size. In that way we can easily determine the difference between expected numbers of offspring between "adjacent" individuals. For example, for MAX = 2.0 and $pop\_size = 50$, the difference between expected numbers of offspring between "adjacent" individuals would be 0.04.

Another possibility is to take a user defined parameter q and define a linear function, e.g.,

$$prob(rank) = q - (rank - 1)r,$$

or a nonlinear function, e.g.,

$$prob(rank) = q(1 - q)^{rank-1}.$$

Both functions return the probability of an individual ranked in position $rank$ ($rank = 1$ means the best individual, $rank = pop\_size$ the worst one) to be selected in a single selection.

Both schemes allow the users to influence the selective pressure of the algorithm. In the case of the linear function, the requirement

$$\sum_{i=1}^{pop\_size} prob(i) = 1$$

implies, that

$$q = r(pop\_size - 1)/2 + 1/pop\_size.$$

If $r = 0$ (and consequently $q = 1/pop\_size$) there is no selection pressure at all: all individuals have the same probability of selection. On the other hand, if $q - (pop\_size - 1)r = 0$, then

$$r = 2/(n(n - 1)), \text{ and } q = 2/n,$$

provide the maximum selective pressure. In other words, if a linear function is selected to provide probabilities for ranked individuals, a single parameter q, which varies between $1/pop\_size$ and $2/pop\_size$ can control the selective pressure of the algorithm. For example, if $pop\_size = 100$, and $q = 0.015$, then $r = q/(pop\_size - 1) = 0.00015151515$ and $prob(1) = 0.015$, $prob(2) = 0.0148484848$, ... , $prob(100) = 0.00000000000000000051$.

For the nonlinear function, the parameter $q \in (0..1)$ does not depend on the population size; larger values of q imply stronger selective pressure of the algorithm. For example, if $q = 0.1$ and $pop\_size = 100$, then $prob(1) = 0.100$, $prob(2) = 0.1 \cdot 0.9 = 0.090$, $prob(3) = 0.1 \cdot 0.9 \cdot 0.9 = .081$, ... , $prob(100) = 0.000003$. Note that

$$\sum_{i=1}^{pop\_size} prob(i) = \sum_{i=1}^{pop\_size} q(1 - q)^{i-1} \approx 1.^2$$

---

[2]It is easy to replace $\approx$ by $=$; it is sufficient to define $prob(i) = c \cdot q(1 - q)^{i-1}$, where $c = \frac{1}{1-(1-q)^{pop\_size}}$.

Such approaches, though shown to improve genetic algorithm behavior in some cases, have some apparent drawbacks. First, they put the responsibility on the user to decide when to use these mechanisms. Second, they ignore the information about the relative evaluations of different chromosomes. Third, they treat all cases uniformly, regardless of the magnitude of the problem. Finally, selection procedures based on ranking violate the Schema Theorem. On the other hand, as shown in some research studies [23], [395], they prevent scaling problems (discussed in the next section), they control better the selective pressure, and (coupled with one-at-a-time reproduction) they give the search a greater focus.

An additional selection method, *tournament selection* [159], combines the idea of ranking in very interesting and efficient way. This method (in a single iteration) selects some number k of individuals and selects the best one from this set of k elements into the next generation. This process is repeated *pop_size* number of times. It is clear, that large values of k increase selective pressure of this procedure; typical value accepted by many applications is $k = 2$ (so-called tournament size). Here it is possible to add a flavor of simulated annealing by considering Boltzmann selection , where two elements, i and j, compete with each other, and the winner is determined accordingly to the formula

$$\frac{1}{1 + e^{\frac{f(i) - f(j)}{T}}},$$

where T is temperature and $f(i)$ and $f(j)$ are values of the objective function for elements i and j, respectively (the formula is for minimization problems).

In [16] Bäck and Hoffmeister discuss categories of selection procedures. They divide selection procedures into *dynamic* and *static* methods — a static selection requires that selection probabilities remain constant between generations (for example, ranking selection), whereas a dynamic selection does not have such a requirement (e.g., proportional selection). Another division of selection procedures is into *extinctive* and *preservative* methods — preservative selection requires non-zero selection probability for each individual, whereas extinctive selection does not. Extinctive selections are further divided into *left* and *right* selections: in left extinctive selection the best individuals are prevented from reproduction in order to avoid premature convergence due to super individuals (right selection does not). Additionally, some selection procedures are *pure* in the sense that parents are allowed to reproduce in one generation only (i.e., the life time of each individual is limited to one generation only regardless of its fitness). We shell return to extinctive, pure selections in Chapter 8, when we discuss evolution strategies and compare them with genetic algorithms. Some selections are *generational* in the sense that the set of parents is fixed until all offspring for the next generation are completely produced; in selections *on-the-fly* an offspring replaces its parent immediately. Some selections are *elitist* in the sense that some (or all) of the parents are allowed to undergo selection with their offspring — we have already seen such selection in the elitist model [82].

In most of the experiments discussed in this volume, we used a new, two–step variation of the basic selection algorithm. However, this modification is not just a new selection mechanism; it can use any of the sampling methods devised so far and is itself designed to decrease the (possible) undesirable influence of some functions' characteristics. It falls into the category of dynamic, preservative, generational, and elitist selection.

The structure of the modified genetic algorithm (modGA) is shown in Figure 4.1. The modification with respect to the classical genetic algorithm is that in the modGA we do not perform the selection step "select $P(t)$ from $P(t-1)$", but rather we select independently r (not necessarily distinct) chromosomes for reproduction and r (distinct) chromosomes to die. These selections are performed with respect to the relative fitness of the strings: a string with a better than average performance has a higher chance to be selected for reproduction; strings with a worse than average performance have higher chances to be selected to die. After the "select-parents" and "select-dead" steps of the modGA are performed, there are three (not necessarily disjoint) groups of strings in the population:

- r (not necessarily distinct) strings to reproduce (parents),

- precisely r strings to die (dead), and

- the remaining strings, called neutral strings.

The number of neutral strings in a generation (at least $pop\_size - 2r$ and at most $pop\_size - r$) depends on the number of selected distinct parents and on the number of overlapping strings in categories "parents" and "dead". Then a new population $P(t + 1)$ is formed, consisting of the $pop\_size - r$ strings (all strings except these selected to die) and r offspring of the r parents.

```
procedure modGA
begin
    t ← 0
    initialize P(t)
    evaluate P(t)
    while (not termination-condition) do
    begin
        t ← t + 1
        select-parents from P(t − 1)
        select-dead from P(t − 1)
        form P(t): reproduce the parents
        evaluate P(t)
    end
end
```

Fig. 4.1. The algorithm modGA

As presented, the algorithm has a potentially problematic step: how to select the r chromosomes to die. Obviously, we wish to perform this selection in such a way that stronger chromosomes have smaller chances of dying. We achieved this by changing the method of forming the new population $P(t + 1)$ to the following one:

step 1: Select r parents from $P(t)$. Each selected chromosome (or rather each of selected copies of some chromosomes) is marked as applicable to exactly one fixed genetic operation.

step 2: Select $pop\_size - r$ distinct chromosomes from $P(t)$ and copy them to $P(t + 1)$.

step 3: Let r parent chromosomes breed to produce exactly r offspring.

step 4: Insert these r new offspring into population $P(t + 1)$.

The above selections (steps 1 and 2) are done according to the chromosomes' fitness (stochastic universal sampling method).

There are a few important differences between different selection routines discussed earlier and the one described above. Firstly, both parent and offspring have a very good chance to be present in a new generation: an above average individual has a good chances to be selected as a parent (step 1) and, in the same time, to be selected in a new population of $pop\_size - r$ elements (step 2). If so, one (or more) of its offspring would take some of the remaining r positions. Secondly, we apply genetic operators on whole individuals as opposed to individual bits (classical mutation). This would provide an uniform treatment of all operators used in evolution program (an evolution program, GENOCOP, uses several genetic operators; see Chapter 7). So, if three operators are used (e.g., mutation, crossover, inversion), some of the parents would undergo mutation, some others crossover, and the rest inversion.

The modified approach (modGA) enjoys similar theoretical properties as the classical genetic algorithm. We can rewrite the growth equation (3.3) from Chapter 3 as:

$$\xi(S, t + 1) \geq \xi(S, t) \cdot p_s(S) \cdot p_g(S), \tag{4.1}$$

where $p_s(S)$ represents the probability of the survival of the schema S and $p_g(S)$ represents the probability of the growth of the schema S. The growth of the schema S happens during the selection stage (growing phase) where several copies of above-average schemata are copied into a new population. The probability $p_g(S)$ of the growth of the schema S, $p_g(S) = eval(S, t)/\overline{F(t)}$, and $p_g(S) > 1$ for better-than-average schemata. Then the selected chromosomes must survive the genetic operators crossover and mutation (shrinking phase). As discussed in Chapter 3, the probability $p_s(S)$ of survival of the schema S,

$$p_s(S) = 1 - p_c \frac{\delta(S)}{m-1} - p_m \cdot o(S) < 1.$$

Formula (4.1) implies that for short, low-order schemata, $p_g(S) \cdot p_s(S) > 1$; because of this, such schemata receive an exponentially increasing number of trials in subsequent generations. The same holds for the modified (modGA) version of the algorithm. The expected number of chromosomes of the schema S in the modGA algorithm is also a product of the number of chromosomes in the old population $\xi(S, t)$, the probability of survival ($p_s(S) < 1$), and the probability of the growth $p_g(S)$ — the only difference is in interpretation of growing and shrinking phases and their relative order. In the modGA version, the shrinking phase is the first one: $n - r$ chromosomes are selected for the new population. The probability of survival is defined as a fraction of chromosomes of the schema S which were not selected to die. The growing phase is next and is manifested in the arrival of r new offspring. The probability of the growth $p_g(S)$ of the schema S is a probability that the schema S expands by a new offspring generated from the r parents. Again, for short, low-order schemata, $p_s(S) \cdot p_g(S) > 1$ holds and such schemata receive exponentially increasing trials in subsequent generations.

One of the ideas of the modGA algorithm is a better utilization of the available storage resource: population size. The new algorithm avoids leaving exact multiple copies of the same chromosome in the new populations (which may still happen by accident by other means but is very unlike). On the other hand, the classical algorithm is quite vulnerable to creation of such multiple copies. Moreover, such multi-occurrences of super individuals create a possibility for a chain reaction: there is an chance for an even larger number of such exact copies in the next population, etc. This way the already limited population size can actually represent only a decreasing number of unique chromosomes. Lower space utilization decreases the performance of the algorithm; note that the theoretical foundations of genetic algorithms assume infinite population size. In the modGA algorithm we may have a number of family members for a chromosome, but all such members are different (by a family we mean offspring of the same parent).

As an example consider a chromosome with an expected value of appearances in $P(t+1)$ equal $p = 3$. Also assume that the classical genetic algorithm has probability of crossover and mutation $p_c = 0.3$ and $p_m = 0.003$, a rather usual scenario. Following the selection, there will be exactly $p = 3$ copies of this chromosome in $P(t+1)$ before reproduction. After reproduction, assuming chromosome length $m = 20$, the expected number of exact copies of this chromosome remaining in $P(t+1)$ will be $p \cdot (1 - p_c - p_m \cdot m) = 1.92$. Therefore, it is safe to say that the next population will have two exact copies of such a chromosome, reducing the number of different chromosomes.

The modification used in the modGA is based on the idea of the crowding factor model [82], where a newly generated chromosome replaces some old one. But the difference is that in the crowding factor model the dying chromosome is selected from those which resemble the new one, whereas in the modGA the dying chromosomes are those with lower fitness.

The modGAs, for small values of the parameter r, belong to a class of Steady State GAs (SSGA) [394], [382]; the main difference between GAs and SSGAs is that in the latter only few members of the population are changed (within each generation). There is also some similarity between the modGA and classifier systems (Chapter 12): a genetic component of a classifier system changes the population as little as possible. In the modGA we can regulate such a change using the parameter r, which determines the number of chromosomes to reproduce and the number of chromosomes to die. In the modGA, $pop\_size - r$ chromosomes are placed in a new population without any change. In particular, for $r = 1$, only one chromosome is replaced in each generation. Recently, Mühlenbein [289] proposed Breeder GAs (BGA), where r best individuals are selected and mated randomly until the number of offspring is equal to the size of the population. The offspring generation replaces the parent population and the best individual found so far remains in the population.

4.2 Characteristics of the function

The modGA algorithm provides a new mechanism for forming a new population from the old one. However, it seems that some additional measures might be helpful in fighting problems related to the characteristic of the function being optimized. Over the years we have seen three basic directions. One of them borrows the simulated annealing technique of varying the system's entropy, (see e.g., [359], where the authors control the rate of population convergence by thermodynamic operators, which use a global temperature parameter).

Another direction is based on allocation of reproductive trials according to rank rather than actual evaluation values (as discussed in the previous section), since ranking automatically introduces a uniform scaling across the population.

The last direction concentrates on trying to fix the function itself by introducing a scaling mechanism. Following Goldberg [154, pp. 122–124] we divide such mechanisms into three categories:

1. *Linear Scaling*. In this method the actual chromosomes' fitness is scaled as

   $$f'_i = a * f_i + b.$$

   The parameters a, b are normally selected so that the average fitness is mapped to itself and the best fitness is increased by a desired multiple of the average fitness. This mechanism, though quite powerful, can introduce negative evaluation values that must be dealt with. In addition, the parameters a, b are normally fixed for the population life and are not problem dependent.

2. *Sigma Truncation*. This method was designed as an improvement of linear scaling both to deal with negative evaluation values and to incorporate

problem dependent information into the mapping itself. Here the new fitness is calculated according to:

$$f_i' = f_i + (\overline{f} - c * \sigma),$$

where c is chosen as a small integer (usually a number from the range 1 and 5) and σ is the population's standard deviation; possible negative evaluations f' are set to zero.

3. *Power Law Scaling*. In this method the initial fitness is taken to some specific power:

$$f_i' = f_i^k,$$

with some k close to one. The parameter k scales the function f; however, in some studies [138] it was concluded that the choice of k should be problem dependent. In the same study the author used $k = 1.005$ to obtain some experimental improvements.

The most noticeable problem associated with the characteristic of the function under consideration involves differences in relative fitness. As an example consider two functions: $f_1(x)$ and $f_2(x) = f_1(x) + const$. Since they are both basically the same (i.e., they share the same optima), one would expect that both can be optimized with similar degree of difficulty. However, if $const \gg \overline{f_1}(x)$, then the function $f_2(x)$ will suffer from (or enjoy) much slower convergence than the function $f_1(x)$. In fact, in the extreme case, the second function will be optimized using a totally random search; such a behavior may be tolerable during the very early life of the population but would be devastating later on. Conversely, $f_1(x)$ might be converging too fast, pushing the algorithm into a local optimum.

In addition, due to the fixed size of the population, the behavior of a GA may be different from run to run — this is caused by errors of finite sampling. Consider a function $f_3(x)$ with a sample $x_i^t \in P(t)$ close to some local optimum and $f(x_i^t)$ much greater than the average fitness $\overline{f}(x^t)$ (i.e., x_i^t is a super individual). Furthermore, assume that there is no x_j^t close to the sought global maximum. This might be the case for a highly non–smooth function. In such a case, there is a fast convergence towards that local optimum. Because of that, the population $P(t+1)$ becomes over–saturated with elements close to that solution, decreasing the chance of a global exploration needed to search for other optima. While such a behavior is permissible at the later evolutionary stages, and even desired at the very final stages, it is quite disturbing at the early ones. Moreover, normally late populations (during late stages of the algorithm) are saturated with chromosomes of similar fitness as all of those are closely related (by the mating processes). Therefore, using the traditional selective techniques the sampling actually becomes random. Such a behavior is exactly the opposite of the most desirable one, where there is a decreased influence of relative chromosomes fitness on the selection process during the initial stages of population life and increased influence at late stages.

One of the best known systems, GENESIS 1.2ucsd, uses two parameters to control the search with respect to the characteristic of the function being optimized: the scaling window and the sigma truncation factor. The system minimizes a function: in such cases usually the evaluation function *eval* returns

$$eval(x) = F - f(x),$$

where F is a constant such that $F > f(x)$ for all x. As discussed earlier, a poor choice of F may have unfortunate effect on the search, moreover, F might be not available a priori. The scaling window W of the GENESIS 1.2ucsd allows the user to control how often the constant F is updated: if $W > 0$, the system sets F to the greatest value of $f(x)$ which has occurred in the last W generations. A value $W = 0$ indicates an infinite window, i.e., $F = \max\{f(x)\}$ over all evaluations. If $W < 0$, the users can use another method discussed earlier: sigma truncation.

It is important also to point out the significance of the termination condition used in the algorithm. The simplest termination condition would check the current generation number; the search is terminated if the total number of generations exceeds a predefined constant. In terms of Figure 0.1 (Introduction), such termination condition is expressed as "$t \geq T$" for some constant T. In many versions of evolution programs, not all individuals need to be re-evaluated: some of them pass from one generation to the next without any alteration. In such cases it might be meaningful (for the sake of comparison with some other, traditional algorithms) to count the number of function evaluations (usually, such a number is proportional to the number of generations) and terminate the search when the number of function evaluations exceeds some predefined constant.

However, the above termination conditions assume user's knowledge on the characteristic of the function, which influence the length of the search. In many instances it is quite difficult to claim that the total number of generations (or function evaluations) should be, say, 10,000. It seems that it would be much better if the algorithm terminates the search, when the chance for a significant improvement is relatively slim.

There are two basic categories of termination conditions, which use the characteristic of the search for making termination decisions. One category is based on the chromosome structure (genotype); the other—on the meaning of a particular chromosome (phenotype). Terminations conditions from the first category measure the convergence of the population by checking the number of converged alleles, where allele is considered converged if some predetermined percentage of the population have the same (or similar—for non-binary representations) value in this allele. If the number of converged alleles exceeds some percentage of total alleles, the search is terminated. Terminations conditions from the second category measure the progress made by the algorithm in a predefined number of generations: if such progress is smaller than some epsilon (which is given as a parameter of the method), the search is terminated.

4.3 Contractive mapping genetic algorithms

The convergence of genetic algorithms is one of the most challenging theoretical issues in the evolutionary computation area. Several researchers explored this problem from different perspectives. Goldberg and Segrest [163] provided a finite Markov chain analysis of genetic algorithm (finite population, reproduction and mutation only). Davis and Principe [80] investigated a possibility of extrapolation of the existing theoretical foundation of the simulated annealing algorithm onto a Markov chain genetic algorithm model. Eiben, Aarts, and Van Hee [98] proposed an abstract genetic algorithm which unifies genetic algorithms and simulated annealing; a Markov chain analysis on a such abstract genetic algorithm is discussed and conditions implying that the evolution process finds an optimum with probability 1 are given. Kingdon [224] investigated starting points, convergence and the class of problems genetic algorithms find hard to solve. The notion of competing schemata is generalized and the probability of convergence of such schemata is given. Several researchers considered also various definitions of deceptive problems [154]. Recently [334] Rudolph proved that a classical genetic algorithm never converges to the global optimum, but modified versions, which maintain the best solution in the population (i.e., elitist model) do.

One possible approach for explaining the convergence properties of genetic algorithm might be based on Banach fixpoint theorem [386]. It provides an intuitive explanation of a convergence of GAs (without elitist model); the only requirement is that there should be an improvement in subsequent populations (not necessarily improvement of the best individual). Banach fixpoint theorem deals with contractive mappings on metric spaces. It states that any such a mapping f has a unique fixpoint, i.e., an element x such that $f(x) = x$. Fixpoint techniques are generally accepted as a powerful tool for defining semantics of computations. For example, the denotational semantics of a program or computation is usually given as the least fixpoint of a continuous mapping defined on a suitable complete lattice. However, unlike in the traditional denotational semantics, we found that these are metric spaces that provide a very simple and natural way to express the semantics of genetic algorithms. Genetic algorithms can be defined as transformations between populations. Suppose now that we are able to find such metric spaces, in which those transformations are contractive. In such a case we are given a semantics of genetic algorithms as fixpoints of the underlying transformations. Since any such transformation has the unique fixpoint, we get the convergence of genetic algorithms as a simple corollary.

Intuitively, a metric space is an ordered pair of a set and a function that allows us to measure the distance between any pair of elements of the set. A mapping f defined on elements of such a set is contractive if the distance between $f(x)$ and $f(y)$ is less[3] than the distance between x and y.

---

[3]Actually, as we shall see later, the term *less* is a bit stronger than usual.

Let us now define the basic notions more formally. Denote the set of real numbers by R. A set S together with a mapping $\delta : S \times S \longrightarrow R$ is a *metric space* if the following conditions are satisfied for any elements $x, y \in S$

- $\delta(x, y) \geq 0$ and $\delta(x, y) = 0$ iff $x = y$

- $\delta(x, y) = \delta(y, x)$

- $\delta(x, y) + \delta(y, z) \geq \delta(x, z)$.

The mapping δ is called a *distance*. We usually denote metric spaces by $\langle S, \delta \rangle$.

Let $\langle S, \delta \rangle$ be a metric space and let $f : S \longrightarrow S$ be a mapping. We shall say that f is *contractive* iff there is a constant $\epsilon \in [0, 1)$ such that for all $x, y \in S$

$$\delta(f(x), f(y)) \leq \epsilon * \delta(x, y).$$

In order to formulate Banach theorem, we have to define the notion of completeness of metric spaces. We say that the sequence $p_0, p_1, \ldots$ of elements of metric space $\langle S, \delta \rangle$ is a *Cauchy sequence* iff for any $\epsilon > 0$ there is k such that for all $m, n > k$, $\delta(p_m, p_n) < \epsilon$. We say that a metric space is *complete* if any Cauchy sequence $p_0, p_1, \ldots$ has a limit $p = \lim_{n \to \infty} p_n$.

We are now ready to formulate the Banach theorem. The proof of the theorem was for the first time given in [25] and can be found in most manuals on topology (e.g., [94], p. 60).

Theorem. [25] Let $\langle S, \delta \rangle$ be a complete metric space and let $f : S \longrightarrow S$ be a contractive mapping. Then there f has a unique fixpoint $x \in S$ such that for any $x_0 \in S$,

$$x = \lim_{i \to \infty} f^i(x_0),$$

where $f^0(x_0) = x_0$ and $f^{i+1}(x_0) = f(f^i(x_0))$.

The Banach theorem has a very intuitive application to the case of genetic algorithms. Namely, if we construct the metric space S in such a way that its elements are populations, then any contractive mapping f has unique fixpoint, which by Banach theorem is obtained by iteration of f applied to an *arbitrary chosen* initial population $P(0)$. Thus if we find a suitable metric space in which genetic algorithms are contracting, then we are able to show the convergence of those algorithms to the same fixpoint independently of the choice of initial population. We shall show that such a construction is possible for a slightly modified genetic algorithm, called Contractive Mapping Genetic Algorithm (CM-GA).

Without any loss of generality we assume that we deal with maximization problems, i.e., problems for which a solution $\overline{x}_i$ is better than solution $\overline{x}_j$ iff $eval(\overline{x}_i) > eval(\overline{x}_j)$.

We assume, that the size of a population *pop_size* $= n$ is fixed; every population consists of n individuals, i.e., $P = \{\overline{x}_1, \ldots, \overline{x}_n\}$. Moreover, let us consider an evaluation function $Eval$ for a population P; for example, we can assume

$$Eval(P) = \frac{1}{n} \sum_{\overline{x}_i \in P} eval(\overline{x}_i),$$

where *eval* returns the 'fitness' of an individual $\overline{x}_i$ from the population P. The set S consists of all possible populations, P, i.e., any vector $\{\overline{x}_1, \ldots, \overline{x}_n\} \in S$.

Now we define a mapping (distance) $\delta : S \times S \longrightarrow R$ and a contractive mapping $f : S \longrightarrow S$ in the metric space $\langle S, \delta \rangle$. The distance δ in the metric space S of populations can be defined as:

$$\delta(P_1, P_2) = \begin{cases} 0 & \text{if } P_1 = P_2 \\ |1 + M - Eval(P_1)| + \\ \quad |1 + M - Eval(P_2)| & \text{otherwise} \end{cases}$$

where M is the upper limit of the *eval* function in the domain of interest, i.e., $eval(\overline{x}) \leq M$ for all individuals $\overline{x}$ (consequently, $Eval(P) \leq M$ for all possible populations P). Indeed,

- $\delta(P_1, P_2) \geq 0$ for any populations P_1 and P_2; moreover, $\delta(P_1, P_2) = 0$ iff $P_1 = P_2$,

- $\delta(P_1, P_2) = \delta(P_2, P_1)$, and

- $\delta(P_1, P_2) + \delta(P_2, P_3) = |1 + M - Eval(P_1)| + |1 + M - Eval(P_2)| + |1 + M - Eval(P_2)| + |1 + M - Eval(P_3)| \geq |1 + M - Eval(P_1)| + |1 + M - Eval(P_3)| = \delta(P_1, P_3)$,

and, consequently, $\langle S, \delta \rangle$ is a metric space.

Moreover, the metric space $\langle S, \delta \rangle$ is complete. This is because for any Cauchy sequence $P_1, P_2, \ldots$ of populations there exist k such that for all $n > k$, $P_n = P_k$. It means that all Cauchy sequences P_i have a limit for $i \to \infty$.[4]

Now we are ready to discuss the contractive mapping, $f : S \longrightarrow S$, which is simply a single iteration of a run[5] of genetic algorithm (see Figure 4.3) provided, that there was an improvement (in terms of function $Eval$) from population $P(t)$ to population $P(t+1)$. In such a case, $f(P(t)) = P(t+1)$. In other words, a t-th iteration of a genetic algorithm would serve as a contractive mapping operator f iff $Eval(P(t)) < Eval(P(t+1))$. If there is no improvement, we do not count such iteration, i.e., we run selection and recombination process again.

The structure of such modified genetic algorithm (Contractive Mapping Genetic Algorithm—CM-GA) is given in Figure 4.2.

The modified iteration of the CM-GA indeed satisfies the requirement for contractive mapping. It is clear that if an iteration $f : P(t) \longrightarrow P(t+1)$ improves a population in terms of function $Eval$, i.e., if

$$Eval(P_1(t)) < Eval(f(P_1(t))) = Eval(P_1(t+1)), \text{ and}$$
$$Eval(P_2(t)) < Eval(f(P_2(t))) = Eval(P_2(t+1)),$$

---

[4]Note that in the case of genetic algorithms we essentially deal with finite metric spaces, as there is only a finite and bounded number of elements of all possible populations. Thus Banach demand as to the completeness of metric spaces is in that case always satisfied. In our case, however, $\langle S, \delta \rangle$ is complete for any set S.

[5]By a run we mean here any observable computation sequence.

```
procedure CM-GA
begin
    t = 0
    initialize P(t)
    evaluate P(t)
    while (not termination-condition) do
    begin contractive mapping f(P(t)) ⟶ P(t + 1)
        t = t + 1
        select P(t) from P(t − 1)
        recombine P(t)
        evaluate P(t)
        if Eval(P(t − 1)) ≥ Eval(P(t))
            then t = t − 1
    end
end
```

Fig. 4.2. Contractive mapping genetic algorithm

then

$$\delta(f(P_1(t)), f(P_2(t))) = |1 + M - Eval(f(P_1(t)))| + |1 + M - Eval(f(P_2(t)))| <$$
$$|1 + M - Eval(P_1(t))| + |1 + M - Eval(P_2(t))| = \delta(P_1(t), P_2(t)).$$

Moreover, as one always deals with a particular implementation of the algorithms the improvement is not less than the smallest real number, given by the implementation.

In summary, the CM-GA satisfies the assumptions of Banach fixpoint theorem: the space of populations $\langle S, \delta \rangle$ is a complete metric space and the iteration $f : P(t) \longrightarrow P(t+1)$ (which improves a population in terms of evaluation function $Eval$) is contractive. Consequently,

$$P^* = \lim_{i \to \infty} f^i(P(0)),$$

i.e., the CM-GA algorithm converges to population P^*, which is a unique fixpoint in the space of all populations.

Obviously, P^* represents the population which yields the global optimum. Note that the $Eval$ function was defined as

$$Eval(P) = \frac{1}{n} \sum_{\overline{x}_i \in P} eval(\overline{x}_i);$$

it means that the fixpoint P^* is achieved when all individuals in this population have the same (global maximum) value. Moreover, P^* does not depend on the initial population, $P(0)$.

An interesting problem appears when the evaluation function $eval$ has more than one maximum. In such a case contractive mapping genetic algorithm does not really define contractive mapping, as for optimal populations P_1, P_2,

$$\delta(f(P_1), f(P_2)) = \delta(P_1, P_2).$$

On the other hand, it can be shown in this case that CM-GA converges to one of possible optimal populations. That follows from the fact that each run of the algorithm converges to an optimal population.

At the first glance the result seems surprising: in case of contractive mapping genetic algorithms the choice of initial population may influence only the convergence speed. The proposed contractive mapping genetic algorithm (on the basis of the Banach theorem) would always converge to the global optimum (in infinite time). However, it is possible that (at some stage of the algorithm) no new population is accepted for a long time and the algorithm loops trying to find a new population $P(t)$. In other words, mutation and crossover operators applied to a particular sub-optimal population are unable to produce "better" population and the algorithm loops trying to perform the next *converging* step. The choice of the distance δ between populations and the evaluation function *Eval* were made just to make things as simple as possible. On the other hand, such choices can have an influence on the convergence speed and seem to be application dependent.

4.4 Genetic algorithms with varying population size

The size of the population is one of the most important choices faced by any user of genetic algorithms and may be critical in many applications. If the population size is too small, the genetic algorithm may converge too quickly; if it is too large, the genetic algorithm may waste computational resources: the waiting time for an improvement might be too long. As we discussed earlier (section 4.1), there are two important issues in the evolution process of the genetic search: population diversity and selective pressure. Clearly, both these factors are influenced by the size of population.

Several researchers have investigated the size of population for genetic algorithms from different perspectives. Grefenstette [169] applied a meta-GA to control parameters of another GA (including populations size and the selection method). Goldberg [151, 153] provides a theoretical analysis of the optimal population size. A study on influence of the control parameters on the genetic search (online performance for function optimization) is presented in [343]. Additional experiments with population size were reported in [206] and [59]. Recently Smith [362] proposed an algorithm which adjusts the population size with respect to the probability of selection error.

In this section we discuss a Genetic Algorithm with Varying Population Size (GAVaPS) [12]. This algorithm does not use any variation of selection mechanism considered earlier (section 4.1), but rather introduces the concept of "age" of a chromosome, which is equivalent to the number of generations the chromosome stays "alive". Thus the age of the chromosome replaces the concept of selection and, since it depends on the fitness of the individual, influences the size of the population at every stage of the process. It seems also that such approach is more "natural" than any selection mechanism considered earlier: after all, the aging process is well-known in all natural environments.

It seems that the proposed method of variable population size is similar to some evolution strategies (Chapter 8) and other methods where offspring compete with parents for survival. However, an important difference is that the in other methods the population size remain constant, whereas the size of the population in GAVaPS varies over time.

Additional motivation for this work was based on the following observation: a few researchers examined a possibility of introducing adaptive probabilities of genetic operators in genetic algorithms [77], [115], [343], [367], [368]; other techniques, like evolutionary strategies [349] already incorporated adaptive probabilities for its operators some time ago (we discuss these aspects briefly in section 4.6 and Chapter 8). It seems reasonable to assume that at different stages of the evolution process different operators would have different significance and the system should be allowed to self-tune their frequencies and scope. The same should be true for population sizes: at different stages of the evolution process different sizes of the population may be 'optimal', thus it is important to experiment with some heuristic rules to tune the size of the population to the current stage of the search.

The GAVaPS algorithm at time t processes a population $P(t)$ of chromosomes. During the 'recombine $P(t)$' step, a new auxiliary population is created (this is a population of offspring). The size of the auxiliary population is proportional to the size of the original population; the auxiliary population contains $AuxPopSize(t) = \lfloor PopSize(t) * \rho \rfloor$ chromosomes (we refer to parameter ρ as a *reproduction ratio*). Each chromosome from the population can be chosen to reproduce (i.e., to place the offspring in the auxiliary population) with equal probability, *independently* of its fitness value. Offspring are created by applying genetic operators (crossover and mutation) to selected chromosomes. Since the selection of the chromosomes does not depend on their fitness values, i.e., there is no selection step as such, we introduce the concept of *age* of the chromosome and its *lifetime* parameter.

The structure of the GAVaPS is shown in Figure 4.3.

The lifetime parameter is assigned once for each chromosome during the evaluation step (either after the initialization for all chromosomes or after the recombination step for members of auxiliary population) and remains constant (for a given chromosome) through the evolution process, i.e., from the birth of the chromosome to its death. It means that for the 'old' chromosomes their lifetime values *are not* re-calculated. The death of a chromosome occurs when its age, i.e., the number of generations the chromosome stays alive (initially set to zero), exceeds its lifetime value. In other words, a chromosome's lifetime determines the number of GAVaPS generations during which the chromosome is kept in the population: after its lifetime expires, the chromosome dies off. Thus the size of the population after single iteration is

$$PopSize(t+1) = PopSize(t) + AuxPopSize(t) - D(t),$$

procedure GAVaPS
begin
  $t = 0$
  initialize $P(t)$
  evaluate $P(t)$
  while (**not** termination-condition) **do**
  begin
    $t = t + 1$
    increase the *age* of each individual by 1
    recombine $P(t)$
    evaluate $P(t)$
    remove from $P(t)$ all individuals
      with *age* greater than their *lifetime*
  end
end

Fig. 4.3. The GAVaPS algorithm

where $D(t)$ is the number of chromosomes which die off during generation t.

There are many possible strategies of assigning lifetime values. Clearly, assigning a constant value (greater than one) independently of any statistics of the search would cause an exponential growth of the population size. Moreover, since there is no selection mechanism as such in the GAVaPS, no selective pressure exists, so assigning a constant value for the lifetime parameter would result in a poor performance of the algorithm. In order to introduce a selective pressure, a more sophisticated lifetime calculation should be performed. The lifetime calculation strategies should (1) reinforce the individuals with above-average fitness, (and consequently, restrict the individuals with below-average fitness), and (2) tune the size of the population to the current stage of the search (in particular, prevent the exponential growth of the population and lower simulation costs). Reinforcement of fit individuals should result in above-average allocation of their offspring in the auxiliary populations. Since there is an equal probability for each individual to undergo the genetic recombination, the expected number of the individual's offspring is proportional to its lifetime value (since the lifetime determines number of generations of keeping the individual in the population). So individuals having above-average fitness values should be granted higher lifetime values. While calculating the lifetime, a state of the genetic search should be taken under consideration. Because of that we use a few measures of the state of the search: $AvgFit$, $MaxFit$ and $MinFit$ represent average, maximal and minimal fitness values, respectively, in the current population, and $AbsFitMax$ and $AbsFitMin$ stand for maximal and minimal fitness values found so far. It should be also noted that the lifetime calculation should be computationally easy in order to spare the computational resources.

Having in mind the above remarks, several lifetime calculation strategies have been implemented and used for the experiments. The lifetime parameter for

the i-th individual ($lifetime[i]$) can be determined by (as in the pr
we assume maximization problems with non-negative evaluatior

(1) proportional allocation:

$$\min(MinLT + \eta\frac{fitness[i]}{AvgFit}, MaxLT)$$

(2) linear allocation:

$$MinLT + 2\eta\frac{fitness[i] - AbsFitMin}{AbsFitMax - AbsFitMin}$$

(3) bi-linear allocation:

$$\begin{cases} MinLT + \eta\frac{fitness[i]-MinFit}{AvgFit-MinFit} & if \ \ AvgFit \geq fitness[i] \\ \frac{1}{2}(MinLT + MaxLT) + \eta\frac{fitness[i]-AvgFit}{MaxFit-AvgFit} & if \ \ AvgFit < fitness[i] \end{cases}$$

where $MaxLT$ and $MinLT$ stand for maximal and minimal allowable lifetime values, respectively (these values are given as the GAVaPS parameters), and $\eta = \frac{1}{2}(MaxLT - MinLT)$.

The first strategy (proportional allocation) has come up from the idea of roulette-wheel selection: the value of lifetime for particular individual is proportional to its fitness (within limits $MinLT$ and $MaxLT$). However, this strategy has a serious drawback — it does not utilize any information about the "objective goodness" of the individual, which can be estimated by relating its fitness to the best value found so far. This observation motivates the linear strategy. In this strategy the lifetime value is calculated accordingly to the individual fitness related to the best value at present. However, if many individuals have their fitness equal or approximately equal to the best value, such strategy results in allocating long lifetime values, thus enlarging the size of the population. Finally, the bi-linear strategy attempts to make a compromise between the first two. It sharpens the difference between lifetime values of nearly-the-best individuals utilizing information about the average fitness value, however also taking into consideration the maximal and minimal fitness values found so far.

The GAVaPS algorithm was tested on the following functions:

G1: $-x\sin(10\pi x) + 1$ $-2.0 \leq x \leq 1.0$
G2: $integer(8x)/8$ $0.0 \leq x \leq 1.0$
G3: $x \cdot sgn(x)$ $-1.0 \leq x \leq 2.0$
G4: $0.5 + \frac{\sin^2\sqrt{x^2+y^2}-0.5}{(1+0.001(x^2+y^2))^2}$ $-100 \leq x,y \leq 100$

The functions were chosen to cover the wide spectrum of possible function types to be optimized. Functions G1 and G4 are multimodal functions with many local maxima. Function G2 cannot be optimized by means of any gradient technique, since there is no gradient information available. Function G3 represents a problem recognized as a "deceptive problem" [154]. While maximizing such function, two directions of growth can easily be recognized, but the boundaries are chosen in such way that only for one of them a global maximum can be obtained. In case of gradient-based techniques with random sampling this should result in frequent finding the local maximum.

GAVaPS performance has been tested and compared to the performance of the Goldberg's Simple Genetic Algorithm (SGA) [154]. Problem coding methods as well as genetic operators were identical for the SGA and GAVaPS (a simple binary coding has been used and two genetic operators: mutation and one-point crossover).

For the experiments we have made the following assumptions. The initial size of any population was 20. In case of the SGA, the size of initial population remained constant through the entire simulation. Reproduction ratio ρ was set to 0.4 (this parameter is meaningless in case of the SGA). Mutation ratio was set to 0.015, and crossover ratio was set to 0.65. The length of chromosomes was 20. Through all our experiments we assumed that minimal and maximal lifetime values were constant and equal to $MaxLT = 7$ and $MinLT = 1$.

To compare SGA with GAVaPS, two parameters have been chosen: cost of the algorithm, represented by *evalnum* (the average of the number of function evaluations over all runs) and performance, represented by *avgmax* (the average of the maximal values found over all runs). Both algorithms have the same termination condition: they terminate if there is no progress in terms of the best value found for consecutive *conv* = 20 generations. Population was initialized at random, and there were 20 independent runs performed. Then, measures of performance and cost were averaged over these 20 runs giving the reported results. While testing the influence of a single parameter on the performance and the cost, the values of the parameters reported above were constant except the one which influence was tested.

Figure 4.4 shows the $PopSize(t)$ and the average fitness of the population for a single GAVaPS run for the function G4 with the bi-linear lifetime calculation (similar observations can be made for other functions and other strategies for allocating lifetime values). The shape of the $PopSize(t)$ curve seems very interesting. At first, when the fitness variation is relatively high, the population size grows. This means, that the GAVaPS makes a wide search for the optima. Once the neighborhood of the optimum is located, the algorithm starts to converge and the population size is reduced. However, there is still a search for an improvement. When a possibility for a better result occurs, another "demographic explosion" takes place, which is followed by another convergence stage. It seems that the GAVaPS incorporates a *self-tuning* process by choosing the population size at each stage of the evolution process.

Figures 4.5–4.6 show the influence of the reproduction ratio on the performance of the GAVaPS. For the SGA, this value has no meaning (since in this case there is a total overlap of the old population by the new one). In case of the GAVaPS, this value strongly influences the simulation cost, which can be decreased by lowering the reproduction ratio, however without loss of accuracy (see relevant values of *avgmax*). Judging from the experiments, it seems that the 'optimal' selection of ρ is approximately 0.4.

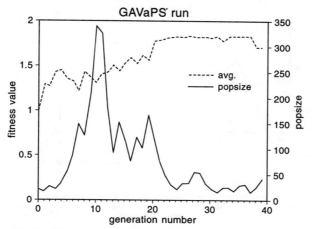

Fig. 4.4. *PopSize(t)* and average fitness of the population for a single GAVaPS run

Figures 4.7–4.8 show the influence of the initial population size on the performance (*avgmax*) and computation cost (*evalnum*) of the algorithms. In the case of the SGA, the population size (for all runs) was constant and equal to its initial value. As expected, for the SGA low values of population size implied low cost and poor performance. Increasing population size at first improves the performance but also increases cost of computations. Then there is a stage of "performance saturation", while cost is still linearly growing. In case of the GAVaPS, the initial population size has in practice no influence on both performance (very good) and cost (reasonable and sufficient for the very good performance).

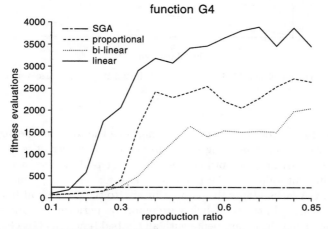

Fig. 4.5. Comparison of the SGA and GAVaPS: reproduction ratio versus number of evaluations

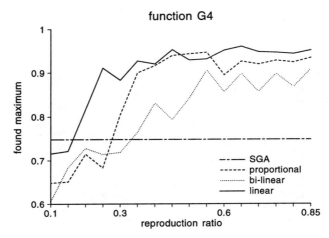

Fig. 4.6. Comparison of the SGA and GAVaPS: reproduction ratio versus average performance

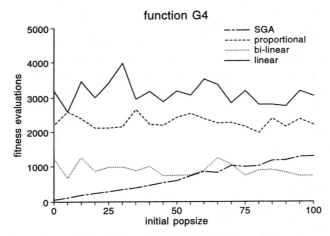

Fig. 4.7. Comparison of the SGA and GAVaPS: initial population size versus number of evaluations

Similar observations can be made by analysing the cost and the performance of both algorithms on remaining functions G1–G3. However, it is important to note that the SGA has the optimal behavior (best performance with minimum cost) for different values of the population size for all four problems. On the other hand, GAVaPS adopts the population size to the problem at hand and the state of the search. In the table below we report on performance and simulation cost obtained from the experiments with all testbed functions G1–G4. The rows 'SGA', 'GAVaPS (1)', 'GAVaPS (2)', and 'GAVaPS (3)' contain the best value found (V) and the number of function evaluations (E) for the SGA and GAVaPS with proportional, linear, and bi-linear lifetime allocations, respectively. The

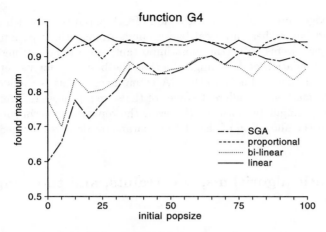

Fig. 4.8. Comparison of the SGA and GAVaPS: initial population size versus average performance

optimal population sizes for the SGA for test cases G1 – G4 were 75, 15, 75, 100, respectively.

Type	Function							
of the	G1		G2		G3		G4	
algorithm	*V*	*E*	*V*	*E*	*V*	*E*	*V*	*E*
SGA	2.814	1467	0.875	345	1.996	1420	0.959	2186
GAVaPS(1)	2.831	1708	0.875	970	1.999	1682	0.969	2133
GAVaPS(2)	2.841	3040	0.875	1450	1.999	2813	0.970	3739
GAVaPS(3)	2.813	1538	0.875	670	1.999	1555	0.972	2106

Table 4.1. Comparison between three strategies

The linear strategy (2) is characterized by the best performance and (unfortunately) the highest cost. On the other hand, the bi-linear strategy (3) is the cheapest one, but the performance is not as good as in the linear case. Finally, the proportional strategy (1) provides the medium performance with the medium cost. It should be noted, that the GAVaPS algorithm with any lifetime allocation strategy (1)–(3) in most test-cases provides better performance than the SGA. The cost of the GAVaPS (in comparison to the SGA) is higher, however, the results of the SGA were reported for the optimal sizes of populations. If, for example, the population size for the SGA in the experiment with function G2 was 75 (instead of the optimal 15), then the SGA simulation cost would be 1035.

The knowledge about the proper selection of GA parameters is still only fragmentary and has rather empirical background. Among these parameters, the population size seems to be the most important, since it has strong influence on the GA simulation cost. It might be that the best way for its setting is to let it to self-tune accordingly to the GA actual needs. This is the idea behind the GAVaPS method: at different stages of the search process different sizes of the population might be optimal. However, the reported experiments are preliminary and the allocation of the lifetime parameter deserves further research.

4.5 Genetic algorithms, constraints, and the knapsack problem

As discussed in the Introduction, the constraint-handling techniques for genetic algorithms can be grouped into a few categories. One way of dealing with candidates that violate the constraints is to generate potential solutions without considering the constraints and then to penalize them by decreasing the "goodness" of the evaluation function. In other words, a constrained problem is transformed to an unconstrained one by associating a penalty with all constraint violations; these penalties are included in the function evaluation. Of course, there are a variety of possible penalty functions which can be applied. Some penalty functions assign a constant as a penalty measure. Other penalty functions depend on the degree of violation: the larger violation is, the greater penalty is imposed (however, the growth of the function can be logarithmic, linear, quadratic, exponential, etc. with respect to the size of the violation).

Additional version of penalty approach is elimination of non-feasible solutions from the population (i.e., application of the most severe penalty: death penalty). This technique was used successfully in evolution strategies (Chapter 8) for numerical optimization problems. However, such approach has its drawbacks. For some problems the probability of generating (by means of standard genetic operators) a feasible solution is relatively small and the algorithm spends a significant amount of time evaluating illegal individuals. Moreover, in this approach non-feasible solutions do not contribute to the gene-pool of any population.

Another category of constraint handling methods is based on application of special repair algorithms to "correct" any infeasible solutions so generated. Again, such repair algorithms might be computationally intensive to run and the resulting algorithm must be tailored to the particular application. Moreover, for some problems the process of correcting a solution may be as difficult as solving the original problem.

The third approach concentrates on the use of special representation mappings (decoders) which guarantee (or at least increase the probability of) the generation of a feasible solution or the use of problem-specific operators which preserve feasibility of the solutions. However, decoders are frequently computationally intensive to run [73], not all constraints can be easily implemented this way, and the resulting algorithm must be tailored to the particular application.

In this section we examine the above techniques on one particular problem: the 0/1 knapsack problem. The problem is easy to formulate, yet, the decision version of it belongs to a family of NP-complete problems. It is an interesting exercise to evaluate the advantages and disadvantages of constraint handling techniques on this particular problem with a single constraint: the conclusions might be applicable to many constrained combinatorial optimization problems. It should be noted, however, that the main purpose of this section is to illustrate the concept of decoders, repair algorithms, and penalty functions (discussed briefly in the Introduction) on one particular example; by no means it is a complete survey of possible methods. For that reason we do not provide the optimum solutions for the test cases: we only make some comparisons between presented methods.

4.5.1 The 0/1 knapsack problem and the test data

There is a variety of knapsack-type problems in which a set of entities, together with their values and sizes, is given, and it is desired to select one or more disjoint subsets so that the total of the sizes in each subset does not exceed given bounds and the total of the selected values is maximized [252]. Many of the problems in this class are NP-hard and large instances of such problems can be approach only by using heuristic algorithms. The problem selected for these experiments is the 0/1 knapsack problem. The task is, for a given set of weights $W[i]$, profits $P[i]$, and capacity C, to find a binary vector $\boldsymbol{x} = \langle x[1], \ldots, x[n] \rangle$, such that

$$\sum_{i=1}^{n} x[i] \cdot W[i] \leq C,$$

and for which

$$\mathcal{P}(\boldsymbol{x}) = \sum_{i=1}^{n} x[i] \cdot P[i]$$

is maximum.

As indicated earlier, in this section we analyse the experimental behavior of a few GA-based algorithms on several sets of randomly generated test problems. Since the difficulty of such problems is greatly affected by the correlation between profits and weights [252], three randomly generated sets of data are considered:

- *uncorrelated*:

  $W[i] := $ (uniformly) random($[1..v]$), and
  $P[i] := $ (uniformly) random($[1..v]$).

- *weakly correlated*:

  $W[i] := $ (uniformly) random($[1..v]$), and
  $P[i] := W[i] + $ (uniformly) random($[-r..r]$),

(if, for some i, $P[i] \leq 0$, such profit value is ignored and the calculations are repeated until $P[i] > 0$).

- *strongly correlated*:

  $W[i] := $ (uniformly) random($[1..v]$), and
  $P[i] := W[i] + r$.

  Higher correlation implies smaller value of the difference:

  $$\max_{i=1..n}\{P[i]/W[i]\} - \min_{i=1..n}\{P[i]/W[i]\};$$

as reported in [252], higher correlation problems have higher expected difficulty.

Data have been generated with the following parameter settings: $v = 10$ and $r = 5$. For the tests we used three data sets of each type containing $n = 100$, 250, and 500 items, respectively. Again, following a suggestion from [252], we have taken under consideration two knapsack types:

- *restrictive knapsack capacity*

  A knapsack with the capacity of $C_1 = 2v$. In this case the optimal solution contains very few items. An area, for which conditions are not fulfilled, occupies almost the whole domain.

- *average knapsack capacity*

  A knapsack with the capacity $C_2 = 0.5 \sum_{i=1}^{n} W[i]$. In this case about half of the items are in the optimal solution.

As reported in [252], further increasing the value of capacity C does not significantly increase the computation times of the classical algorithms.

4.5.2 Description of the algorithms

Three types of algorithms were implemented and tested: algorithms based on penalty functions ($A_p[i]$, where i is the index of a particular algorithm in this class), algorithms based on repair methods ($A_r[i]$), and algorithms based on decoders ($A_d[i]$). We discuss these three categories of algorithms in turn.

Algorithms $A_p[i]$

In all algorithms in this category a binary string of the length n represents a solution $\boldsymbol{x}$ to the problem: the i-th item is selected for the knapsack iff $x[i] = 1$.

The fitness $eval(\boldsymbol{x})$ of each string is determined as:

$$eval(\boldsymbol{x}) = \sum_{i=1}^{n} x[i] \cdot P[i] - Pen(\boldsymbol{x}),$$

where penalty function $Pen(\boldsymbol{x})$ is zero for all feasible solutions $\boldsymbol{x}$, i.e., solutions such that $\sum_{i=1}^{n} x[i] \cdot W[i] \leq C$, and is greater than zero, otherwise.

There are many possible strategies for assigning the penalty value. Here, three cases only were considered, where the growth of the penalty function is logarithmic, linear, and quadratic with respect to the degree of violation, respectively:

- $A_p[1]$: $Pen(\boldsymbol{x}) = \log_2(1 + \rho \cdot (\sum_{i=1}^{n} x[i] \cdot W[i] - C))$,

- $A_p[2]$: $Pen(\boldsymbol{x}) = \rho \cdot (\sum_{i=1}^{n} x[i] \cdot W[i] - C)$,

- $A_p[3]$: $Pen(\boldsymbol{x}) = (\rho \cdot (\sum_{i=1}^{n} x[i] \cdot W[i] - C))^2$.

In all three cases, $\rho = \max_{i=1..n}\{P[i]/W[i]\}$.

Algorithms $A_r[i]$

As in the previous category of algorithms, a binary string of the length n represents a solution to the problem $\boldsymbol{x}$: the i-th item is selected for the knapsack iff $x[i] = 1$.

The fitness $eval(\boldsymbol{x})$ of each string is determined as:

$eval(\boldsymbol{x}) = \sum_{i=1}^{n} x'[i] \cdot P[i]$,

where vector $\boldsymbol{x}'$ is a repaired version of the original vector $\boldsymbol{x}$.

There are two interesting aspects here. First, we may consider different repair methods. Second, some percentage of repaired chromosomes may replace the original chromosomes in the population. Such replacement rate may vary from 0% to 100%; recently Orvosh and Davis [301] reported so-call 5% rule which states that if replacing original chromosomes with a 5% probability, the performance of the algorithm is better than if replacing with any other rate (in particular, it is better than with 'never replacing' or 'always replacing' strategies).

We have implemented and tested two different repair algorithms. Both algorithms are based on the same procedure, shown in Figure 4.9.

> **procedure repair $(\boldsymbol{x})$**
> **begin**
>   knapsack-overfilled := false
> $\boldsymbol{x}' := \boldsymbol{x}$
>   **if** $\sum_{i=1}^{n} x'[i] \cdot W[i] > C$
>   **then** knapsack-overfilled := true
>   **while** (knapsack-overfilled) **do**
>   **begin**
> $i :=$ **select** an item from the knapsack
>   remove the selected item from the knapsack:
>   i.e., $x'[i] := 0$
>   **if** $\sum_{i=1}^{n} x'[i] \cdot W[i] \le C$
>   **then** knapsack-overfilled := false
>   **end**
> **end**

Fig. 4.9. The repair procedure

The two repair algorithms considered here differ only in selection procedure **select**, which chooses an item for removal from the knapsack:

- $A_r[1]$ (random repair). The procedure **select** selects a random element from the knapsack.

- $A_r[2]$ (greedy repair). All items in the knapsack are sorted in the decreasing order of their profit to weight ratios. The procedure **select** chooses always the last item (from the list of available items) for deletion.

Algorithms $A_d[i]$

A possible decoder for the knapsack problem is based on integer representation.[6] Here we used the ordinal representation (see Chapter 10 for more details on this representation) of selected items. Each chromosome is a vector of n integers; the i-th component of the vector is an integer in the range from 1 to $n - i + 1$. The ordinal representation references a list L of items; a vector is decoded by selecting appropriate item from the current list. For example, for a list of items $L = (1, 2, 3, 4, 5, 6)$, the vector $\langle 4, 3, 4, 1, 1, 1 \rangle$ is decoded as the following sequence of items: 4, 3, 6 (since 6 is the 4-th element on the current list after removal of 4 and 3), 1, 2, and 5. Clearly, in this method a chromosome can be interpreted as a strategy of incorporating items into the solution. Additionally, one-point crossover applied to any two feasible parents would produce a feasible offspring. A mutation operator is defined in a similar way as for the binary representation: if the i-th gene undergoes mutation, it takes a value random value (uniform distribution) from the range $[1..n - i + 1]$. The decoding algorithm is presented in Figure 4.10.

The two algorithms based on decoding techniques considered here differ only in the procedure **build**:

- $A_d[1]$ (random decoding). In this algorithm the procedure **build** creates a list L of items such that the order of items on the list corresponds to the order of items in the input file (which is random).

- $A_d[2]$ (greedy decoding). The procedure **build** creates a list L of items in the decreasing order of their profit to weight ratios. The decoding of the vector x is done on the basis of the sorted sequence (there are some similarities with the $A_r[2]$ method). For example, $x[i] = 23$ is interpreted as the 23-rd item (in the decreasing order of the profit to weight ratios) on the current list L.

4.5.3 Experiments and results

In all experiments the population size was constant and equal to 100. Also, probabilities of mutation and crossover were fixed: 0.05 and 0.65, respectively.

---

[6]There are, of course, other possibilities. We can stay, for example, with binary representation, and interpret a string from the left to the right (i.e., from $i = 1$ to n) in the following fashion: take the i-th item if (1) $i = 1$, and (2) there is a room in the knapsack for this item. Such interpretation always results in a feasible solution. Moreover, if items are sorted by the profit to weight ratio, a solution of all 1's correspond to the solution by the greedy algorithm. An example of such decoder is given in section 15.3, part I.

```
procedure decode (x)
begin
    build a list L of items
    i := 1
    WeightSum := 0
    ProfitSum := 0
    while i ≤ n do
    begin
        j := x[i]
        remove the j-th item from the list L
        if WeightSum + W[j] ≤ C then
        begin
            WeightSum := WeightSum + Weight[j]
            ProfitSum := ProfitSum + Profit[j]
        end
        i := i + 1
    end
end
```

Fig. 4.10. The decoding procedure for the ordinal representation

As a performance measure of the algorithm we have collected the best solution found within 500 generations. It has been empirically verified that after such number of generations no improvement has been observed. The results reported in the Table 4.2 are mean values of the 25 experiments. The exact solutions are not listed there; the table compares only the relative effectiveness of different algorithms. Note, that the data files were unsorted (arbitrary sequences of items, not related to their $P[i]/W[i]$ ratios). The capacity types C_1 and C_2 stand for restrictive and average capacities (section 2), respectively.

Results for the methods $A_r[1]$ and $A_r[2]$ have been obtained using the 5% repair rule. We have also examined whether the 5% rule works for the 0/1 knapsack problem (the rule was discovered during experiments on two other combinatorial problems: network design problem and graph coloring problem [301]). For the purpose of comparison we have chosen the test data sets with the weak correlation between weights and profits. All parameters settings were fixed and the value of the repair ratio varied from 0% to 100%. We have observed no influence of the 5% rule on performance of the genetic algorithm. The results (algorithm $A_r[2]$) have been collected in the Table 4.3.

The main conclusions drawn from the experiments can be summarized as follows:

- Penalty functions $A_p[i]$ (for all i) do not produce feasible results on problems with restrictive knapsack capacity (C_1). This is the case for any number of items ($n = 100$, 250, and 500) and any correlation.

Correl.	No. of items	Cap. type	method						
			$A_p[1]$	$A_p[2]$	$A_p[3]$	$A_r[1]$	$A_r[2]$	$A_d[1]$	$A_d[2]$
none	100	C_1	*	*	*	62.9	94.0	63.5	59.4
		C_2	398.1	341.3	342.6	344.6	371.3	354.7	353.3
	250	C_1	*	*	*	62.6	135.1	58.0	60.4
		C_2	919.6	837.3	825.5	842.2	894.4	867.4	857.5
	500	C_1	*	*	*	63.9	156.2	61.0	61.4
		C_2	1712.2	1570.8	1565.1	1577.4	1663.2	1602.8	1597.0
weak	100	C_1	*	*	*	39.7	51.0	38.2	38.4
		C_2	408.5	327.0	328.3	330.1	358.2	333.6	332.3
	250	C_1	*	*	*	43.7	74.0	42.7	44.7
		C_2	920.8	791.3	788.5	798.4	852.1	804.4	799.0
	500	C_1	*	*	*	44.5	93.8	43.2	44.5
		C_2	1729.0	1531.8	1532.0	1538.6	1624.8	1548.4	1547.1
strong	100	C_1	*	*	*	61.6	90.0	59.5	59.5
		C_2	741.7	564.5	564.4	566.5	577.0	576.2	576.2
	250	C_1	*	*	*	65.5	117.0	65.5	64.0
		C_2	1631.9	1339.5	1343.4	1345.8	1364.4	1366.4	1359.0
	500	C_1	*	*	*	67.5	120.0	67.1	64.1
		C_2	3051.6	2703.8	2700.8	2709.5	2748.1	2738.0	2744.0

Table 4.2. Results of experiments; symbol '*' means, that no valid solution has been found within given time constraints

- Judging only from the results of the experiments on problems with average knapsack capacity (C_2), the algorithm $A_p[1]$ based on logarithmic penalty function is a clear winner: it outperforms all other techniques on all cases (uncorrelated, weakly and strongly correlated, with $n = 100$, 250, and 500 items). However, as mentioned earlier, it fails on problems with restrictive capacity.

- Judging only from the results of the experiments on problems with restrictive knapsack capacity (C_1), the repair method $A_r[2]$ (greedy repair) outperforms all other methods on all test cases.

These results are quite intuitive. In the case of restrictive knapsack capacity, only very small fraction of possible subsets of items constitute feasible solutions; consequently, most penalty methods would fail. This is generally the case for many other combinatorial problems: the smaller the ratio between feasible part of the search space and the whole search space, the harder it is for the penalty function methods to provide feasible results. This was already observed in [332], where the authors wrote:

"On sparse problems, [harsh penalty functions] seldom found solutions. The solutions which it did occasionally manage to find were poor. The reason for this is clear: with a sparse feasible region, an initial population is very unlikely to contain any solutions. Since [the harsh penalty function] made no distinction between infeasible

correlation	weak					
no. of items	100		250		500	
capacity type	C_1	C_2	C_1	C_2	C_1	C_2
repair ratio						
0	94.0	371.3	134.1	895.0	158.0	1649.1
5	94.0	371.7	135.1	891.4	155.8	1648.5
10	94.0	370.2	135.1	889.5	157.3	1640.3
15	94.0	368.3	135.3	895.4	156.6	1646.2
20	94.0	372.0	135.6	905.7	155.8	1643.6
25	94.0	370.2	135.1	894.0	155.8	1644.0
30	94.0	367.2	135.1	895.7	157.3	1648.0
35	94.0	370.3	136.1	896.5	156.3	1643.3
40	94.0	368.6	134.3	886.5	156.1	1648.4
45	94.0	369.0	135.3	891.5	156.6	1649.0
50	94.0	371.7	134.6	891.0	156.1	1641.5
55	94.0	371.3	135.0	895.7	157.0	1647.0
60	94.0	369.6	135.0	894.0	156.1	1645.0
65	94.0	370.0	135.1	893.2	156.6	1642.8
70	94.0	367.6	135.0	893.4	156.1	1640.9
75	94.0	367.7	135.3	895.7	157.1	1648.1
80	94.0	368.2	134.3	898.5	155.8	1648.5
85	94.0	364.7	135.6	897.4	156.1	1646.2
90	94.0	368.7	134.3	885.2	156.1	1648.9
95	94.0	371.2	135.0	890.5	155.3	1642.3
100	94.0	370.2	134.6	901.0	156.3	1646.1

Table 4.3. Influence of the repair ratio on the performance of the algorithm $A_r[2]$

solutions, the GA wandered around aimlessly. If through a lucky mutation or strange crossover, an offspring happened to land in the feasible region, this child would become a *super-individual*, whose genetic material would quickly dominate the population and premature convergence would ensue".

On the other hand, the repair algorithms perform quite well. In the case of average knapsack capacity, logarithmic penalty function method (i.e., small penalties) is superior: it is interesting to note that the size of the problem does not influence the conclusions.

As indicated earlier, these experiments do not provide a complete picture yet; many additional experiments are planned in the near future. In the category of penalty functions, it might be interesting to experiment with additional formulae. All considered penalties were of the form $Pen(\boldsymbol{x}) = f(\boldsymbol{x})$, where f was logarithmic, linear, and quadratic. Another possibility would be to experiment with $Pen(\boldsymbol{x}) = a + f(\boldsymbol{x})$ for some constant a. This would provide a minimum penalty for any infeasible vector. Further, it might be interesting to experiment with *dynamic* penalty functions, where their values depend on additional parameters, like the generation number (this was done in [267] for numerical optimization for continuous variables) or characteristic of the search (see [360], where dynamic penalty functions were used for the facility layout problem: as better feasible and infeasible solutions are found, the penalty imposed on a given infeasible solution will change). Also, it seems worthwhile to experiment with *self-adaptive* penalty functions. After all, probabilities of applied operators might be adaptive (as in evolution strategies); some initial experiments indicate that adaptive population sizes may have some merit (section 4.4); so the idea of adaptive penalty functions deserves some attention. In its simplest version, a penalty coefficient would be part of the solution vector and undergo all genetic (random) changes (as opposed the idea of dynamic penalty functions, where such penalty coefficient is changed on regular basis as a function of, for example, generation number).

It is also possible to experiment with many repair schemes, including other heuristics than the ratio of the profit and the weight. Also, it might be interesting to combine the penalty methods with repair algorithms: infeasible solutions for algorithms $A_p[i]$ could have been repaired into feasible ones.

In the category of decoders it would be necessary to experiment with different (integer) representations (as it was done for the traveling salesman problem — Chapter 10): adjacency representation (with alternating-edges crossover, subtour-chunks crossover, or heuristic crossover), or path representation (with the PMX, OX, and CX crossovers, or even the edge recombination crossover). It would be interesting to compare the usefulness of these representations and operators for the 0/1 knapsack problem (as it was done for the traveling salesman problem and scheduling [370]). It is quite possible, that some new problem-specific crossover would provide with the best results.

4.6 Other ideas

In the previous sections of this chapter we have discussed some issues connected with removing possible errors in sampling mechanisms. The basic aim of that research was to enhance genetic search; in particular, to fight the premature convergence of GAs. During the last few years there have been some other research efforts in the quest for better schema processing, but using different approaches. In this section we discuss some of them.

The first direction is related to the genetic operator, crossover. This operator was inspired by a biological process; however, it has some drawbacks. For example, assume there are two high performance schemata:

$$S_1 = (0\ 0\ 1 \star \star \star \star \star \star \star \star 0\ 1) \text{ and}$$
$$S_2 = (\star \star \star \star 1\ 1 \star \star \star \star \star \star \star).$$

There are also two strings in a population, v_1 and v_2, matched by S_1 and S_2, respectively:

$$v_1 = (\textbf{001}0001101\textbf{001}) \text{ and}$$
$$v_2 = (1110\textbf{11}0001000).$$

Clearly, the crossover cannot combine certain combinations of features encoded on chromosomes; it is impossible to get a string to be matched by a schema

$$S_3 = (0\ 0\ 1 \star 1\ 1 \star \star \star \star \star 0\ 1),$$

since the first schema will be destroyed.

Additional argument against the classical, one-point crossover is in some asymmetry between mutation and crossover: a mutation depends on the length of the chromosome and crossover does not. For example, if the probability of mutation is $p_m = 0.01$, and the length of a chromosome is 100, the expected number of mutated bits within the chromosome is one. If the length of a chromosome is 1000, the expected number of mutated bits within the chromosome is ten. On the other hand, in both cases, one-point crossover combines two strings on the basis of one cross site, regardless the length of the strings.

Some researchers (see, e.g., [104] or [382]) have experimented with other crossovers. For example, two-point crossover selects two crossing sites and chromosomal material is swapped between them. Clearly, strings v_1 and v_2 may produce a pair of offspring

$$v'_1 = (\mathbf{001}|01100|01001) \text{ and}$$
$$v'_2 = (111|00011|01000),$$

where v'_1 is matched by

$$S_3 = (0\ 0\ 1 \star 1\ 1 \star \star \star \star \star 0\ 1),$$

which was not possible with one-point crossover.

Similarly, there are schemata that two-point crossover cannot combine. We can experiment then with multi-point crossover [104], a natural extension of two-point crossover. Note, however, that since multi-point crossover must alternate segments (obtained after cutting a chromosome into s pieces) between the two parents, the number of segments must be even, i.e., a multi-point crossover is not a natural extension of single-point crossover.

Schaffer and Morishima [341] experimented with a crossover which adapts the distribution of its crossover points by the same processes already in place (survival of the fittest and recombination). This was done by introducing special marks into string representation. These marks keep track of the sites in the string where crossover occurred. The hope was that the crossover sites would undergo an evolution process: if a particular site produces poor offspring, the site dies off (and vice versa). Experiments indicated [341] that adaptive crossover performed as well or better than a classical GA for a set of test problems. Spears [367] experimented with adapting a particular crossover (two crossovers: one-point and uniform crossovers were considered) by extending a chromosomal representation by additional bit.

Some researchers [104] experimented with other crossovers: segmented crossover and shuffle crossover. Segmented crossover is a variant of the multi-point crossover, which allows the number of crossover points to vary. The (fixed) number of crossover points (or segments) is replaced by a segment switch rate. This rate specifies the probability that a segment will end at any point in the string. For example, if the segment switch rate is $s = 0.2$, then, starting from the beginning of a segment, the chances of terminating this segment at each bit are 0.2. In other words, with the segment switch rate $s = 0.2$, the expected number of segments will be $m/5$, however, unlike multi-point crossover, the number of crossover points will vary. Shuffle crossover should be perceived as an additional mechanism which can be applied with other crossovers. It is independent of the number of crossover points. The shuffle crossover (1) randomly shuffles the bit positions of the two strings in tandem, (2) crosses the strings, i.e., exchanges segments between crossover points, and (3) unshuffles the strings.

A further generalization of one-point, two-point, and multi-point crossover is the uniform crossover [366], [382]: for each bit in the first offspring it decides (with some probability p) which parent will contribute its value in that position. The second offspring would receive the bit from the other parent.

For example, for $p = 0.5$ (0.5-uniform crossover), the strings

$$v_1 = (0010001101001) \text{ and}$$
$$v_2 = (1110110001000)$$

may produce the following pair of offspring:

$$v_1' = (0_1 0_1 1_2 0_1 1_2 1_2 0_2 1_1 0_1 1_2 0_1 0_1 0_2) \text{ and}$$
$$v_2' = (1_2 1_2 1_1 0_2 0_1 0_1 1_1 0_2 0_2 1_1 0_2 0_2 1_1),$$

where subscripts 1 and 2 indicate the parent (vectors v_1 and v_2, respectively) for a given bit. If $p = 0.1$ (0.1-uniform crossover), two strings v_1 and v_2 may produce

$$v_1' = (0_1 0_1 1_1 0_1 1_2 0_1 1_1 1_1 0_1 1_2 0_1 0_1 1_1) \text{ and}$$
$$v_2' = (1_2 1_2 1_2 0_2 0_1 1_2 0_2 0_2 0_2 1_1 0_2 0_2 0_2).$$

Since the uniform crossover exchanges bits rather than segments, it can combine features regardless their relative location. For some problems [382] this ability outweighs the disadvantage of destroying building blocks. However, for other problems, the uniform crossover was inferior to two-point crossover. Syswerda [382] compared theoretically the 0.5-uniform crossover with one-point and two-point crossovers. Spears and De Jong [366] provided an analysis of p-uniform crossovers, i.e., crossovers which involve on average $m \cdot p$ crossover points.

Eshelman [104] reports on several experiments for various crossover operators. The results indicate that the 'loser' is one-point crossover; however, there is no clear winner. A general comment on the above experiments is that each of these crossovers is particularly useful for some classes of problems and quite poor for other problems. This strengthens our idea of problem dependent operators leading us towards evolution programs.

Mühlenbein and Voigt [290] investigated the properties of a new recombination operator, called *gene pool recombination*, where the genes are randomly picked from the gene pool defined by the selected parents. An interesting aspect of this operator is that it allows so-called *orgies*: several parents in producing an offspring. Such a multi-parent crossover was also investigated by Eiben et al. [100], where several *gene scanning techniques* (which produce a single offspring out of several parents) are considered. Renders and Bersini [324] experimented with *simplex crossover* for numerical optimization problems; this crossover involves computing centroid of group of parents and moving from the worst individual beyond the centroid point. Also, scatter search techniques [142] propose the use of multiple parents.

Several researchers looked also at the effect of control parameters of a GA (population size, probabilities of operators) on the performance of the system. Grefenstette [169] applied a meta-GA to control parameters of another GA. Goldberg [153] provides a theoretical analysis of the optimal population size. A complete study on influence of the control parameters on the genetic search (online performance for function optimization) is presented in [343]. The results suggest that (1) mutation plays a stronger role than has previously been admitted (mutation is regarded as a background operator), (2) the importance of the crossover rate is smaller than expected, (3) the search strategy based on selection and mutation only might be a powerful search procedure even without the assistance of crossover (like evolution strategies, presented in Chapter 8). However, GAs still lack good heuristics to determine good values for their parameters: there is no single setting which would be universal for all considered problems. It seems that finding good values for the GA parameters is still more an art than a science.

Until this chapter we have discussed two basic genetic operators: crossover (one-point, two-point, uniform, etc.) and mutation, which were applied to individuals or single bits at some fixed rates (probabilities of crossover p_c and mutation p_m). As we have seen in the running example of Chapter 2, it was possible to apply crossover and mutation to the same individual (e.g., v_{13}). In fact, these two basic operators can be viewed as one recombination operator, the "crossover and mutation" operator, since both operations can be applied to individuals at the same time. One possibility in experimenting with genetic operators is to make them independent: one or the other of these operators will be applied during the reproduction event, but not both [78]. There are a few advantages of such separation. Firstly, a mutation will not be applied to the result of the crossover operator any longer, making the whole process conceptually simpler. Secondly, it is easier to extend a list of genetic operators by adding new ones: such a list may consist of several, problem dependent operators. This is precisely the idea behind evolution programming: there are many problem dependent "genetic operators", which are applied to individual data structures. Recall also, that for evolution programs presented in this book, we have developed a special selection routine modGA (previous section) which facilitates the above idea. Moreover, we can go further. Each operator may have its own fitness

which would undergo some evolution process as well. Operators are selected and
applied randomly, however, according to their fitness. This idea is not new and
has been around for some time [77], [115], [78], but it is getting a new meaning
and significance in the evolution programming technique.

Another interesting direction in search for a better schema processing, mentioned already in the previous chapter in connection with a deception problem,
was proposed recently [155], [159]: a messy genetic algorithm (mGA).

The mGAs differ from classical GAs in many ways: representation, operators, size of population, selection, and phases of the evolution process. We
discuss them briefly in turn.

First of all, each bit of a chromosome is tagged with its name (number) —
the same trick we used discussing inversion operator in the previous chapter.
Additionally, strings are of variable length and there is no requirement for a
string to have full gene complements. A string may have redundant and even
contradictory genes. For example, the following strings are legitimate chromosomes in mGA:

$$v_1 = ((7,1)(1,0)),$$
$$v_2 = ((3,1)(9,0)(3,1)(3,1),(3,1)),$$
$$v_3 = ((2,1)(2,0)(4,1)(5,0)(6,0)(7,1)(8,1)).$$

The first number in each parenthesis indicate a position, the second number the
value of the bit. Thus the first string, v_1, specifies two bits: bit 1 on the 7th
position and bit 0 on the 1st position.

To evaluate such strings, we have to deal with overspecification (string v_3,
where two bits are specified on the 2nd position) and underspecification (all
three vectors are underspecified, assuming 9 bit positions) problems. Overspecification can be handled in many ways; for example, some voting procedure
(deterministic or probabilistic) can be used, or a positional precedence. Underspecification is harder to deal with, and we refer the interested reader to the
source information [155], [158], [159].

Clearly, variable-length, overspecified or underspecified strings would influence the operators used. Simple crossover is replaced by two (even simpler)
operators: *splice* and *cut*. The splice operator concatenates two selected strings
(with the specified splice probability). For example, splicing strings v_1 with v_2
we get

$$v_4 = ((7,1)(1,0)(3,1)(9,0)(3,1)(3,1),(3,1)).$$

The cut operator cuts (with some cut probability) the selected string at a position determined randomly along its length. For example, cutting string v_3 at
position 4, we get

$$v_5 = ((2,1)(2,0)(4,1)(5,0)) \text{ and}$$
$$v_6 = ((6,0)(7,1)(8,1)).$$

In addition, there is an unchanged mutation operator, which changes 0 to 1 (or
vice versa) with some specified probability.

There are some other differences between GA and mGA. Messy genetic algorithms (for reliable selection regardless of function scaling) use a form of tournament selection [155]; they also divide the evolution process into two phases (the first phase selects building blocks, and only in the second phase are genetic operators invoked), and the population size changes in the process.

Messy genetic algorithms were tested on several deceptive functions with very good results [155], [158]. As stated by Goldberg [155]:

> "A difficult test function has been designed, and in two sets of experiments the mGA is able to find its global optimum. [...] In all runs on both sets of experiments, the mGA converges to the test function global optimum. By contrast, a simple GA using a random ordering of the string is able to get only 25% of the subfunctions correct."

and in [158]:

> "Because mGAs can converge in these worst-case problems, it is believed that they will find global optima in all other problems with bounded deception. Moreover, mGAs are structured to converge in computational time that grows only as a polynomial function of the number of decision variables on a serial machine and as a logarithmic function of the number of decision variables on a parallel machine. Finally, mGAs are a practical tool that can be used to climb a function's ladder of deception, providing useful and relatively inexpensive intermediate results along the way."

There were also some other attempts to enhance genetic search. A modification of GAs, called Delta Coding, was proposed recently by Whitley et al. [400]. Schraudoph and Belew [347] proposed a Dynamic Parameter Encoding (DPE) strategy, where the precision of the encoded individual is dynamically adjusted. These algorithms are discussed later in the book (Chapter 8).

Part II
Numerical Optimization

.

5. Binary or Float?

> There were rules in the
> monastery, but the Master
> always warned against
> the tyranny of the law.
>
> Anthony de Mello, *One Minute Wisdom*

As discussed in the previous chapter, there are some problems that GA applications encounter that sometimes delay, if not prohibit, finding the optimal solutions with the desired precision. One of the implications of these problems was premature convergence of the entire population to a non–global optimum (Chapter 4); other consequences include inability to perform fine local tuning and inability to operate in the presence of nontrivial constraints (Chapters 6 and 7).

The binary representation traditionally used in genetic algorithms has some drawbacks when applied to multidimensional, high-precision numerical problems. For example, for 100 variables with domains in the range $[-500, 500]$ where a precision of six digits after the decimal point is required, the length of the binary solution vector is 3000. This, in turn, generates a search space of about 10^{1000}. For such problems genetic algorithms perform poorly.

The binary alphabet offers the maximum number of schemata per bit of information of any coding [154] and consequently the bit string representation of solutions has dominated genetic algorithm research. This coding also facilitates theoretical analysis and allows elegant genetic operators. But the 'implicit parallelism' result does not depend on using bit strings [9] and it may be worthwhile to experiment with large alphabets and (possibly) new genetic operators. In particular, for parameter optimization problems with variables over continuous domains, we may experiment with real-coded genes together with special "genetic" operators developed for them.

In [157] Goldberg wrote:

> "The use of real-coded or floating-point genes has a long, if controversial, history in artificial genetic and evolutionary search schemes, and their use as of late seems to be on the rise. This rising usage has been somewhat surprising to researchers familiar with fundamental

genetic algorithm (GA) theory ([154], [188]), because simple analyses seem to suggest that enhanced schema processing is obtained by using alphabets of low cardinality, a seemingly direct contradiction of empirical findings that real codings have worked well in a number of practical problems."

In this chapter we describe the results of experiments with various modifications of genetic operators on floating point representation. The main objective behind such implementations is (in line with the principle of evolution programming) to move the genetic algorithm closer to the problem space. Such a move forces, but also allows, the operators to be more problem specific — by utilizing some specific characteristics of real space. For example, this representation has the property that two points close to each other in the representation space must also be close in the problem space, and vice versa. This is not generally true in the binary approach, where the distance in a representation is normally defined by the number of different bit positions. However, it is possible to reduce such discrepancy by using Gray coding.

The procedures for converting a binary number $b = \langle b_1, \ldots, b_m \rangle$ into Gray code number $g = \langle g_1, \ldots, g_m \rangle$ and vice versa are given in Figure 5.1; the parameter m denotes the number of bits in these representations.

procedure Binary-to-Gray
begin
$g_1 = b_1$
for $k = 2$ **to** m **do**
$g_k = b_{k-1} \, XOR \, b_k$
end

procedure Gray-to-Binary
begin
$value = g_1$
$b_1 = value$
for $k = 2$ **to** m **do**
begin
if $g_k = 1$ **then** $value = NOTvalue$
$b_k = value$
end
end

Fig. 5.1. The Binary-to-Gray and Gray-to-Binary procedures

Table 5.1 lists the first 16 binary numbers together with the corresponding Gray codes.

Note that the Gray coding representation has the property that any two points next to each other in the problem space differ by one bit only. In other words, an increase of one step in the parameter value corresponds to a change

Binary	Gray
0000	0000
0001	0001
0010	0011
0011	0010
0100	0110
0101	0111
0110	0101
0111	0100
1000	1100
1001	1101
1010	1111
1011	1110
1100	1010
1101	1011
1110	1001
1111	1000

Table 5.1. Binary and Gray codes

of a single bit in the code. Note also that there are other equivalent procedures for converting binary and Gray codes. For example (case of $m = 4$), a pair of matrices:

$$A = \begin{bmatrix} 1 & 0 & 0 & 0 \\ 1 & 1 & 0 & 0 \\ 0 & 1 & 1 & 0 \\ 0 & 0 & 1 & 1 \end{bmatrix} \qquad A^{-1} = \begin{bmatrix} 1 & 0 & 0 & 0 \\ 1 & 1 & 0 & 0 \\ 1 & 1 & 1 & 0 \\ 1 & 1 & 1 & 1 \end{bmatrix}$$

provides the following transformations:

$$g = Ab \text{ and } b = A^{-1}g,$$

where multiplication operations are done modulo 2.

However, we use a floating point representation as it is conceptually closest to the problem space and also allows for an easy and efficient implementation of closed and dynamic operators (see also Chapters 6 and 7). Subsequently, we empirically compared a binary and floating point implementations using various new operators on many test cases. In this chapter, we illustrate the differences between binary and float representations on one typical test case of a dynamic control problem. This is a linear-quadratic problem, which is a particular case of a problem we use in the next chapter (together with two other dynamic control problems) to illustrate the progress of our evolution program in connection with premature convergence and local fine tuning. As expected, the results are better than those from binary representation. The same conclusion was also reached by other researchers, e.g., [78], [408].

5.1 The test case

For experiments we have selected the following dynamic control problem:

$$\min\left(x_N^2 + \sum_{k=0}^{N-1}(x_k^2 + u_k^2)\right),$$

subject to

$$x_{k+1} = x_k + u_k, \ \ k = 0, 1, \dots, N-1,$$

where x_0 is a given initial state, $x_k \in R$ is a state, and $\boldsymbol{u} \in R^N$ is the sought control vector. The optimal value can be analytically expressed as

$$J^* = K_0 x_0^2,$$

where K_k is the solution of the Riccati equation:

$$K_k = 1 + K_{k+1}/(1 + K_{k+1}) \text{ and } K_N = 1.$$

During the experiments a chromosome represented a vector of the control states $\boldsymbol{u}$. We have also assumed a fixed domain $\langle -200, 200 \rangle$ for each u_i (actual solutions fall within this range for the class of tests performed). For all subsequent experiments we used $x_0 = 100$ and $N = 45$, i.e., a chromosome $\boldsymbol{u} = \langle u_0, \dots, u_{44} \rangle$, having the optimal value $J^* = 16180.4$.

5.2 The two implementations

For the study we have selected two genetic algorithm implementations differing only by representation and applicable genetic operators, and equivalent otherwise. Such an approach gave us a better basis for a more direct comparison. Both implementations used the same selective mechanism: stochastic universal sampling [23].

5.2.1 The binary implementation

In the binary implementation each element of a chromosome vector was coded using the same number of bits. To facilitate fast run-time decoding, each element occupied its own word (in general it occupied more than one if the number of bits per element exceeded the word size, but this case is an easy extension) of memory: this way elements could be accessed as integers, which removed the need for binary to decimal decoding. Then, each chromosome was a vector of N words, which equals the number of elements per chromosome (or a multiple of such for cases where multiple words were required to represent a desired number of bits).

The precision of such an approach depends (for a fixed domain size) on the number of bits actually used and equals $(UB - LB)/(2^n - 1)$, where UB and LB are domain bounds and n is the number of bits per one element of a chromosome.

5.2.2 The floating point implementation

In the floating point (FP) implementation each chromosome vector was coded as a vector of floating point numbers, of the same length as the solution vector. Each element was forced to be within the desired range, and the operators were carefully designed to preserve this requirement.

The precision of such an approach depends on the underlying machine, but is generally much better than that of the binary representation. Of course, we can always extend the precision of the binary representation by introducing more bits, but this considerably slows down the algorithm (see section 5.4).

In addition, the FP representation is capable of representing quite large domains (or cases of unknown domains). On the other hand, the binary representation must sacrifice precision with an increase in domain size, given fixed binary length. Also, in the FP representation it is much easier to design special tools for handling nontrivial constraints: this is discussed fully in Chapter 7.

5.3 The experiments

The experiments were conducted on a DEC3100 workstation. All results presented here represent the average values obtained from 10 independent runs. During all experiments the population size was kept fixed at 60, and the number of iterations was set at 20,000. Unless otherwise stated, the binary representation used $n = 30$ bits to code one variable (one element of the solution vector), making $30 \cdot 45 = 1350$ bits for the whole vector.

Because of possible differences in interpretation of the mutation operator, we accepted the probability of chromosomes' update as a fair measure for comparing the floating point and binary representations. All experimental values were obtained from runs with the operators set to achieve the same such rate; therefore, some number of iterations can be approximately treated interchangeably with the same number of function evaluations.

5.3.1 Random mutation and crossover

In this part of the experiment we ran both implementations with operators which are equivalent (at least for the binary representation) to the traditional ones.

Binary

The binary implementation used traditional operators of mutation and crossover. However, to make them more similar to those of the FP implementation, we allowed crossover only between elements. The probability of crossover was fixed at 0.25, while the probability of mutation varied to achieve the desired rate of chromosome update (shown in Table 5.2).

implementation	Probability of chromosome's update				
	0.6	0.7	0.8	0.9	0.95
Binary, p_m	0.00047	0.00068	0.00098	0.0015	0.0021
FP, p_m	0.014	0.02	0.03	0.045	0.061

Table 5.2. Probabilities of chromosome's update versus mutation rates

FP

The crossover operator was quite analogous to that of the binary implementation (split points between float numbers) and applied with the same probability (0.25). The mutation, which we call random, applies to a floating point number rather that to a bit; the result of such mutation is a random value from the domain $\langle LB, UB \rangle$.

Results

implementation	Probability of chromosome's update					standard deviation
	0.6	0.7	0.8	0.9	0.95	
Binary	42179	46102	29290	52769	30573	31212
FP	46594	41806	47454	69624	82371	11275

Table 5.3. Average results as a function of probability of chromosome's update

The results (Table 5.3) are slightly better for the binary case; however, it is rather difficult to judge them better as all fell well away from the optimal solution (16180.4). Moreover, an interesting pattern emerged that showed the FP implementation to be more stable, with much lower standard deviation.

In addition, it is interesting to note that the above experiment was not quite fair for the FP representation; its random mutation behaves "more" randomly than that of the binary implementation, where changing a random bit doesn't imply producing a totally random value from the domain. As an illustration let us consider the following question: what is the probability that after mutation an element will fall within $\delta\%$ of the domain range (400, since the domain is $\langle -200, 200 \rangle$) from its old value? The answer is:

FP: Such probability clearly falls in the range $\langle \delta, 2 \cdot \delta \rangle$. For example, for $\delta = 0.05$ it is in $\langle 0.05, 0.1 \rangle$.

Binary: Here we need to consider the number of low-order bits that can be safely changed. Assuming $n = 30$ as an element length and m as the length of permissible change, m must satisfy $m \leq n + \log_2 \delta$. Since m

is an integer, then $m = \lfloor n + \log_2 \delta \rfloor = 25$ and the sought probability is $m/n = 25/30 = 0.833$, a quite different number.

Therefore, we will try to design a method of compensating for this drawback in the following subsection.

5.3.2 Non-uniform mutation

In this part of the experiments we ran, in addition to the operators discussed in section 5.3.1, a special dynamic mutation operator aimed at both improving single-element tuning and reducing the disadvantage of random mutation in the FP implementation. We call it a *non-uniform mutation*; a full discussion of this operator is presented in the next two chapters.

FP
The new operator is defined as follows: if $s_v^t = \langle v_1, \ldots, v_m \rangle$ is a chromosome (t is the generation number) and the element v_k was selected for this mutation, the result is a vector $s_v^{t+1} = \langle v_1, \ldots, v_k', \ldots, v_m \rangle$, where

$$v_k' = \begin{cases} v_k + \triangle(t, UB - v_k) & \text{if a random digit is 0,} \\ v_k - \triangle(t, v_k - LB) & \text{if a random digit is 1,} \end{cases}$$

and LB and UB are lower and upper domain bounds of the variable v_k. The function $\triangle(t, y)$ returns a value in the range $[0, y]$ such that the probability of $\triangle(t, y)$ being close to 0 increases as t increases. This property causes this operator to search the space uniformly initially (when t is small), and very locally at later stages; thus increasing the probability of generating the new number closer to its successor than a random choice. We have used the following function:

$$\triangle(t, y) = y \cdot \left(1 - r^{(1-\frac{t}{T})^b}\right),$$

where r is a random number from $[0..1]$, T is the maximal generation number, and b is a system parameter determining the degree of dependency on iteration number (we used $b = 5$).

Binary
To be more than fair to the binary implementation, we modeled the dynamic operator into its space, even though it was introduced mainly to improve the FP mutation. Here, it is analogous to that of the FP, but with a differently defined v_k':

$$v_k' = mutate(v_k, \nabla(t, n)),$$

where $n = 30$ is the number of bits per one element of a chromosome; $mutate(v_k, pos)$ means: mutate value of the k-th element on pos bit (0 bit is the least significant), and

$$\nabla(t, n) = \begin{cases} \lfloor \Delta(t, n) \rfloor & \text{if a random digit is 0,} \\ \lceil \Delta(t, n) \rceil & \text{if a random digit is 1,} \end{cases}$$

with the b parameter of Δ adjusted appropriately if similar behavior is desired (we used $b = 1.5$).

Results

We repeated similar experiments to those of section 5.3.1 using also the non–uniform mutations applied at the same rate as the previously defined mutations.

implementation	Probability of chromosome's update		standard deviation
	0.8	0.9	
Binary	35265	30373	40256
FP	20561	26164	2133

Table 5.4. Average results as a function of probability of chromosome's update

Now the FP implementation shows a better average performance (Table 5.4). In addition, again the binary's results were more unstable. However, it is interesting to note here that despite its high average, the binary implementation produced the two single best results for this round (16205 and 16189).

5.3.3 Other operators

In this part of the experiment we decided to implement and use as many additional operators as could be easily defined in both representation spaces.

Binary

In addition to those previously described we implemented a multi–point crossover, and also allowed for crossovers within bits of an element. The multi–point operator had the probability of application to a single element controlled by a system parameter (set at 0.3).

FP

Here we also implemented a similar multi-point crossover. In addition, we implemented single and multi-point arithmetical crossovers; they average values of two elements rather that exchange them, at selected points. Such operators have the property that each element of the new chromosomes is still within the original domain. More details of these operators are provided in the next two chapters.

Results

Here the FP implementation shows an outstanding superiority (Table 5.5); even though the best results are not so very different, only the FP was consistent in achieving them.

implementation	Probability of chromosome's update			standard deviation	Best
	0.7	0.8	0.9		
Binary	23814	19234	27456	6078	16188.2
FP	16248	16798	16198	54	16182.1

Table 5.5. Average results as a function of probability of chromosome's update

5.4 Time performance

Many complain about the high time complexity of GAs on nontrivial problems. In this section we compare the time performance of both implementations. The results presented in Table 5.6 are those for runs of section 5.3.3.

implementation	Number of elements (N)				
	5	15	25	35	45
Binary	1080	3123	5137	7177	9221
FP	184	398	611	823	1072

Table 5.6. CPU time (sec) as a function of number of elements

Table 5.6 compares CPU time for both implementations on varying number of elements in the chromosome. The FP version is much faster, even for the moderate 30 bits per variable in the binary implementation. For large domains and high precision the total length of the chromosome grows, and the relative difference would expand as further indicated in Table 5.7.

5.5 Conclusions

The conducted experiments indicate that the floating point representation is faster, more consistent from run to run, and provides a higher precision (especially with large domains where binary coding would require prohibitively long representation). At the same time its performance can be enhanced by special

implementation	Number of bits per binary element					
	5	10	20	30	40	50
Binary	4426	5355	7438	9219	10981	12734
FP	1072 (constant)					

Table 5.7. CPU time (sec) as a function of number of bits per element; $N = 45$

operators to achieve high (even higher than that of the binary representation) accuracy. In addition, the floating point representation, as intuitively closer to the problem space, is easier for designing other operators incorporating problem specific knowledge. This is especially essential in handling nontrivial, problem–specific constraints (Chapter 7).

These conclusions are in accordance with the reasons of the users of genetic-evolutionary techniques who prefer floating point representation given in [157]: (1) comfort with one-gene-one-variable correspondence, (2) avoidance of Hamming cliffs and other artifacts of mutation operating on bit strings treated as unsigned binary integers, (3) fewer generations to population conformity.

At this stage the reader is encouraged to run a few experiments (see Appendix D; exercise 3). Select a few test functions (e.g., take some functions from Appendix B) and experiment with three GA-based systems with binary, Gray, and floating point representations. For the first two systems, use GENESIS 1.2ucsd ; for the third one — use GENOCOP (Chapter 7).

6. Fine Local Tuning

Genetic algorithms display inherent difficulties in performing local search for numerical applications. Holland suggested [188] that the genetic algorithm should be used as a preprocessor to perform the initial search, before turning the search process over to a system that can employ domain knowledge to guide the local search. As observed in [170]:

> "Like natural genetic systems, GAs progress by virtue of changing the distribution of high performance substructures in the overall population; individual structures are not the focus of attention. Once the high performance regions of the search space are identified by a GA, it may be useful to invoke a local search routine to optimize the members of the final population."

Local search requires the utilization of schemata of higher order and longer defining length than those suggested by the Schema Theorem. Additionally, there are problems where the domains of parameters are unlimited, the number of parameters is quite large, and high precision is required. These requirements imply that the length of the (binary) solution vector is quite significant (for 100 variables with domains in the range [−500, 500], where the precision of six digits after the decimal point is required, the length of the binary solution vector is 3000). As mentioned in the previous chapter, for such problems the performance of genetic algorithms is quite poor.

To improve the fine local tuning capabilities of a genetic algorithm, which is a must for high precision problems, we designed a special mutation operator whose performance is quite different from the traditional one. Recall that a traditional mutation changes one bit of a chromosome at a time; therefore, such a change uses only local knowledge — only the bit undergoing mutation is known. Such a bit, if located in the left portion of a sequence coding a variable, is very significant to the absolute magnitude of the mutation effect on the variable. On the other hand, bits far to the right of such a sequence have quite a smaller influence while mutated. We decided to use such positional global knowledge in the following way: as the population ages, bits located further to the right of each sequence coding one variable get higher probability of being mutated, while those on the left have such a probability decreasing. In other words, such a mutation causes global search of the search space at the beginning of the iterative process, but an increasingly local exploitation later on. We call this a non-uniform mutation and discuss it later in the chapter. First, we discuss the problems used for a test bed for this new operator.

6.1 The test cases

In general, the task of designing and implementing algorithms for the solution of optimal control problems is a difficult one. The highly touted dynamic programming is a mathematical technique that can be used in variety of contexts, particularly in optimal control [37]. However, this algorithm breaks down on problems of moderate size and complexity, suffering from what is called "the curse of dimensionality" [33].

Optimal control problems are quite difficult to deal with numerically. Some numerical dynamic optimization programs available for general users are typically offspring of the static packages [50] and they do not use dynamic-optimization specific methods. Thus the available programs do not make an explicit use of the Hamiltonian, transversality conditions, etc. On the other hand, if they did use the dynamic-optimization specific methods, they would be even more difficult for a layman to handle.

On the other hand, to the best of the author's knowledge, it is only recently that GAs have been applied to optimal control problems in a systematic way [273], [271]. We believe that previous GA implementations were too weak to deal with problems where high precision was required. In this chapter we present our modification of a GA designed to enhance its performance. We show the quality and applicability of the developed system by a comparative study of some dynamic optimization problems. Later, the system evolved to include typical constraints for such optimization problems — this is discussed in the next chapter (section 7.2). As a reference for these test cases (as well as many other experiments discussed later in the book), we use a standard computational package used for solving such problems: the Student Version of General Algebraic Modeling System with MINOS optimizer [50]. We will refer to this package in the rest of the book as GAMS.

Three simple discrete-time optimal control models (frequently used in applications of optimal control) have been chosen as test problems for the evolution program: the linear-quadratic problem, the harvest problem, and the (discretized) push-cart problem. We discuss them in turn.

6.1.1 The linear-quadratic problem

The first test problem is a one-dimensional linear-quadratic model:

$$\min q \cdot x_N^2 + \sum_{k=0}^{N-1} (s \cdot x_k^2 + r \cdot u_k^2), \tag{6.1}$$

subject to

$$x_{k+1} = a \cdot x_k + b \cdot u_k, \quad k = 0, 1, \ldots, N-1, \tag{6.2}$$

where x_0 is given, a, b, q, s, r are given constants, $x_k \in R$, is the state and $u_k \in R$ is the control of the system.

The value for the optimal performance of (6.1) subject to (6.2) is

$$J^* = K_0 x_0^2, \tag{6.3}$$

where K_k is the solution of the Riccati equation:

$$K_k = s + ra^2 K_{k+1}/(r + b^2 K_{k+1}), \quad K_N = q. \tag{6.4}$$

In the sequel, the problem (6.1) subject to (6.2) will be solved for the sets of the parameters displayed in Table 6.1.

Case	N	x_0	s	r	q	a	b
I	45	100	1	1	1	1	1
II	45	100	10	1	1	1	1
III	45	100	1000	1	1	1	1
IV	45	100	1	10	1	1	1
V	45	100	1	1000	1	1	1
VI	45	100	1	1	0	1	1
VII	45	100	1	1	1000	1	1
VIII	45	100	1	1	1	0.01	1
IX	45	100	1	1	1	1	0.01
X	45	100	1	1	1	1	100

Table 6.1. Ten test cases

In the experiments the value of N was set at 45 as this was the largest horizon for which a comparative numerical solution from GAMS was still achievable.

6.1.2 The harvest problem

The harvest problem is defined as:

$$\max \sum_{k=0}^{N-1} \sqrt{u_k}, \tag{6.5}$$

subject to the equation of growth,

$$x_{k+1} = a \cdot x_k - u_k, \tag{6.6}$$

and one equality constraint,

$$x_0 = x_N, \tag{6.7}$$

where initial state x_0 is given, a is a constant, and $x_k \in R$ and $u_k \in R^+$ are the state and the (nonnegative) control, respectively.

The optimal value J^* of (6.5) subject to (6.6) and (6.7) is:

$$J^* = \sqrt{\frac{x_0 \cdot (a^N - 1)^2}{a^{N-1} \cdot (a - 1)}}. \tag{6.8}$$

Problem (6.5) subject to (6.6) and (6.7) will be solved for $a = 1.1$, $x_0 = 100$, and the following values of $N = 2, 4, 10, 20, 45$.

6.1.3 The push-cart problem

The push-cart problem is to maximize the total distance $x_1(N)$ traveled in a given time (a unit, say), minus the total effort. The system is second order:

$$x_1(k + 1) = x_2(k) \tag{6.9}$$

$$x_2(k + 1) = 2x_2(k) - x_1(k) + \frac{1}{N^2} u(k), \tag{6.10}$$

and the performance index to be maximized is:

$$x_1(N) - \frac{1}{2N} \sum_{k=0}^{N-1} u^2(k). \tag{6.11}$$

For this problem the optimal value of index (6.11) is:

$$J^* = \frac{1}{3} - \frac{3N - 1}{6N^2} - \frac{1}{2N^3} \sum_{k=0}^{N-1} k^2. \tag{6.12}$$

The push-cart problem will be solved for different values $N = 5, 10, 15, 20, 25, 30, 35, 40, 45$. Note that different N correspond to the number of discretization periods (of an equivalent continuous problem) rather than to the actual length of the optimization horizon which will be assumed as one.

6.2 The evolution program for numerical optimization

The evolution program we have built for numerical optimization problems is based on the floating point representation, and some new (specialized) genetic operators; we discuss them in turn.

6.2.1 The representation

In floating point representation each chromosome vector is coded as a vector of floating point numbers of the same length as the solution vector. Each element is initially selected as to be within the desired domain, and the operators are carefully designed to preserve this constraint (there is no such problem in the binary representation, but the design of the operators is rather simple; we do not see that as a disadvantage; on the other hand, it provides for other advantages mentioned below).

The precision of such an approach depends on the underlying machine, but is generally much better than that of the binary representation. Of course, we can always extend the precision of the binary representation by introducing more bits, but this considerably slows down the algorithm, as discussed in the previous chapter.

In addition, the floating point representation is capable of representing quite large domains (or cases of unknown domains). On the other hand, the binary representation must sacrifice precision with an increase in domain size, given fixed binary length. Also, in the floating point representation it is much easier to design special tools for handling nontrivial constraints: this is discussed fully in next chapter.

6.2.2 The specialized operators

The operators we use are quite different from the classical ones, as they work in a different space (real valued). However, because of intuitive similarities, we will divide them into the standard classes, mutation and crossover. In addition, some operators are non-uniform, i.e., their action depends on the age of the population.

Mutation group:

- **uniform mutation**, defined similarly to that of the classical version: if $x_i^t = \langle v_1, \ldots, v_n \rangle$ is a chromosome, then each element v_k has exactly equal chance of undergoing the mutative process. The result of a single application of this operator is a vector $\langle v_1, \ldots, v'_k, \ldots, v_n \rangle$, with $1 \leq k \leq n$, and v'_k a random value from the domain of the corresponding parameter $domain_k$.

- **non-uniform mutation** is one of the operators responsible for the fine tuning capabilities of the system. It is defined as follow: The non-uniform mutation operator was defined as follows: if $s_v^t = \langle v_1, \ldots, v_m \rangle$ is a chromosome and the element v_k was selected for this mutation (domain of v_k is $[l_k, u_k]$), the result is a vector $s_v^{t+1} = \langle v_1, \ldots, v'_k, \ldots, v_m \rangle$, with $k \in \{1, \ldots, n\}$, and

$$v'_k = \begin{cases} v_k + \triangle(t, u_k - v_k) & \text{if a random digit is } 0, \\ v_k - \triangle(t, v_k - l_k) & \text{if a random digit is } 1, \end{cases}$$

where the function $\triangle(t,y)$ returns a value in the range $[0, y]$ such that the probability of $\triangle(t,y)$ being close to 0 increases as t increases. This property causes this operator to search the space uniformly initially (when t is small), and very locally at later stages. We have used the following function:

$$\triangle(t,y) = y \cdot \left(1 - r^{(1-\frac{t}{T})^b}\right),$$

where r is a random number from $[0..1]$, T is the maximal generation number, and b is a system parameter determining the degree of non-uniformity. Figure 6.1 displays the value of $\triangle$ for two selected times; this picture clearly indicates the behavior of the operator.

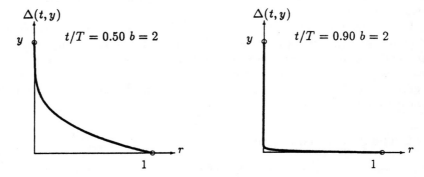

Fig. 6.1. $\triangle(t, y)$ for two selected times

Moreover, in addition to the standard way of applying mutation we have some new mechanisms: e.g., non-uniform mutation is also applied to a whole solution vector rather than a single element of it, causing the whole vector to be slightly slipped in the space.

Crossover group:

- **simple crossover**, defined in the usual way, but with the only permissible split points between v's, for a given chromosome x.

- **arithmetical crossover** is defined as a linear combination of two vectors: if s_v^t and s_w^t are to be crossed, the resulting offspring are $s_v^{t+1} = a \cdot s_w^t + (1-a) \cdot s_v^t$ and $s_w^{t+1} = a \cdot s_v^t + (1-a) \cdot s_w^t$. This operator can use a parameter a which is either a constant (uniform arithmetical crossover), or a variable whose value depends on the age of population (non-uniform arithmetical crossover).

Here again we have some new mechanisms to apply these operators; e.g., the arithmetical crossover may be applied either to selected elements of two vectors or to the whole vectors.

6.3 Experiments and results

In this section we present the results of the evolution program for the optimal control problems. For all test problems, the population size was fixed at 70, and the runs were made for 40,000 generations. For each test case we have made three random runs and reported the best results; it is important to note, however, that the standard deviations of such runs were almost negligibly small. The vectors $\langle u_0, \ldots, u_{N-1} \rangle$ were initialized randomly (but within a desired domains). Tables 6.2, 6.3, and 6.4 report the values found along with intermediate results at some generation intervals. For example, the values in column "10,000" indicate the partial result after 10,000 generations, while running 40,000. It is important to note that such values are worse than those obtained while running only 10,000 generation, due to the nature of some genetic operators. In the next section we compare these results with the exact solutions and solutions obtained from the computational package GAMS.

Case	\multicolumn{7}{c}{Generations}	Factor
	1	100	1,000	10,000	20,000	30,000	40,000	
I	17904.4	3.87385	1.73682	1.61859	1.61817	1.61804	1.61804	10^4
II	13572.3	5.56187	1.35678	1.11451	1.09201	1.09162	1.09161	10^5
III	17024.8	2.89355	1.06954	1.00952	1.00124	1.00102	1.00100	10^7
IV	15082.1	8.74213	4.05532	3.71745	3.70811	3.70162	3.70160	10^4
V	5968.42	12.2782	2.69862	2.85524	2.87645	2.87571	2.87569	10^5
VI	17897.7	5.27447	2.09334	1.61863	1.61837	1.61805	1.61804	10^4
VII	2690258	18.6685	7.23567	1.73564	1.65413	1.61842	1.61804	10^4
VIII	123.942	72.1958	1.95783	1.00009	1.00005	1.00005	1.00005	10^4
IX	7.28165	4.32740	4.39091	4.42524	4.31021	4.31004	4.31004	10^5
X	9971341	148233	16081.0	1.48445	1.00040	1.00010	1.00010	10^4

Table 6.2. Evolution program for the linear-quadratic problem (6.1)–(6.2)

Note, that the problem (6.5)–(6.7) has the final state constrained. It differs from the problem (6.1)–(6.2) in the sense that not every randomly initialized vector $\langle u_0, \ldots, u_{N-1} \rangle$ of positive real numbers generates an admissible sequence x_k (see condition (6.6)) such that $x_0 = x_N$, for given a and x_0. In our evolution program, we have generated a random sequence of $u_0, \ldots, u_{N-2}$, and have set $u_{N-1} = a \cdot x_{N-1} - x_N$. For negative u_{N-1}, we have discarded the sequence and repeated the initialization process (we discuss this process in detail in the next chapter, section 7.2). The same difficulty occurred during the reproduction process. An offspring (after some genetic operations) need not satisfy the constraint $x_0 = x_N$. In such a case we replaced the last component of the offspring vector u using the formula $u_{N-1} = a \cdot x_{N-1} - x_N$. Again, if u_{N-1} turns out to be negative, we do not introduce such offspring into the new population.

It is the only test problem considered in this chapter which includes a non-trivial constraint. We discuss the general issue of constrained optimal control problems in the next chapter.

	Generations						
N	1	100	1,000	10,000	20,000	30,000	40,000
2	6.3310	6.3317	6.3317	6.3317	6.3317	6.3317	6.331738
4	12.6848	12.7127	12.7206	12.7210	12.7210	12.7210	12.721038
8	25.4601	25.6772	25.9024	25.9057	25.9057	25.9057	25.905710
10	32.1981	32.5010	32.8152	32.8209	32.8209	32.8209	32.820943
20	65.3884	68.6257	73.1167	73.2372	73.2376	73.2376	73.237668
45	167.1348	251.3241	277.3990	279.0657	279.2612	279.2676	279.271421

Table 6.3. Evolution program for the harvest problem (6.5)–(6.7)

	Generations						
N	1	100	1,000	10,000	20,000	30,000	40,000
5	-3.008351	0.081197	0.119979	0.120000	0.120000	0.120000	0.120000
10	-5.668287	-0.011064	0.140195	0.142496	0.142500	0.142500	0.142500
15	-6.885241	-0.012345	0.142546	0.150338	0.150370	0.150370	0.150371
20	-7.477872	-0.126734	0.149953	0.154343	0.154375	0.154375	0.154377
25	-8.668933	-0.015673	0.143030	0.156775	0.156800	0.156800	0.156800
30	-12.257346	-0.194342	0.123045	0.158241	0.158421	0.158426	0.158426
35	-11.789546	-0.236753	0.110964	0.159307	0.159586	0.159592	0.159592
40	-10.985642	-0.235642	0.072378	0.160250	0.160466	0.160469	0.160469
45	-12.789345	-0.342671	0.072364	0.160913	0.161127	0.161152	0.161152

Table 6.4. Evolution program for the push-cart problem (6.9)–(6.11)

6.4 Evolution program versus other methods

In this section we compare the above results with the exact solutions as well as those obtained from the computational package GAMS.

6.4.1 The linear-quadratic problem

Exact solutions of the problems for the values of the parameters specified in Table 6.1 have been obtained using formulae (6.3) and (6.4).

To highlight the performance and competitiveness of the evolution program, the same test problems were solved using GAMS. The comparison may be regarded as not totally fair for the evolution program since GAMS is based on search methods particularly appropriate for linear-quadratic problems. Thus the problem (6.1)–(6.2) must be an easy case for this package. On the other hand, if for these test problems the evolution program proved to be competitive, or close to, there would be an indication that it should behave satisfactorily in general. Table 6.5 summarizes the results, where columns D refer to the percentage of the relative error.

	Exact solution	Evolution Program		GAMS	
Case	value	value	D	value	D
I	16180.3399	16180.3928	0.000%	16180.3399	0.000%
II	109160.7978	109161.0138	0.000%	109160.7978	0.000%
III	10009990.0200	10010041.3789	0.000%	10009990.0200	0.000%
IV	37015.6212	37016.0426	0.000%	37015.6212	0.000%
V	287569.3725	287569.4357	0.000%	287569.3725	0.000%
VI	16180.3399	16180.4065	0.000%	16180.3399	0.000%
VII	16180.3399	16180.3784	0.000%	16180.3399	0.000%
VIII	10000.5000	10000.5000	0.000%	10000.5000	0.000%
IX	431004.0987	431004.4182	0.000%	431004.0987	0.000%
X	10000.9999	10001.0038	0.000%	10000.9999	0.000%

Table 6.5. Comparison of solutions for the linear-quadratic problem

As shown above, the performance of GAMS for the linear-quadratic problem is perfect. However, this was not at all the case for the second test problem.

6.4.2 The harvest problem

To begin with, none of the GAMS solutions was identical with the analytical one. The difference between the solutions increased with the optimization horizon as shown in Table 6.6, and for $N > 4$ the system failed to find any value.

It appears that GAMS is sensitive to non-convexity of the optimizing problem and to the number of variables. Even adding another constraint to the problem $(u_{k+1} > 0.1 \cdot u_k)$ to restrict the feasibility set so that the GAMS algorithm does not "lose itself"[1] has not helped much (see column "GAMS+"). As this column shows, for sufficiently long optimization horizons there is no chance to obtain a satisfactory solution from GAMS.

6.4.3 The push-cart problem

For the push-cart problem both GAMS and the evolution program produce very good results (Table 6.7). However, it is interesting to note the relationship between the times different search algorithms need to complete the task.

For most optimization programs, the time necessary for an algorithm to converge to the optimum depends on the number of decision variables. This relationship for dynamic programming is exponential ("curse of dimensionality"). For the search methods (like GAMS) it is usually "worse than linear".

---

[1]This is "unfair" from the point of view of the genetic algorithm which works without such help.

N	Exact solution	GAMS		GAMS+		Genetic Alg	
		value	D	value	D	value	D
2	6.331738	4.3693	30.99%	6.3316	0.00%	6.3317	0.000%
4	12.721038	5.9050	53.58%	12.7210	0.00%	12.7210	0.000%
8	25.905710	*		18.8604	27.20%	25.9057	0.000%
10	32.820943	*		22.9416	30.10%	32.8209	0.000%
20	73.237681	*		*		73.2376	0.000%
45	279.275275	*		*		279.2714	0.001%

Table 6.6. Comparison of solutions for the harvest problem. The symbol * means that the GAMS failed to report a reasonable value

	Exact solution	GAMS		GA	
N	value	value	D	value	D
5	0.120000	0.120000	0.000%	0.120000	0.000 %
10	0.142500	0.142500	0.000%	0.142500	0.000 %
15	0.150370	0.150370	0.000%	0.150370	0.000 %
20	0.154375	0.154375	0.000%	0.154375	0.000 %
25	0.156800	0.156800	0.000%	0.156800	0.000 %
30	0.158426	0.158426	0.000%	0.158426	0.000 %
35	0.159592	0.159592	0.000%	0.159592	0.000 %
40	0.160469	0.160469	0.000%	0.160469	0.000 %
45	0.161152	0.161152	0.000%	0.161152	0.000 %

Table 6.7. Comparison of solutions for the push-cart problem

Table 6.8 reports the number of iterations the evolution program needed to obtain an exact solution (with six decimal place rounding), the time needed for that, and the total time for all 40,000 iterations (for unknown exact solution we cannot determine the precision of the current solution). Also, the time for GAMS is given. Note that GAMS was run on PC Zenith z-386/20 and the evolution program on a DEC–3100 station.

It is clear that the evolution program is much slower than GAMS: there is a difference in absolute values of CPU time as well as computers used. However, let us compare not the times needed for both systems to complete their calculations, but rather their growth rates of the time as a function of the size of the problem. Figure 6.2 show the growth rate of the time needed to obtain the result for the evolution program and GAMS.

These graphs are self-explanatory: although the evolution program is generally slower, its close to linear growth rate is much better than that of GAMS (which is at least quadratic). Similar results hold for the linear-quadratic problem and the harvest problem.

N	No. of iterations needed	Time needed (CPU sec)	Time for 40,000 iterations (CPU sec)	Time for GAMS (CPU sec)
5	6234	65.4	328.9	31.5
10	10231	109.7	400.9	33.1
15	19256	230.8	459.8	36.6
20	19993	257.8	590.8	41.1
25	18804	301.3	640.4	47.7
30	22976	389.5	701.9	58.2
35	23768	413.6	779.5	68.0
40	25634	467.8	850.7	81.3
45	28756	615.9	936.3	95.9

Table 6.8. Time performance of evolution program and GAMS for the push-cart problem (6.9)–(6.11): number of iterations needed to obtain the result with precision of six decimal places, time needed for that number of iterations, time needed for all 40,000 iterations

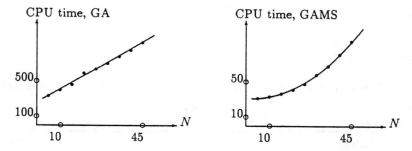

Fig. 6.2. Time as a function of problem size (N).

6.4.4 The significance of non-uniform mutation

It is interesting to compare these results with the exact solutions as well as those obtained from another GA, exactly the same but without the *non-uniform mutation* on. Table 6.9 summarizes the results; columns labeled D indicate the relative errors in percents.

The genetic algorithm using the *non-uniform mutation* clearly outperforms the other one with respect to the accuracy of the found optimal solution; while the enhanced GA rarely erred by more than a few thousandths of one percent, the other one hardly ever beat one percent. Moreover, it also converged much faster to that solution.

	Exact solution	GA w/ non-uniform mutation		GA w/o non-uniform mutation	
Case	value	value	D	value	D
I	16180.3399	16180.3939	0.000%	16234.3233	0.334%
II	109160.7978	109163.0278	0.000%	113807.2444	4.257%
III	10009990.0200	10010391.3989	0.004%	10128951.4515	1.188%
IV	37015.6212	37016.0806	0.001%	37035.5652	0.054%
V	287569.3725	287569.7389	0.000%	298214.4587	3.702%
VI	16180.3399	16180.6166	0.002%	16238.2135	0.358%
VII	16180.3399	16188.2394	0.048%	17278.8502	6.786%
VIII	10000.5000	10000.5000	0.000%	10000.5000	0.000%
IX	431004.0987	431004.4092	0.000%	431610.9771	0.141%
X	10000.9999	10001.0045	0.000%	10439.2695	4.380%

Table 6.9. Comparison of solutions for the linear-quadratic dynamic control problem

As an illustration of the *non-uniform mutation*'s effect on the evolutionary process check Figure 6.3; the new mutation causes quite an increase in the number of improvements observed in the population at the end of the population's life. Moreover, a smaller number of such improvements prior to that time, together with an actually faster convergence, clearly indicates a better overall search.

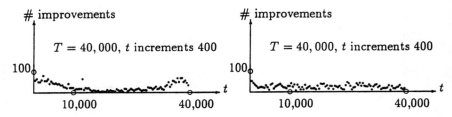

Fig. 6.3. Number of improvements on case I of the linear-quadratic dynamic control problem

6.5 Conclusions

In this chapter we have studied a new operator, a non-uniform mutation, to improve the fine local tuning capabilities of a GA. The experiments were successful on the three discrete-time optimal control problems which were selected as test cases. In particular, the results were encouraging because the closeness of the numerical solutions to the analytical ones was satisfying. Additionally, the computation effort was reasonable (for the 40,000 generations, a few minutes of CPU time on a CRAY Y-MP and up to 15 minutes on a DEC–3100 station).

The numerical results were compared with those obtained from a search-based computational package (GAMS). While the evolution program gave us results comparable with the analytic solutions for all test problems, GAMS failed for one of them. The developed evolution program displayed some qualities not always present in the other (gradient-based) systems:

- The optimization function for the evolution program need not be continuous. At the same time some optimization packages will not accept such functions at all.

- Some optimization packages are all-or-nothing propositions: the user has to wait until the program completes. Sometimes it is not possible to get partial (or approximate) results at some early stages. Evolution programs give the users additional flexibility, since the user can monitor the "state of the search" during the run time and make appropriate decisions. In particular, the user can specify the computation time (s)he is willing to pay for (longer time provides better precision in the answer).

- The computational complexity of evolution programs grows at a linear rate; most of other search methods are very sensitive to the length of the optimization horizon. As usual, we can easily improve the performance of the system using parallel implementations; often this is difficult for other optimization methods.

Recently there have been many interesting developments in the area of evolutionary algorithms to enhance their fine local tuning capabilities. These include Delta Coding algorithm, Dynamic Parameter Encoding, ARGOT strategy, IRM strategies, extension of evolutionary programming, granularity evolution, , and interval genetic algorithms. A short description of these attempts is provided in the last section of Chapter 8, after discussion of series of GENOCOP systems (Chapter 7) and evolution strategies (Chapter 8).

There are some other activities which (more or less directly) aim at the fine local tuning capabilities. These include research of Arabas et al. [11], where adaptive intermediate and uniform crossovers for evolution strategies (see Chapter 8) were introduced. On the other hand, Hinterding [184] experimented with mutation of genes (which correspond to variables) as opposed to mutation of bits. From this perspective the authors analysed the significance of coding (Gray versus binary), granularity, and the frequency of such gene mutations.

There were also interesting results reported by Srinivas and Patnaik [368], who experimented with adaptive probabilities of mutation and crossover to maintain diversity of the population (and sustaining the convergence capacity of the algorithm). In this approach the probabilities of these operators were varied depending on the fitness values of the solutions: 'good' solutions are protected and 'poor' solutions are disrupted. More precisely,

$$p_c = \begin{cases} k_1 \cdot (f_{max} - f')/(f_{max} - \overline{f}) & if \ \ f' \leq \overline{f} \\ k_3 & otherwise, \end{cases}$$

and

$$p_m = \begin{cases} k_2 \cdot (f_{max} - f)/(f_{max} - \overline{f}) & if \ f \leq \overline{f} \\ k_4 & otherwise, \end{cases}$$

where k_1 and k_2 are positive constants (not greater than one), f_{max} and $\overline{f}$ denote the maximum and the average values of the fitness function f in the current population, respectively, f denotes the value of the fitness function for a given solution, and f' — the larger value (for two solutions selected for crossover). Note, that (1) the value of $f_{max} - \overline{f}$ is essential in the above formulae; it is also important in measuring the convergence of the algorithm; (2) p_c and p_m are zeros for the solution with the maximum fitness; (3) $p_c = k_1$ and $p_m = k_2$ for a solution with $f = \overline{f}$; and (4) $p_c = k_3$ and $p_m = k_4$ for a below average solutions. For the proper choice of values k_1, k_2, k_3, and k_4, and experimental results, see [368].

7. Handling Constraints

A traveler in quest of the divine
asked the Master how to distinguish
a true teacher from a false one when
he got back to his own land.

Said the Master, 'A good teacher
offers practice; a bad one offers
theories.'

'But how shall I know good practice
from bad?'

'In the same way that the farmer
knows good cultivation from bad.'

Anthony de Mello, *One Minute Wisdom*

The general nonlinear programming problem $\mathcal{NLP}$ is to find $\boldsymbol{x}$ so as to

optimize $f(\boldsymbol{x})$, $\boldsymbol{x} = (x_1, \ldots, x_q) \in R^q$,

subject to $p \geq 0$ equations:

$c_i(\boldsymbol{x}) = 0$, $i = 0, \ldots, p$,

and $m - p \geq 0$ inequalities:

$c_i(\boldsymbol{x}) \leq 0$, $i = p + 1, \ldots, m$.

There is no known method of determining the global maximum (or minimum) to the general nonlinear programming problem. Only if the objective function f and the constraints c_i satisfy certain properties, the global optimum can sometimes be found. Several algorithms were developed for unconstrained problems (e.g., direct search method, gradient method) and constrained problems (these algorithms usually are classified as indirect and direct methods). An indirect method attacks the problem by extracting one or more linear problems from the original one, whereas a direct method tries to determine successive search points. This is usually done by converting the original problem into unconstrained one for which gradient methods are applied with some modifications [387]. Despite the active research and progress in global optimization in recent years [114], it is probably fair to say that no efficient solution procedure is in sight for the general nonlinear problems $\mathcal{NLP}$. As stated in [172]:

"It's unrealistic to expect to find one general $\mathcal{NLP}$ code that's going to work for every kind of nonlinear model. Instead, you should try to select a code that fits the problem you are solving. If your problem doesn't fit in any category except 'general', or if you insist on a globally optimal solution (except when there is no chance of encountering multiple local optima), you should be prepared to have to use a method that boils down to exhaustive search, i.e., you have an intractable problem."

There are many other problems connected with traditional optimization techniques. For example, most proposed methods are local in scope, they depend on the existence of derivatives, and they are insufficiently robust in discontinuous, vast multimodal, or noisy search spaces. It is important then to investigate other (heuristic) methods, which, for many real world problems, may prove very useful.

In this chapter we discuss several methods which were developed in connection with the nonliner programming problem. We start with a description of GENOCOP, a system developed for convex search spaces. In further sections we survey other evolutionary approaches for nonlinear programming problem and describe two other systems, GENOCOP II and GENOCOP III.

7.1 An evolution program: the GENOCOP system

Many researchers [78, 408] investigated GAs based on floating point representation. But the optimization problems they considered were defined on a search space $\mathcal{D} \subseteq R^q$, where $\mathcal{D} = \prod_{k=1}^{q} \langle l_k, r_k \rangle$, i.e., each variable x_k was restricted to a given interval $\langle l_k, r_k \rangle$ $(1 \leq k \leq q)$. Yet it seems important to include other constraints into the considerations; as stated in [66]:

"A little observation and reflection will reveal that all optimization problems of the real world are, in fact, constrained problems. Suppose one has an expression for the output of a chemical reaction vessel in which some set of chemical reactions are taking place and one wishes to maximize the output. It is also necessary to take into account material balance restrictions on the reactans and products, the laws governing flow of materials into and out of the reaction vessel, and other conditions. All of these are additional constraints on the variables of the function to be optimized."

In a constrained optimization problem, the geometric shape of the set of solutions in R^q is perhaps the most crucial characteristic of the problem, with respect to the degree of difficulty that is likely to be encountered in attempting to solve the problem [66]. There is only one special type of set—a convex set—for which a significant amount of theory has been developed.

In this section we are concerned with the following optimization problem:

optimize $f(x_1, \ldots, x_q) \in R$,

where $(x_1, \ldots, x_q) \in \mathcal{D} \subseteq R^q$ and $\mathcal{D}$ is a *convex* set.

The domain $\mathcal{D}$ is defined by ranges of variables ($l_k \leq x_k \leq r_k$ for $k = 1, \ldots, q$) and by a set of constraints $\mathcal{C}$. From the convexity of the set $\mathcal{D}$ it follows that for each point in the search space $(x_1, \ldots, x_q) \in \mathcal{D}$ there exists a feasible range $\langle left(k), right(k) \rangle$ of a variable x_k ($1 \leq k \leq q$), where other variables x_i ($i = 1, \ldots, k-1, k+1, \ldots, q$) remain fixed. In other words, for a given $(x_1, \ldots, x_k, \ldots, x_q) \in \mathcal{D}$:

$$y \in \langle left(k), right(k) \rangle \text{ iff } (x_1, \ldots, x_{k-1}, y, x_{k+1}, \ldots, x_q) \in \mathcal{D},$$

where all x_i's ($i = 1, \ldots, k-1, k+1, \ldots, q$) remain constant. We assume also that the ranges $\langle left(k), right(k) \rangle$ can be efficiently computed.

For example, if $\mathcal{D} \subseteq R^2$ is defined as:

$$-3 \leq x_1 \leq 3,$$
$$0 \leq x_2 \leq 8,$$
$$\text{and } x_1^2 \leq x_2 \leq x_1 + 4,$$

then for a given point $(2, 5) \in \mathcal{D}$:

$$left(1) = 1, \; right(1) = \sqrt{5},$$
$$left(2) = 4, \; right(2) = 6.$$

This means that the first component of the vector $(2, 5)$ can vary from 1 to $\sqrt{5}$ (while $x_2 = 5$ remains constant) and the second component of this vector can vary from 4 to 6 (while $x_1 = 2$ remains constant).

Of course, if the set of constraints $\mathcal{C}$ is empty, then the search space $\mathcal{D} = \prod_{k=1}^{q} \langle l_k, r_k \rangle$ is convex; additionally $left(k) = l_k$, $right(k) = r_k$ for $k = 1, \ldots, q$.

The above property is a basis for all mutation operators: if the x_k variable is to be mutated, the range of the mutation is $\langle left(k), right(k) \rangle$; consequently, an offspring is always feasible.

An additional property of convex search spaces guarantees that for any two points $\boldsymbol{x}_1$ and $\boldsymbol{x}_2$ in the solution space $\mathcal{D}$, the linear combination $a\boldsymbol{x}_1 + (1 - a)\boldsymbol{x}_2$, where $a \in [0, 1]$, is also a point in $\mathcal{D}$. This property is important for implementation of arithmetical crossover.

We consider a particular class of optimization problems which are defined on a convex domain; these problems can be formulated as follows:
Optimize a function $f(x_1, x_2, \ldots, x_q)$, subject to the following sets of linear constraints:

1. Domain constraints: $l_i \leq x_i \leq u_i$ for $i = 1, 2, \ldots, q$. We write $\boldsymbol{l} \leq \boldsymbol{x} \leq \boldsymbol{u}$, where $\boldsymbol{l} = \langle l_1, \ldots, l_q \rangle$, $\boldsymbol{u} = \langle u_1, \ldots, u_q \rangle$, $\boldsymbol{x} = \langle x_1, \ldots, x_q \rangle$.

2. Equalities: $A\boldsymbol{x} = \boldsymbol{b}$, where $\boldsymbol{x} = \langle x_1, \ldots, x_q \rangle$, $A = (a_{ij})$, $\boldsymbol{b} = \langle b_1, \ldots, b_p \rangle$, $1 \leq i \leq p$, and $1 \leq j \leq q$ (p is the number of equations).

3. Inequalities: $C\boldsymbol{x} \leq \boldsymbol{d}$, where $\boldsymbol{x} = \langle x_1, \ldots, x_q \rangle$, $C = (c_{ij})$, $\boldsymbol{d} = \langle d_1, \ldots, d_m \rangle$, $1 \leq i \leq m$, and $1 \leq j \leq q$ (m is the number of inequalities).

This formulation is general enough to handle a large class of standard Operations Research optimization problems with linear constraints and any objective function. The example considered later, the nonlinear transportation problem, is one of many problems in this class.

The developed system (GENOCOP, for GEnetic algorithm for Numerical Optimization for COnstrained Problems) provides a way of handling constraints that is both general and problem independent. It combines some of the ideas seen in the previous approaches, but in a totally new context. The main idea behind this approach lies in (1) an elimination of the equalities present in the set of constraints, and (2) careful design of special genetic operators, which guarantee to keep all chromosomes within the constrained solution space. This can be done very efficiently for linear constraints and, while we do not claim these results extend easily to nonlinear constraints, the former class contains many interesting optimization problems.

In some optimization techniques, such as linear programming, equality constraints are welcome since it is known that the optimum, if it exists, is situated at the surface of the convex set. Inequalities are converted to equalities by the addition of slack variables, and the solution method proceeds by moving from vertex to vertex, around the surface.

In contrast, for a method that generates solutions randomly, such equality constraints are a nuisance. In GENOCOP they are eliminated at the start, together with an equal number of problem variables; this action removes also part of the space to be searched. The remaining constraints, in the form of linear inequalities, form a convex set which must be searched for a solution. The convexity of the search space ensures that linear combinations of solutions yield solutions without needing to check the constraints—a property used throughout this approach. The inequalities can be used to generate bounds for any given variable: such bounds are dynamic as they depend on the values of the other variables and can be efficiently computed.

Suppose the equality constraint set is represented in matrix form:

$$A\boldsymbol{x} = \boldsymbol{b}.$$

We assume there are p independent equality equations (there are easy methods to verify this), i.e., there are p variables $x_{i_1}, x_{i_2}, \ldots, x_{i_p}$ ($\{i_1, \ldots, i_p\} \subseteq \{1, 2, \ldots, q\}$) which can be determined in terms of the other variables. These can, therefore, be eliminated from the problem statement, as follows.

We can split the array A vertically into two arrays A_1 and A_2, such that the j-th column of the matrix A belong to A_1 iff $j \in \{i_1, \ldots, i_p\}$. Thus A_1^{-1} exists. Similarly, we split matrix C and vectors $\boldsymbol{x}, \boldsymbol{l}, \boldsymbol{u}$ (ie. $\boldsymbol{x}^1 = \langle x_{i_1}, \ldots, x_{i_p} \rangle$, $\boldsymbol{l}_1 = \langle l_{i_1}, \ldots, l_{i_p} \rangle$, and $\boldsymbol{u}_1 = \langle u_{i_1}, \ldots, u_{i_p} \rangle$). Then

$$A_1\boldsymbol{x}^1 + A_2\boldsymbol{x}^2 = \boldsymbol{b}.$$

and it is easily seen that

$$x^1 = A_1^{-1}b - A_1^{-1}A_2x^2.$$

Using the above rule, we can eliminate the variables $x_{i_1}, \ldots, x_{i_p}$ replacing them by a linear combination of remaining variables. However, each variable x_{i_j} $(j = 1, 2, \ldots, p)$ is constrained additionally by a domain constraint: $l_{i_j} \leq x_{i_j} \leq u_{i_j}$. Eliminating all variables x_{i_j} leads us to introduce a new set of inequalities:

$$l_1 \leq A_1^{-1}b - A_1^{-1}A_2x^2 \leq u_1,$$

which is added to the original set of inequalities.

The original set of inequalities,

$$Cx \leq d,$$

can be represented as

$$C_1x^1 + C_2x^2 \leq d.$$

This can be transformed into

$$C_1(A_1^{-1}b - A_1^{-1}A_2x^2) + C_2x^2 \leq d.$$

So, after eliminating the p variables $x_{i_1}, \ldots, x_{i_p}$, the final set of constraints consists of the following inequalities only:

1. original domain constraints: $l_2 \leq x^2 \leq u_2$,

2. new inequalities: $l_1 \leq A_1^{-1}b - A_1^{-1}A_2x^2 \leq u_1$,

3. original inequalities (after removal of x^1 variables): $(C_2 - C_1A_1^{-1}A_2)x^2 \leq d - C_1A_1^{-1}b$.

7.1.1 An example

Let us start with an example and assume we wish to optimize a function of six variables:

$$f(x_1, x_2, x_3, x_4, x_5, x_6),$$

subject to the following constraints:

$$2x_1 + x_2 + x_3 = 6,$$
$$x_3 + x_5 - 3x_6 = 10,$$
$$x_1 + 4x_4 = 3,$$
$$x_2 + x_5 \leq 120,$$
$$-40 \leq x_1 \leq 20, \ 50 \leq x_2 \leq 75,$$
$$0 \leq x_3 \leq 10, \ 5 \leq x_4 \leq 15,$$
$$0 \leq x_5 \leq 20, \ -5 \leq x_6 \leq 5.$$

We can take advantage of the presence of three independent equations and express four variables as functions of the remaining three:

$$x_1 = 3 - 4x_4,$$
$$x_2 = -10 + 8x_4 + x_5 - 3x_6,$$
$$x_3 = 10 - x_5 + 3x_6.$$

Thus we have reduced the original problem to the optimization problem of a function of three variables x_4, x_5, and x_6:

$$g(x_4, x_5, x_6) = f((3 - 4x_4), (-10 + 8x_4 + x_5 - 3x_6),$$
$$(10 - x_5 + 3x_6), x_4, x_5, x_6),$$

subject to the following constraints (inequalities only):

$$-10 + 8x_4 + 2x_5 - 3x_6 \leq 120, \text{ (original } x_2 + x_5 \leq 120),$$
$$50 \leq 3 - 4x_4 \leq 20, \text{ (original } 50 \leq x_1 \leq 20),$$
$$-20 \leq -10 + 8x_4 + x_5 - 3x_6 \leq 75, \text{ (original } -20 \leq x_2 \leq 75),$$
$$0 \leq 10 - x_5 + 3x_6 \leq 10, \text{ (original } 0 \leq x_3 \leq 10),$$
$$5 \leq x_4 \leq 15, 0 \leq x_5 \leq 20, \text{ and } -5 \leq x_6 \leq 5.$$

These can be reduced further; for example the second and fifth inequalities can be replaced by a single one:

$$5 \leq x_4 \leq 10.75.$$

Such transformation completes the first step of our algorithm: elimination of equalities. The resulting search space is, of course, convex. As discussed earlier, from the convexity of the search space it follows that for each feasible point (x_1, x_2, x_3) there exists a feasible range $\langle left(k), right(k) \rangle$ of a variable x_k ($1 \leq k \leq 3$), where the other two variables are fixed. For example, for the feasible space defined above and for a given feasible point $(x_4, x_5, x_6) = (10, 8, 2)$:

$$left(1) = 7.25, right(1) = 10.375,$$
$$left(2) = 6, right(2) = 11,$$
$$left(3) = 1, right(3) = 2.666,$$

($left(1)$ and $right(1)$ are ranges of the first component of the vector $(10, 8, 2)$, i.e., of the variable x_4, etc.). This means that the first component of the vector $(10, 8, 2)$ can vary from 7.25 to 10.375 (while $x_5 = 8$ and $x_6 = 2$ remain constant), the second component of this vector can vary from 6 to 11 (while $x_4 = 10$ and $x_6 = 2$ remain constant), and the third component of this vector can vary from 1 to 2.666 (while $x_4 = 10$ and $x_5 = 8$ remain constant).

The GENOCOP system tries to locate an initial (feasible) solution by sampling the feasible region. If some predefined number of trials is unsuccessful, the system would prompt the user for a feasible initial point. The initial population consists of identical copies of such an initial point (whether generated or provided by the user).

There are several operators in the GENOCOP system which proved to be useful on many test problems. We discuss them in turn in the next subsection.

7.1.2 Operators

In this subsection we describe six genetic operators based on floating point representation, which were used in the modified version of the GENOCOP system. The first three are unary operators (category of mutation), the other three are binary (various types of crossovers). We discuss them in turn.

Uniform mutation

This operator requires a single parent x and produces a single offspring x'. The operator selects a random component $k \in (1, \ldots, q)$ of the vector $x = (x_1, \ldots, x_k, \ldots, x_q)$ and produces $x' = (x_1, \ldots, x'_k, \ldots, x_q)$, where x'_k is a random value (uniform probability distribution) from the range $\langle left(k), right(k) \rangle$.

The operator plays an important role in the early phases of the evolution process as the solutions are allowed to move freely within the search space. In particular, the operator is essential in the case where the initial population consists of multiple copies of the same (feasible) point. Such situations may occur quite often in constrained optimization problems where users specify the starting point for the process. Moreover, such a single starting point (apart from its drawbacks) has a powerful advantage: it allows for developing an iterative process, where the next iteration starts from the best point of the previous iteration. This very technique was used in a development of a system to handle nonlinear constraints in spaces that were not necessarily convex (GENOCOP II).

Also, in the later phases of an evolution process the operator allows possible movement away from a local optimum in the search for a better point.

Boundary mutation

This operator requires also a single parent x and produces a single offspring x'. The operator is a variation of the uniform mutation with x'_k being either $left(k)$ or $right(k)$, each with equal probability.

The operator is constructed for optimization problems where the optimal solution lies either on or near the boundary of the feasible search space. Consequently, if the set of constraints $\mathcal{C}$ is empty, and the bounds for variables are quite wide, the operator is a nuisance. But it can prove extremely useful in the presence of constraints. A simple example demonstrates the utility of this operator. The example is a linear programming problem; in such a case we know that the global solution lies on the boundary of the search space.

Example 7.1.

Let us consider the following test case [387]:

$$\text{maximize } f(x_1, x_2) = 4x_1 + 3x_2,$$

subject to the following constraints:

$$2x_1 + 3x_2 \leq 6,$$
$$-3x_1 + 2x_2 \leq 3,$$
$$2x_1 + x_2 \leq 4, \text{ and}$$
$$0 \leq x_i \leq 2, \ i = 1, 2.$$

num is $(x_1, x_2) = (1.5, 1.0)$, and $f(1.5, 1.0) = 9.0$.
tility of this operator in optimizing the above problem,
un with all operators functioning and another ten ex-
indary mutation. The system with boundary mutation
ium easily in all runs, on average within 32 generations,
operator, even in 100 generations the *best* point found
$= (1.501, 0.997)$ and $f(x) = 8.996$ (the worst point was
h $f(x) = 8.803$).

Non-uniform mutation

This is the (unary) operator responsible for the fine tuning capabilities of the
system. It is defined as follows. For a parent x, if the element x_k was selected
for this mutation, the result is $x' = \langle x_1, \ldots, x'_k, \ldots, x_q \rangle$, where

$$x'_k = \begin{cases} x_k + \triangle(t, right(k) - x_k) & \text{if a random binary digit is 0} \\ x_k - \triangle(t, x_k - left(k)) & \text{if a random binary digit is 1} \end{cases}$$

The function $\triangle(t, y)$ returns a value in the range $[0, y]$ such that the probability
of $\triangle(t, y)$ being close to 0 increases as t increases (t is the generation number).
This property causes this operator to search the space uniformly initially (when
t is small), and very locally at later stages. We have used the following function:

$$\triangle(t, y) = y \cdot r \cdot (1 - \frac{t}{T})^b,$$

where r is a random number from $[0..1]$, T is the maximal generation number,
and b is a system parameter determining the degree of non-uniformity.

Arithmetical crossover

This binary operator is defined as a linear combination of two vectors: if x_1
and x_2 are to be crossed, the resulting offspring are $x'_1 = a \cdot x_1 + (1 - a) \cdot x_2$
and $x'_2 = a \cdot x_2 + (1 - a) \cdot x_1$. This operator uses a random value $a \in [0..1]$,
as it always guarantees closure ($x'_1, x'_2 \in \mathcal{D}$). Such a crossover was called a
guaranteed average crossover [77] (when $a = 1/2$); *intermediate crossover* [18];
linear crossover [408]; and *arithmetical crossover* [268, 269].

The importance of arithmetical crossover is illustrated by the following example.

Example 7.2.

Let us consider the following problem [114]:

$$\text{minimize } f(x_1, x_2, x_3, x_4, x_5) = -5sin(x_1)sin(x_2)sin(x_3)sin(x_4)sin(x_5) + \\ -sin(5x_1)sin(5x_2)sin(5x_3)sin(5x_4)sin(5x_5),$$

where

$$0 \leq x_i \leq \pi, \text{ for } 1 \leq i \leq 5.$$

The known global solution is $(x_1, x_2, x_3, x_4, x_5) = (\pi/2, \pi/2, \pi/2, \pi/2, \pi/2)$, and $f(\pi/2, \pi/2, \pi/2, \pi/2, \pi/2) = -6$.

It appears that the system without arithmetical crossover has slower convergence. After 50 generations the average value of the best point (out of 10 runs) was -5.9814, and the average value of the best point after 100 generations was -5.9966. In the same time, these averages for the system with arithmetical crossover were -5.9930 and -5.9996, respectively.

Moreover, an interesting pattern that emerged showed that the system with arithmetical crossover was more stable, with much lower standard deviation of the best solutions (obtained in ten runs).

Simple crossover

This binary operator is defined as follows: if $\boldsymbol{x}_1 = (x_1, \ldots, x_q)$ and $\boldsymbol{x}_2 = (y_1, \ldots, y_q)$ are crossed after the k-th position, the resulting offspring are: $\boldsymbol{x}_1' = (x_1, \ldots, x_k, y_{k+1}, \ldots, y_q)$ and $\boldsymbol{x}_2' = (y_1, \ldots, y_k, x_{k+1}, \ldots, x_q)$. Such an operator may produce offspring outside the domain $\mathcal{D}$. To avoid this, we use the property of convex spaces that there exists $a \in [0, 1]$ such that

$$\boldsymbol{x}_1' = \langle x_1, \ldots, x_k, y_{k+1} \cdot a + x_{k+1} \cdot (1 - a), \ldots, y_q \cdot a + x_q \cdot (1 - a) \rangle$$

and

$$\boldsymbol{x}_2' = \langle y_1, \ldots, y_k, x_{k+1} \cdot a + y_{k+1} \cdot (1 - a), \ldots, x_q \cdot a + y_q \cdot (1 - a) \rangle$$

are feasible.

The only remaining question to be answered is how to find the largest a to obtain the greatest possible information exchange. The simplest method would start with $a = 1$ and, if at least one of the offspring does not belong to $\mathcal{D}$, decreases a by some constant $\frac{1}{\rho}$. After ρ attempts $a = 0$ and both offspring are in $\mathcal{D}$ since they are identical to their parents. The necessity for such maximal decrement is small in general and decreases rapidly over the life of the population.

It seems that the merits of simple crossover are the same as of arithmetical crossover (for experiments, we have used the problem from the test case #6; see next subsection). The results showed that the system without simple crossover was even less stable than the system without arithmetical crossover; in this case the standard deviation of the best solutions obtained in ten runs was much higher. Also, the worst solution obtained in 100 generations had a value of -5.9547, which was much worse than the worst solution obtained with all operators (-5.9982) or the worst solution obtained without arithmetical crossover (-5.9919).

Heuristic crossover

This operator [408] is a unique crossover for the following reasons: (1) it uses values of the objective function in determining the direction of the search, (2) it produces only one offspring, and (3) it may produce no offspring at all.

The operator generates a single offspring $\boldsymbol{x}_3$ from two parents $\boldsymbol{x}_1$ and $\boldsymbol{x}_2$ according to the following rule:

$$x_3 = r \cdot (x_2 - x_1) + x_2,$$

where r is a random number between 0 and 1, and the parent x_2 is not worse than x_1, i.e., $f(x_2) \geq f(x_1)$ for maximization problems and $f(x_2) \leq f(x_1)$ for minimization problems.

It is possible for this operator to generate an offspring vector which is not feasible. In such a case another random value r is generated and another offspring created. If after w attempts no new solution meeting the constraints is found, the operator gives up and produces no offspring.

It seems that heuristic crossover contributes to the precision of the solution found; its major responsibilities are (1) fine local tuning, and (2) search in the most promising direction.

7.1.3 Testing GENOCOP

In order to evaluate the GENOCOP method, a set of test problems have been carefully selected to illustrate the performance of the algorithm and to indicate that it has been successful in practice. The eight test cases, which include quadratic, nonlinear, and discontinuous functions with several linear constraints, are discussed below.

All runs of the system were performed on SUN SPARC station 2. We used the following parameters for all experiments:

> $pop\_size = 70$, $k = 28$ (number of parents in each generation; classification step), and $b = 2$ (coefficient for non-uniform mutation).

(The third version of the GENOCOP system is available from ftp.uncc.edu, directory coe/evol, file GENOCOP3.0.tar.Z). For each test case we run the GENOCOP ten times. For all problems, the number of generations T was 500 or 1000 (with exception of the test case #6, where the system was run for 10,000 generations). The eight test cases and the results of the GENOCOP system are reported in the following subsections.

Test Case #1
The problem [114] is

$$\text{minimize } f(x, y) = -10.5x_1 - 7.5x_2 - 3.5x_3 - 2.5x_4 - 1.5x_5 - 10y - 0.5\sum_{i=1}^{5} x_i^2,$$

subject to:

$$6x_1 + 3x_2 + 3x_3 + 2x_4 + x_5 \leq 6.5, \quad 10x_1 + 10x_3 + y \leq 20,$$
$$0 \leq x_i \leq 1, \qquad\qquad\qquad 0 \leq y.$$

The global solution is $(x^*, y^*) = (0, 1, 0, 1, 1, 20)$, and $f(x^*, y^*) = -213$.

GENOCOP found solutions that were very close to the optimum in all ten runs; a typical discovered optimum point was:

(0.000000, 1.000000, 0.000000, 0.999999, 1.000000, 20.000000),

for which the value of the objective function is equal to -213.0. A single run of 1000 iterations took 21 sec of CPU time.

Test Case #2
The problem [186] is

$$\text{minimize } f(\boldsymbol{x}) = \sum_{j=1}^{10} x_j \left(c_j + \ln \frac{x_j}{x_1 + \ldots + x_{10}} \right),$$

subject to:

$$x_1 + 2x_2 + 2x_3 + x_6 + x_{10} = 2, \quad x_4 + 2x_5 + x_6 + x_7 = 1,$$
$$x_3 + x_7 + x_8 + 2x_9 + x_{10} = 1, \quad x_i \geq 0.000001, \ (i = 1, \ldots, 10),$$

where

$c_1 = -6.089$; $c_2 = -17.164$; $c_3 = -34.054$; $c_4 = -5.914$; $c_5 = -24.721$; $c_6 = -14.986$; $c_7 = -24.100$; $c_8 = -10.708$; $c_9 = -26.662$; $c_{10} = -22.179$;

The previously best known solution [186] was

$$\boldsymbol{x}^* = (.01773548, .08200180, .8825646, .0007233256, .4907851,$$
$$.0004335469, .01727298, .007765639, .01984929, .05269826),$$

and $f(\boldsymbol{x}^*) = -47.707579$.

GENOCOP found points with better value than the one above in all ten runs; the best solution found was

$$\boldsymbol{x}^* = (.04034785, .15386976, .77497089, .00167479, .48468539,$$
$$.00068965, .02826479, .01849179, .03849563, .10128126),$$

for which the value of the objective function is equal to -47.760765. A single run of 1000 iterations took 56 sec of CPU time.

Test Case #3
The problem [114] is

$$\text{minimize } f(\boldsymbol{x}, \boldsymbol{y}) = 5x_1 + 5x_2 + 5x_3 + 5x_4 - 5\sum_{i=1}^{4} x_i^2 - \sum_{i=1}^{9} y_i,$$

subject to:

$$2x_1 + 2x_2 + y_6 + y_7 \leq 10, \quad 2x_1 + 2x_3 + y_6 + y_8 \leq 10,$$
$$2x_2 + 2x_3 + y_7 + y_8 \leq 10, \quad -8x_1 + y_6 \leq 0,$$
$$-8x_2 + y_7 \leq 0, \quad -8x_3 + y_8 \leq 0,$$
$$-2x_4 - y_1 + y_6 \leq 0, \quad -2y_2 - y_3 + y_7 \leq 0,$$
$$-2y_4 - y_5 + y_8 \leq 0, \quad 0 \leq x_i \leq 1, \ i = 1, 2, 3, 4,$$
$$0 \leq y_i \leq 1, \ i = 1, 2, 3, 4, 5, 9, \quad 0 \leq y_i, \ i = 6, 7, 8.$$

The global solution is $(\boldsymbol{x}^*, \boldsymbol{y}^*) = (1, 1, 1, 1, 1, 1, 1, 1, 1, 3, 3, 3, 1)$, and $f(\boldsymbol{x}^*, \boldsymbol{y}^*) = -15$.

GENOCOP found the optimum in all ten runs; a typical optimum point found was:

$$(1.000000, 1.000000, 1.000000, 1.000000, 0.999995, 1.000000, 0.999999,$$
$$1.000000, 1.000000, 2.999984, 2.999995, 2.999995, 0.999999),$$

for which the value of the objective function is equal to -14.999965. A single run of 1000 iterations took 41 sec of CPU time.

Test Case #4
The problem [113] is

$$\text{maximize } f(\boldsymbol{x}) = \frac{3x_1 + x_2 - 2x_3 + 0.8}{2x_1 - x_2 + x_3} + \frac{4x_1 - 2x_2 + x_3}{7x_1 + 3x_2 - x_3},$$

subject to:

$$
\begin{array}{ll}
x_1 + x_2 - x_3 \le 1, & -x_1 + x_2 - x_3 \le -1, \\
12x_1 + 5x_2 + 12x_3 \le 34.8, & 12x_1 + 12x_2 + 7x_3 \le 29.1, \\
-6x_1 + x_2 + x_3 \le -4.1, & 0 \le x_i,\ i = 1, 2, 3.
\end{array}
$$

The global solution is $\boldsymbol{x}^* = (1, 0, 0)$, and $f(\boldsymbol{x}^*) = 2.471428$.

GENOCOP found the optimum in all ten runs; a single run of 500 iterations took 10 sec of CPU time.

Test Case #5
The problem[114] is

$$\text{minimize } f(\boldsymbol{x}) = x_1^{0.6} + x_2^{0.6} - 6x_1 - 4x_3 + 3x_4,$$

subject to:

$$
\begin{array}{ll}
-3x_1 + x_2 - 3x_3 = 0, & x_1 + 2x_3 \le 4, \\
x_2 + 2x_4 \le 4, & x_1 \le 3, \\
x_4 \le 1, & 0 \le x_i,\ i = 1, 2, 3, 4.
\end{array}
$$

The best known global solution is $\boldsymbol{x}^* = (\frac{4}{3}, 4, 0, 0)$, and $f(\boldsymbol{x}^*) = -4.5142$.

GENOCOP found this point in all ten runs; a single run of 500 iterations took 9 sec of CPU time.

Test Case #6
The problem[64] is

$$\text{minimize } f(\boldsymbol{x}) = 100(x_2 - x_1^2)^2 + (1 - x_1)^2 + 90(x_4 - x_3^2)^2 +$$
$$+ (1 - x_3)^2 + 10.1((x_2 - 1)^2 + (x_4 - 1)^2) + 19.8(x_2 - 1)(x_4 - 1),$$

subject to:

$$-10.0 \le x_i \le 10.0,\ i = 1, 2, 3, 4.$$

The global solution is $\boldsymbol{x}^* = (1, 1, 1, 1)$, and $f(\boldsymbol{x}^*) = 0$.

GENOCOP approached the optimum quite closely in all ten runs; a typical optimum point found was:

$$(x_1, x_2, x_3, x_4) = (1.000044, 1.000087, 0.999954, 0.999909),$$

for which the value of the objective function is equal to 0.00000001. A single run of 10,000 iterations took 159 sec of CPU time.

Test Case #7

The problem [114] is

$$\text{minimize } f(x, \boldsymbol{y}) = 6.5x - 0.5x^2 - y_1 - 2y_2 - 3y_3 - 2y_4 - y_5,$$

subject to:

$$x + 2y_1 + 8y_2 + y_3 + 3y_4 + 5y_5 \leq 16,$$
$$-8x - 4y_1 - 2y_2 + 2y_3 + 4y_4 - y_5 \leq -1,$$
$$2x + 0.5y_1 + 0.2y_2 - 3y_3 - y_4 - 4y_5 \leq 24,$$
$$0.2x + 2y_1 + 0.1y_2 - 4y_3 + 2y_4 + 2y_5 \leq 12,$$
$$-0.1x - 0.5y_1 + 2y_2 + 5y_3 - 5y_4 + 3y_5 \leq 3,$$
$$y_3 \leq 1, \ y_4 \leq 1, \text{ and } y_5 \leq 2,$$
$$x \geq 0, \ y_i \geq 0, \text{ for } 1 \leq i \leq 5.$$

The global solution is $(x, \boldsymbol{y}^*) = (0, 6, 0, 1, 1, 0)$, and $f(x, \boldsymbol{y}^*) = -11$.

GENOCOP found the optimum in all runs; a single run of 1000 iterations took 23 sec of CPU time.

Test Case #8

The problem was constructed from three separate problems [186] in the following way:

$$\text{minimize } f(\boldsymbol{x}) = \begin{cases} f_1 = x_2 + 10^{-5}(x_2 - x_1)^2 - 1.0 & \text{if } 0 \leq x_1 < 2 \\ f_2 = \frac{1}{27\sqrt{3}}((x_1 - 3)^2 - 9)x_2^3 & \text{if } 2 \leq x_1 < 4 \\ f_3 = \frac{1}{3}(x_1 - 2)^3 + x_2 - \frac{11}{3} & \text{if } 4 \leq x_1 \leq 6 \end{cases}$$

subject to:

$$x_1/\sqrt{3} - x_2 \geq 0,$$
$$-x_1 - \sqrt{3}x_2 + 6 \geq 0,$$
$$0 \leq x_1 \leq 6, \text{ and } x_2 \geq 0.$$

The function f has three global solutions:

$$\boldsymbol{x}_1^* = (0, 0), \ \boldsymbol{x}_2^* = (3, \sqrt{3}), \text{ and } \boldsymbol{x}_3^* = (4, 0),$$

in all cases $f(\boldsymbol{x}_i^*) = -1$ ($i = 1, 2, 3$).

We made three separate experiments. In experiment k ($k = 1, 2, 3$) all functions f_i except f_k were increased by 0.5. As a result, the global solution for the first experiment was $\boldsymbol{x}_1^* = (0, 0)$, the global solution for the second experiment was $\boldsymbol{x}_2^* = (3, \sqrt{3})$, and the global solution for the third experiment was $\boldsymbol{x}_3^* = (4, 0)$.

GENOCOP found global optima in all runs in all three cases; a single run of 500 iterations took 9 sec of CPU time.

Summary

As demonstrated earlier, GENOCOP worked very well for many experimental test problems with linear constraints. However, it is not clear how to generalize GENOCOP to handle nonlinear constraints (i.e., problems in class $\mathcal{NP}$). Some sets of nonlinear constraints can still yield a convex search spaces—the property important for many operators (all mutations, arithmetical crossover). However, even in this case the process of finding the ranges $left(k)$ and $right(k)$ might be hard computationally. Another possibility would be to cover the search space by a (not necessarily disjoint) family of convex subspaces and run GENOCOP on each of these. Again, the method would still have many computational problems.

For the above reasons, we looked at traditional optimization methods for further inspiration. Two particular methods (described in the following sections) were selected for 'mating' with the GENOCOP system.

7.2 Nonlinear optimization: GENOCOP II

In this section we discuss a new hybrid system, GENOCOP II, to solve nonlinear programming problems. The concept of the system is based on ideas taken from recent developments in the area of optimization [13] combined with iterative execution of GENOCOP (in this section we refer to GENOCOP as GENOCOP I to distinguish it from GENOCOP II).

Calculus-based methods assume that the objective function $f(\boldsymbol{x})$ and all constraints are twice continuously differentiable functions of $\boldsymbol{x}$. The general approach of most methods is to transform the nonlinear problem $\mathcal{NLP}$ into a sequence of solvable subproblems. The amount of work involved in a subproblem varies considerably among methods. These methods require explicit (or implicit) second derivative calculations of the objective (or transformed) function, which in some methods can be ill-conditioned and cause the algorithm to fail.

During the last 30 years there has been considerable research directed toward nonlinear optimization problems and progress has been made in theory and practice [114]. Several approaches have been developed in this area, including: the sequential quadratic penalty function [51], [13], recursive quadratic programming method [40], penalty trajectory methods [291], and the SOLVER method [112].

One of these approaches, the sequential quadratic penalty function method, was used as the main idea behind the GENOCOP II system. The method replaces a problem $\mathcal{NLP}$ by the problem $\mathcal{NLP}'$:

$$\text{optimize } F(\boldsymbol{x}, r) = f(\boldsymbol{x}) + \frac{1}{2r}\overline{C}^T\overline{C},$$

where $r > 0$ and $\overline{C}$ is a vector of all active constraints $c_1, \ldots, c_\ell$.

Fiacco and McCormick [253] have shown that the solutions of $\mathcal{NLP}$ and $\mathcal{NLP}'$ are identical in the limit as $r \longrightarrow 0$. It was thought that $\mathcal{NLP}'$ could then be solved simply by minimizing $F(x, r)$ for a sequence of decreasing positive values of r by Newton's method [111]. This hope, however, was short-lived, because minimizing $F(x, r)$ proved to be extremely inefficient for the smaller values of r; it was shown by Murray [291] that this was due to the Hessian matrix of $F(x, r)$ becoming increasingly ill-conditioned as $r \longrightarrow 0$. As there seemed to be no obvious way of overcoming this problem, the method gradually fell into disuse. More recently, Broyden and Attia [52], [51] offered a method of overcoming the numerical difficulties associated with the simple quadratic penalty function. The computation of the search direction does not require the solution of any system of linear equations, and can thus be expected to require less work than is needed for some other algorithms. The method also provides an automatic technique for calculating the initial value for the parameter r and its successive values [51].

The above technique together with the existing system GENOCOP I was used to construct a new system, GENOCOP II. The structure of GENOCOP II is given in Figure 7.1.

procedure GENOCOP II
begin
    $t \longleftarrow 0$
    split the set of constraints C into
        $C = L \cup N_e \cup N_i$
    select a starting point x_s
    set the set of active constraints, A to
        $A \longleftarrow N_e \cup V$
    set penalty $\tau \longleftarrow \tau_0$
    while (not termination-condition) **do**
    begin
        $t \longleftarrow t + 1$
        execute GENOCOP I for the function
          $F(x, \tau) = f(x) + \frac{1}{2\tau}\overline{A}^T\overline{A}$
          with linear constraints L
          and the starting point x_s
        save the best individual x^*:
          $x_s \longleftarrow x^*$
        update A:
          $A \leftarrow A - S \cup V,$
        decrease penalty τ:
          $\tau \leftarrow g(\tau, t)$
    end
  end

Fig. 7.1. The structure of GENOCOP II

There are several steps of the algorithm in the first phase of its execution (before it enters the *while* loop). The parameter t (which counts the number of iterations of the algorithm, i.e., the number of times the algorithm GENOCOP I is applied) is initialized to zero. The set of all constraints C is divided into three subsets: linear constraints L, nonlinear equations N_e and nonlinear inequalities N_i. A starting point x_s (which need not be feasible) for the following optimization process is selected (or a user is prompted for it). The set of active constraints A consists initially of elements of N_e and set $V \subseteq N_i$ of violated constraints from N_i. A constraint $c_j \in N_i$ is violated at point x iff $c_j(x) > \delta$ $(j = p+1, \ldots, m)$, where δ is a parameter of the method. Finally, the initial penalty coefficient of the system τ is set to τ_0 (a parameter of the method).

In the main loop of the algorithm we apply GENOCOP I to optimize a modified function

$$F(x, \tau) = f(x) + \frac{1}{2\tau}\overline{A}^T\overline{A}$$

with linear constraints L. Note that the initial population for GENOCOP I consists of *pop_size* identical copies (of the initial point for the first iteration and of the best saved point for subsequent ones); several mutation operators introduce diversity in the population at early stages of the process. When GENOCOP I converges, its best individual x^* is saved and used later as the starting point x_s for the next iteration. However, the next iteration is executed with a decreased value of the penalty parameter $(\tau \leftarrow g(\tau, t))$ and a new set of active constraints A:

$$A \leftarrow A - S \cup V,$$

where S and V are subsets of N_i satisfied and violated by x^*, respectively. Note that the decrease of τ results in an increase of the penalty.

The mechanism of the algorithm is illustrated on the following example. The problem is to

$$\text{minimize } f(x) = x_1 \cdot x_2^2,$$

subject to one nonlinear constraint:

$$c_1 : \quad 2 - x_1^2 - x_2^2 \geq 0.$$

The known global solution is $x^* = (-0.816497, -1.154701)$, and $f(x^*) = -1.088662$. The starting feasible point is $x_0 = (-0.99, -0.99)$. After the first iteration of GENOCOP II (A is empty) the system converged to $x_1 = (-1.5, -1.5)$, $f(x_1) = -3.375$. The point x_1 violates the constraint c_1, which becomes active. The point x_1 is used as the starting point for the second iteration. The second iteration ($\tau = 10^{-1}$, $A = \{c_1\}$) resulted in $x_2 = (-0.831595, -1.179690)$, $f(x_2) = -1.122678$. The point x_2 is used as the starting point for the third iteration. The third iteration ($\tau = 10^{-2}$, $A = \{c_1\}$) resulted in $x_3 = (-0.815862, -1.158801)$, $f(x_3) = -1.09985$. The sequence of points x_t (where $t = 4, 5, \ldots$ is the iteration number of the algorithm) approaches the optimum.

In order to evaluate the method of GENOCOP II, a set of test problems have been selected to indicate the performance of the algorithm and to illustrate that it has been successful in practice. The five test cases include quadratic, nonlinear, and discontinuous functions with several nonlinear constraints.

All runs of the system were performed on SUN SPARC station 2. We used the following parameters for GENOCOP I in all experiments:

$pop\_size = 70$, $k = 28$ (number of parents in each generation), $b = 6$ (coefficient for non-uniform mutation), $\delta = 0.01$ (parameter which determines whether or not a constraint is active).

In most cases, the initial penalty coefficient τ_0 was set at 1 (i.e., $g(\tau, 0) = 1$); additionally, $g(\tau, t) = 10^{-1} \cdot g(\tau, t - 1)$.

GENOCOP II was executed ten times for each test case. For most problems, the number of generations necessary for GENOCOP I to converge was 1000 (more difficult problems required a larger number of iterations). We did not report the computational times for these test cases, because we do not have full implementation of GENOCOP II yet. The actions of the system were simulated by executing its external loop in manual fashion: when GENOCOP I converges, the best point is incorporated as the starting point for the next iteration, the constraints are checked for their activity status, and the evaluation function is adjusted accordingly.

Test Case #1
The problem (taken from [186]) is

$$\text{minimize } f(\boldsymbol{x}) = 100(x_2 - x_1^2)^2 + (1 - x_1)^2,$$

subject to nonlinear constraints:

$$c_1 : \quad x_1 + x_2^2 \geq 0,$$
$$c_2 : \quad x_1^2 + x_2 \geq 0,$$

and bounds:

$$-0.5 \leq x_1 \leq 0.5, \text{ and } x_2 \leq 1.0.$$

The known global solution is $\boldsymbol{x}^* = (0.5, 0.25)$, and $f(\boldsymbol{x}^*) = 0.25$. The starting feasible point is $\boldsymbol{x}_0 = (0, 0)$.

GENOCOP II found the exact optimum in one iteration, as none of the nonlinear constraints are active at the optimum.

Test Case #2
The problem (taken from [113]) is

$$\text{minimize } f(x, y) = -x - y,$$

subject to nonlinear constraints:

$$c_1 : \quad y \leq 2x^4 - 8x^3 + 8x^2 + 2,$$
$$c_2 : \quad y \leq 4x^4 - 32x^3 + 88x^2 - 96x + 36,$$

and bounds:

$$0 \le x \le 3 \text{ and } 0 \le y \le 4.$$

The known global solution is $x^* = (2.3295, 3.1783)$, and $f(x^*) = -5.5079$. The starting feasible point is $x_0 = (0,0)$. The feasible region is almost disconnected.

GENOCOP II approached the optimum very closely at the 4th iteration. The progress of the system is shown in Table 7.1.

Iteration number	The best point	Active constraints
0	(0,0)	none
1	(3,4)	c_2
2	(2.06, 3.98)	c_1, c_2
3	(2.3298, 3.1839)	c_1, c_2
4	(2.3295, 3.1790)	c_1, c_2

Table 7.1. Progress of GENOCOP II on test case #2; for iteration 0 the best point is the starting point

Test Case #3

The problem (taken from [113]) is

$$\text{minimize } f(x) = (x_1 - 10)^3 + (x_2 - 20)^3,$$

subject to nonlinear constraints:

$$c_1 : \quad (x_1 - 5)^2 + (x_2 - 5)^2 - 100 \ge 0,$$
$$c_2 : \quad -(x_1 - 6)^2 - (x_2 - 5)^2 + 82.81 \ge 0,$$

and bounds:

$$13 \le x_1 \le 100 \text{ and } 0 \le x_2 \le 100.$$

The known global solution is $x^* = (14.095, 0.84296)$, and $f(x^*) = -6961.81381$ (see figure 7.2). The starting point, which is not feasible, is $x_0 = (20.1, 5.84)$.

GENOCOP II approached the optimum very closely at the 12th iteration. The progress of the system is shown in Table 7.2.

Test Case #4

The problem (taken from [39]) is

$$\text{minimize } f(x_1, x_2) = 0.01x_1^2 + x_2^2,$$

subject to nonlinear constraints:

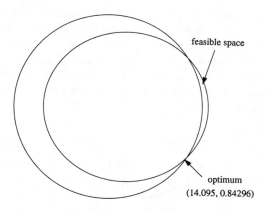

Fig. 7.2. A feasible space for test case #3

Iteration number	The best point	Active constraints
0	$(20.1, 5.84)$	c_1, c_2
1	$(13.0, 0.0)$	c_1, c_2
2	$(13.63, 0.0)$	c_1, c_2
3	$(13.63, 0.0)$	c_1, c_2
4	$(13.73, 0.16)$	c_1, c_2
5	$(13.92, 0.50)$	c_1, c_2
6	$(14.05, 0.75)$	c_1, c_2
7	$(14.05, 0.76)$	c_1, c_2
8	$(14.05, 0.76)$	c_1, c_2
9	$(14.10, 0.87)$	c_1, c_2
10	$(14.10, 0.86)$	c_1, c_2
11	$(14.10, 0.85)$	c_1, c_2
12	$(14.098, 0.849)$	c_1, c_2

Table 7.2. Progress of GENOCOP II on test case #3; for iteration 0 the best point is the starting point

$$c_1 : x_1 x_2 - 25 \geq 0,$$
$$c_2 : x_1^2 + x_2^2 - 25 \geq 0,$$

and bounds:

$$2 \leq x_1 \leq 50 \text{ and } 0 \leq x_2 \leq 50.$$

The global solution is $x^* = (\sqrt{250}, \sqrt{2.5}) = (15.811388, 1.581139)$, and $f(x^*) = 5.0$. The starting point (not feasible) is $x_0 = (2, 2)$.

It is interesting to note that the standard cooling scheme (i.e., $g(\tau, t) = 10^{-1} \cdot g(\tau, t-1)$) did not produce good results; however, when the cooling process was slowed down (i.e., $g(\tau, 0) = 5$ and $g(\tau, t) = 2^{-1} \cdot g(\tau, t-1)$), the system approached optimum easily (Table 7.3). This, of course, leads to some questions about how to control temperature for a given problem: this is one of the topics for future research.

Iteration number	The best point	Active constraints
0	(2,2)	c_1, c_2
1	(3.884181, 3.854748)	c_1
2	(15.805878, 1.581057)	c_1
3	(15.811537, 1.580996)	c_1

Table 7.3. Progress of GENOCOP II on test case #4; for iteration 0 the best point is the starting point

Test Case #5

The final test problem (taken from [186]) is

$$\text{minimize } f(x) = (x_1 - 2)^2 + (x_2 - 1)^2,$$

subject to a nonlinear constraint:

$$c_1 : \quad -x_1^2 + x_2 \geq 0,$$

and a linear constraint:

$$x_1 + x_2 \leq 2.$$

The global solution is $x^* = (1, 1)$ and $f(x^*) = 1$. The starting (feasible) point is $x_0 = (0, 0)$.

GENOCOP II approached the optimum very closely at the 6th iteration. The progress of the system is shown in Table 7.4.

Summary

There are several interesting points connected with the above method. First, like any other GA-based method, it does not require any implicit (or explicit) calculations of the gradient or Hessian matrix of the objective function and constraints. Consequently, the method does not suffer from the ill-conditioned Hessian problem usually associated with some calculus-based methods.

It should be noted that any genetic algorithm can be used in place of GENO-COP I for the inner loop of GENOCOP II. In such a case all constraints (linear

Iteration number	The best point	Active constraints
0	$(0, 0)$	c_1
1	$(1.496072, 0.503928)$	c_1
2	$(1.020873, 0.979127)$	c_1
3	$(1.013524, 0.986476)$	c_1
4	$(1.002243, 0.997757)$	c_1
5	$(1.000217, 0.999442)$	c_1
6	$(1.000029, 0.999971)$	c_1

Table 7.4. Progress of GENOCOP II on test case #5; for iteration 0 the best point is the starting point

and nonlinear) should be considered for placement in the set of active constraints A (the elements of L should be distributed between N_e and N_i). However, such a method is much slower and less effective: for efficiency reasons, it is much better to process linear constraints separately (as done in GENOCOP I).

7.3 Other techniques

During the last two years several methods were proposed for handling constraints by genetic algorithms for numerical optimization problems. Most of them are based on the concept of penalty functions, which penalize infeasible solutions, i.e.,[1]

$$eval(\boldsymbol{x}) = \begin{cases} f(\boldsymbol{x}), & \text{if } \boldsymbol{x} \text{ is feasible} \\ f(\boldsymbol{x}) + penalty(\boldsymbol{x}), & \text{otherwise,} \end{cases}$$

where $penalty(\boldsymbol{x})$ is zero, if no violation occurs, and is positive, otherwise. In most methods a set of functions f_j $(1 \leq j \leq m)$ is used to construct the penalty; the function f_j measures the violation of the j-th constraint in the following way:

$$f_j(\boldsymbol{x}) = \begin{cases} \max\{0, g_j(\boldsymbol{x})\}, & \text{if } 1 \leq j \leq q \\ |h_j(\boldsymbol{x})|, & \text{if } q+1 \leq j \leq m. \end{cases}$$

However, these methods differ in many important details as to how the penalty function is designed and applied to infeasible solutions. In the following subsections we discuss them in turn; the methods are sorted in decreasing order of parameters they require.

---

[1]In the rest of the section we assume minimization problems.

Method #1

This method was proposed by Homaifar et al. [195]. The method assumes that for every constraint we establish a family of intervals which determine an appropriate penalty coefficient. It works as follows:

- for each constraint, create several (ℓ) levels of violation,

- for each level of violation and for each constraint, create a penalty coefficient R_{ij} ($i = 1, 2, \ldots, \ell$, $j = 1, 2, \ldots, m$); higher levels of violation require larger values of this coefficient.

- start with a random population of individuals (feasible or infeasible),

- evolve the population; evaluate individuals using the formula

$$eval(\boldsymbol{x}) = f(\boldsymbol{x}) + \sum_{j=1}^{m} R_{ij} f_j^2(\boldsymbol{x}).$$

The weakness of the method is in the number of parameters: for m constraints the method requires m parameters to establish number of intervals for each constraint (in [195], these parameters are the same for all constraints equal to $\ell = 4$), plus ℓ parameters for each constraint (i.e., $\ell \times m$ parameters in total; these parameters represent boundaries of intervals or levels of violation), plus ℓ parameters for each constraint ($\ell \times m$ parameters in total; these parameters represent the penalty coefficients R_{ij}). So the method requires $m(2\ell + 1)$ parameters in total to handle m constraints. In particular, for $m = 5$ constraints and $\ell = 4$ levels of violation, we need to set 45 parameters! Clearly, the results are parameter dependent. It is quite likely that for a given problem there exists a unique optimal set of parameters for which the system returns feasible near-optimum solution, but it might be quite hard to find it.

Method #2

The second method was proposed by Joines and Houck [210]. As opposed to the previous method, the authors assumed dynamic penalties. Individuals are evaluated (at the iteration t) by the following formula:

$$eval(\boldsymbol{x}) = f(\boldsymbol{x}) + (C \times t)^\alpha \sum_{j=1}^{m} f_j^\beta(\boldsymbol{x}),$$

where C, α and β are constants. A reasonable choice for these parameters (reported in [210]) is $C = 0.5$, $\alpha = \beta = 2$. The method requires a much smaller number (independent of the number of constraints) of parameters than the first method. Also, instead of defining several levels of violation, the pressure on infeasible solutions is increased due to the $(C \times t)^\alpha$ component of the penalty term: towards the end of the process (for high values of the generation number t), this component assumes large values.

Method #3

The third method was proposed by Schoenauer and Xanthakis [346]; it works as follows:

- start with a random population of individuals (feasible or infeasible),

- set $j = 1$ (j is a constraint counter),

- evolve this population with $eval(\boldsymbol{x}) = f_j(\boldsymbol{x})$, until a given percentage of the population (so-called flip threshold ϕ) is feasible for this constraint,[2]

- set $j = j + 1$,

- the current population is the starting point for the next phase of the evolution, where $eval(\boldsymbol{x}) = f_j(\boldsymbol{x})$. During this phase, points that do not satisfy one of the 1st, 2nd, ... , or $(j-1)$-th constraint are eliminated from the population. The stop criterion is again the satisfaction of the j-th constraint by the flip threshold percentage ϕ of the population.

- if $j < m$, repeat the last two steps, otherwise ($j = m$) optimize the objective function, i.e., $eval(\boldsymbol{x}) = f(\boldsymbol{x})$, rejecting infeasible individuals.

The method requires a linear order of all constraints which are processed in turn. It is unclear what is the influence of the order of constraints on the results of the algorithm; our experiments indicated that different orders provide different results (different in the sense of the total running time and precision).

In total, the method requires 3 parameters: the sharing factor σ, the flip threshold ϕ, and a particular order of constraints. The method is very different from the previous two methods, and, in general, is different from other penalty approaches, since it considers only one constraint at the time. Also, in the last step of the algorithm the method optimizes the objective function f itself without any penalty component.

Method #4

Let us denote the GENOCOP II system as the method #4. As discussed earlier, this is the only method described here which distinguishes between linear and nonlinear constraints. The algorithm maintains feasibility of all linear constraints using a set of closed operators, which convert a feasible solution (feasible in terms of linear constraints only) into another feasible solution. At every iteration the algorithm considers active constraints only; the pressure on infeasible solutions is increased due to the decreasing values of temperature τ.

The method has an additional unique feature: it starts from a single point.[3] Consequently, it is relatively easy to compare this method with other classical optimization methods whose performance is tested (for a given problem) from some starting point.

The method requires a starting and 'freezing' temperatures, τ_0 and τ_f, respectively, and a cooling scheme to decrease temperature τ. Standard values (reported in [267]) are $\tau_0 = 1$, $\tau_{i+1} = 0.1 \cdot \tau_i$, with $\tau_f = 0.000001$.

---

[2]The method suggests the use of a sharing scheme (to maintain diversity of the population).

[3]This feature, however, is not essential. The only important requirement is that the next population contains the best individual from the previous population.

Method #5
The fifth method was developed by Powell and Skolnick [313]. The method is a classical penalty method with one notable exception. Each individual is evaluated by the formula:

$$eval(\boldsymbol{x}) = f(\boldsymbol{x}) + r \sum_{j=1}^{m} f_j(\boldsymbol{x}) + \lambda(t, \boldsymbol{x}),$$

where r is a constant; however, there is also a component $\lambda(t, \boldsymbol{x})$. This is an additional iteration dependent function which influences the evaluations of infeasible solutions. The point is that the method distinguishes between feasible and infeasible individuals by adopting an additional heuristic rule (suggested earlier in [332]): for any feasible individual $\boldsymbol{x}$ and any infeasible individual $\boldsymbol{y}$, $eval(\boldsymbol{x}) < eval(\boldsymbol{y})$, i.e., any feasible solution is better than any infeasible one. This can be achieved in many ways; one possibility is to set

$$\lambda(t, \boldsymbol{x}) = \begin{cases} 0, & \text{if } \boldsymbol{x} \in \mathcal{F} \\ \max\{0, \max_{x \in \mathcal{F}}\{f(\boldsymbol{x})\} - \\ \quad \min_{x \notin \mathcal{F}}\{f(\boldsymbol{x}) + r \sum_{j=1}^{m} f_j(\boldsymbol{x})\}\}, & \text{otherwise} \end{cases}$$

where $\mathcal{F}$ denotes the feasible part of the search space. In other words, infeasible individuals have increased penalties: their values cannot be better than the value of the worst feasible individual (i.e., $\max_{x \in \mathcal{F}}\{f(\boldsymbol{x})\}$).[4]

Method #6
The final method rejects infeasible individuals (death penalty); the method has been used by evolution strategies [18], evolutionary programming adopted for numerical optimization [117], and simulated annealing.

7.3.1 Five test cases

In the selection process of the following five test cases we took into account (1) the type of the objective function, (2) the number of variables, (3) the number of constraints, (4) the types of constraints, (5) the number of active constraints at the optimum, and (6) the ratio ρ between the sizes of the feasible search space and the whole search space. We do not make any claims on the completeness of the proposed set of these test cases G1–G5; however, it may constitute a handy collection for preliminary tests for other constraint handling methods.

Test Case #1
The problem [114] is to minimize a function:

$$G1(\boldsymbol{x}) = 5x_1 + 5x_2 + 5x_3 + 5x_4 - 5 \sum_{i=1}^{4} x_i^2 - \sum_{i=5}^{13} x_i,$$

subject to

---

[4]Powell and Skolnick achieved the same result by mapping evaluations of feasible solutions into the interval $(-\infty, 1)$ and infeasible solutions into the interval $(1, \infty)$. For ranking and tournament selections this implementational difference is not important.

$$2x_1 + 2x_2 + x_{10} + x_{11} \leq 10, \ 2x_1 + 2x_3 + x_{10} + x_{12} \leq 10,$$
$$2x_2 + 2x_3 + x_{11} + x_{12} \leq 10, \ -8x_1 + x_{10} \leq 0, \quad -8x_2 + x_{11} \leq 0,$$
$$-8x_3 + x_{12} \leq 0, \quad -2x_4 - x_5 + x_{10} \leq 0, \quad -2x_6 - x_7 + x_{11} \leq 0,$$
$$-2x_8 - x_9 + x_{12} \leq 0, \quad 0 \leq x_i \leq 1, \ i = 1, \ldots, 9,$$
$$0 \leq x_i \leq 100, \ i = 10, 11, 12, \quad 0 \leq x_{13} \leq 1.$$

The problem has 9 linear constraints; the function $G1$ is quadratic with its global minimum at

$$\boldsymbol{x}^* = (1, 1, 1, 1, 1, 1, 1, 1, 1, 3, 3, 3, 1),$$

where $G1(\boldsymbol{x}^*) = -15$. Six (out of nine) constraints are active at the global optimum (all except the following three: $-8x_1 + x_{10} \leq 0$, $-8x_2 + x_{11} \leq 0$, $-8x_3 + x_{12} \leq 0$).

Test Case #2

The problem [186] is to minimize a function:

$$G2(\boldsymbol{x}) = x_1 + x_2 + x_3,$$

where

$$1 - 0.0025(x_4 + x_6) \geq 0, \quad 1 - 0.0025(x_5 + x_7 - x_4) \geq 0,$$
$$1 - 0.01(x_8 - x_5) \geq 0, \quad x_1 x_6 - 833.33252 x_4 - 100 x_1 + 83333.333 \geq 0,$$
$$x_2 x_7 - 1250 x_5 - x_2 x_4 + 1250 x_4 \geq 0, \quad x_3 x_8 - 1250000 - x_3 x_5 + 2500 x_5 \geq 0,$$
$$100 \leq x_1 \leq 10000, \quad 1000 \leq x_i \leq 10000, \ i = 2, 3, \quad 10 \leq x_i \leq 1000,$$
$$i = 4, \ldots, 8.$$

The problem has 3 linear and 3 nonlinear constraints; the function $G2$ is linear and has its global minimum at

$$\boldsymbol{x}^* = (579.3167, 1359.943, 5110.071, 182.0174,$$
$$295.5985, 217.9799, 286.4162, 395.5979),$$

where $G2(\boldsymbol{x}^*) = 7049.330923$. All six constraints are active at the global optimum.

Test Case #3

The problem [186] is to minimize a function:

$$G3(\boldsymbol{x}) = (x_1 - 10)^2 + 5(x_2 - 12)^2 + x_3^4 + 3(x_4 - 11)^2 +$$
$$10x_5^6 + 7x_6^2 + x_7^4 - 4x_6 x_7 - 10x_6 - 8x_7,$$

where

$$127 - 2x_1^2 - 3x_2^4 - x_3 - 4x_4^2 - 5x_5 \geq 0,$$
$$282 - 7x_1 - 3x_2 - 10x_3^2 - x_4 + x_5 \geq 0,$$
$$196 - 23x_1 - x_2^2 - 6x_6^2 + 8x_7 \geq 0,$$
$$-4x_1^2 - x_2^2 + 3x_1 x_2 - 2x_3^2 - 5x_6 + 11x_7 \geq 0$$
$$-10.0 \leq x_i \leq 10.0, \ i = 1, \ldots, 7.$$

The problem has 4 nonlinear constraints; the function $G3$ is nonlinear and has its global minimum at

$$\boldsymbol{x}^* = (2.330499, 1.951372, -0.4775414,$$
$$4.365726, -0.6244870, 1.038131, 1.594227),$$

where $G3(\boldsymbol{x}^*) = 680.6300573$. Two (out of four) constraints are active at the global optimum (the first and the last one).

Test Case #4
The problem [186] is to minimize a function:

$$G4(\boldsymbol{x}) = e^{x_1 x_2 x_3 x_4 x_5},$$

subject to

$$x_1^2 + x_2^2 + x_3^2 + x_4^2 + x_5^2 = 10, \quad x_2 x_3 - 5x_4 x_5 = 0,$$
$$x_1^3 + x_2^3 = -1, \quad -2.3 \leq x_i \leq 2.3, \ i = 1, 2, \ -3.2 \leq x_i \leq 3.2,$$
$$i = 3, 4, 5.$$

The problem has 3 nonlinear equations; nonlinear function $G4$ has its global minimum at

$$\boldsymbol{x}^* = (-1.717143, 1.595709, 1.827247, -0.7636413, -0.7636450),$$

where $G4(\boldsymbol{x}^*) = 0.0539498478$.

Test Case #5
The problem [186] is to minimize a function:

$$G5(\boldsymbol{x}) = x_1^2 + x_2^2 + x_1 x_2 - 14x_1 - 16x_2 + (x_3 - 10)^2 + 4(x_4 - 5)^2 + (x_5 - 3)^2 +$$
$$2(x_6 - 1)^2 + 5x_7^2 + 7(x_8 - 11)^2 + 2(x_9 - 10)^2 + (x_{10} - 7)^2 + 45,$$

where

$$105 - 4x_1 - 5x_2 + 3x_7 - 9x_8 \geq 0,$$
$$-10x_1 + 8x_2 + 17x_7 - 2x_8 \geq 0,$$
$$8x_1 - 2x_2 - 5x_9 + 2x_{10} + 12 \geq 0,$$
$$-3(x_1 - 2)^2 - 4(x_2 - 3)^2 - 2x_3^2 + 7x_4 + 120 \geq 0,$$
$$-5x_1^2 - 8x_2 - (x_3 - 6)^2 + 2x_4 + 40 \geq 0,$$
$$-x_1^2 - 2(x_2 - 2)^2 + 2x_1 x_2 - 14x_5 + 6x_6 \geq 0,$$
$$-0.5(x_1 - 8)^2 - 2(x_2 - 4)^2 - 3x_5^2 + x_6 + 30 \geq 0,$$
$$3x_1 - 6x_2 - 12(x_9 - 8)^2 + 7x_{10} \geq 0,$$
$$-10.0 \leq x_i \leq 10.0, \quad i = 1, \ldots, 10.$$

The problem has 3 linear and 5 nonlinear constraints; the function $G5$ is quadratic and has its global minimum at

$$\boldsymbol{x}^* = (2.171996, 2.363683, 8.773926, 5.095984, 0.9906548,$$
$$1.430574, 1.321644, 9.828726, 8.280092, 8.375927),$$

where $G5(\boldsymbol{x}^*) = 24.3062091$. Six (out of eight) constraints are active at the global optimum (all except the last two).

Summary

All test cases are summarized in Table 7.5; for each test case (TC) we list number n of variables, type of the function f, the ratio ρ between sizes of the feasible and the whole search space, the number of constraints of each category (linear inequalities LI, nonlinear equations NE and inequalities NI), and the number a of active constraints at the optimum.

TC	n	Type of f	ρ	LI	NE	NI	a
#1	13	quadratic	0.0111%	9	0	0	6
#2	8	linear	0.0010%	3	0	3	6
#3	7	polynomial	0.5121%	0	0	4	2
#4	5	nonlinear	0.0000%	0	3	0	3
#5	10	quadratic	0.0003%	3	0	5	6

Table 7.5. Summary of five test cases. The ratio ρ was determined experimentally by generating 1,000,000 random points from the search space and checking whether they are feasible. LI, NE, and NI represent the number of linear inequalities, and nonlinear equations and inequalities, respectively.

7.3.2 Experiments

In all experiments we assumed floating point representation, nonlinear ranking selection, Gaussian mutation, arithmetical and heuristic crossovers; the probabilities of all operators were set at 0.08, and the population size was 70. For all methods the system was run for 5,000 generations.

The results are summarized in Tables 7.6 and 7.7, which report (for each method) the best (row b), median (row m), and the worst (row w) result (out of 10 independent runs) and numbers (row c) of violated constraints at the median solution: the sequence of three numbers indicate the number of violations with violation amount between 1.0 and 10.0, between 0.1 and 1.0, and between 0.001 and 0.1, respectively (a sequence of three zeros indicates a feasible solution). If at least one constraint was violated by more than 10.0 (in terms of functions f_j), the solution was considered as 'not meaningful'. In some cases it was hard to determine "the best" solution due to a relationship between the objective value and the number of violated constraints; the tables report the smallest objective value (for the best solution); consequently, some values are "better" than the value at the global minimum.

It is difficult to provide a complete analysis of all six methods on the basis of five test cases, however, it seems that method #1 can provide good results only if violation levels and penalty coefficients R_{ij} are tuned to the problem

TC	Exact opt.		Method #1	Method #2	Method #3
#1	−15.000	b	−15.002	−15.000	−15.000
		m	−15.002	−15.000	−15.000
		w	−15.001	−14.999	−14.998
		c	0, 0, 4	0, 0, 0	0, 0, 0
#2	7049.331	b	2282.723	3117.242	7485.667
		m	2449.798	4213.497	8271.292
		w	2756.679	6056.211	8752.412
		c	0, 3, 0	0, 3, 0	0, 0, 0
#3	680.630	b	680.771	680.787	680.836
		m	681.262	681.111	681.175
		w	689.660	682.798	685.640
		c	0, 0, 1	0, 0, 0	0, 0, 0
#4	0.054	b	0.084	0.059	
		m	0.955	0.812	*
		w	1.000	2.542	
		c	0, 0, 0	0, 0, 0	
#5	24.306	b	24.690	25.486	
		m	29.258	26.905	—
		w	36.060	42.358	
		c	0, 1, 1	0, 0, 0	

Table 7.6. Experimental results. For each method (#1, #2, and #3) we report the best (b), median (m), and the worst (w) result (out of 10 independent runs) and the number (c) of violated constraints at the median solution: the sequence of three numbers indicate the number of violations by more than 1.0, more than 0.1, and more than 0.001, respectively. The symbols '∗' and '—' stand for 'the method was not applied to this test case' and 'the results were not meaningful', respectively.

(e.g., our arbitrary choice of 4 violation levels with penalties of 100, 200, 500, and 1,000, respectively, worked well for the test cases #1 and #3, whereas it did not work well for other test cases, where some other values for these parameters are required). Also, these violation levels and penalty coefficients did not prevent the system from violating 3 constraints in the test case #2, since the 'reward' was too large to resist (i.e., the optimum value *outside* the feasible region is 2,100 for $x = (100.00, 1000.00, 1000.00, 128.33, 447.95, 336.07, 527.85, 578.08)$ with relatively small penalties). In all test cases (except test case #4) the method returned solutions which were infeasible by a relatively small margin, which is an interesting characteristic of this method.

Method #2 provided better results than the previous method for almost all test cases: for test case #1 (where all returned solutions were feasible), test cases #2 and #4 (where constraint violations were much smaller), and test case

Method #4	Method #5	Method #6	Method #6(f)
-15.000	-15.000		-15.000
-15.000	-15.000	—	-14.999
-15.000	-14.999		-13.616
0, 0, 0	0, 0, 0		0, 0, 0
7377.976	2101.367		7872.948
8206.151	2101.411	—	8559.423
9652.901	2101.551		8668.648
0, 0, 0	1, 2, 0		0, 0, 0
680.642	680.805	680.934	680.847
680.718	682.682	681.771	681.826
680.955	685.738	689.442	689.417
0, 0, 0	0, 0, 0	0, 0, 0	0, 0, 0
0.054	0.067		
0.064	0.091	*	*
0.557	0.512		
0, 0, 0	0, 0, 0		
18.917	17.388		25.653
24.418	22.932	—	27.116
44.302	48.866		32.477
0, 1, 0	1, 0, 0		0, 0, 0

Table 7.7. Continuation of Table 7.6; experimental results for methods #4, #5, and #6. The results for method #6 report also the results of experiments (method #6(f)) where the initial population was feasible.

#3 (the standard deviation of results was much smaller). On the other hand, method #2 seems to provide too strong penalties: often the factor $(C \times t)^\alpha$ grows too fast to be useful. The system has little chance of escaping from local optima: in most experiments the best individual was found in early generations. It is also worth mentioning that this method gave very good results for test cases #1 and #5, where the objective functions were quadratic.

Method #3 was not applied to test case #4, and it did not give meaningful results for test case #5, which has a very small ratio ρ (the smallest apart from the test case #4). Clearly, the method is quite sensitive to the size of the feasible part of the search space. Also, some additional experiments indicated that the order of constraints to be considered influenced the results in a significant way. On the other hand, the method performed very well for test cases #1 and #3, and for test case #2 it gave reasonable results.

Method #4 performed very well for the test cases #1, #3 and #4, where it provided the best results. It also gave a reasonable performance in test case #2 (where linear constraints were responsible for the failure of the methods

#1 and #2). However, for test case #5 the method gave quite poor results in comparison with methods #1, #2, and #6(f); it seems the linear constraints of this problem prevented the system from moving closer to the optimum. This is an interesting example of the damaging effect of limiting the population to the feasible (with respect to linear constraints) region only. Additional experiments indicated that the method is very sensitive to the cooling scheme. For example, the results of the test case #5 were improved in a significant way for different cooling scheme ($\tau_{i+1} = 0.01 \cdot \tau_i$).

Method #5 had difficulties in locating a feasible solution for test case #2: similarly to methods #1 and #2 the algorithm settled for infeasible solution. In all other test cases the method gave a stable, reasonable performance, returning feasible solutions (test cases #1, #3, and #5), or slightly infeasible solutions (test case #4). Additional experiments (not reported in the tables) included runs of the method #5 with a feasible initial population. For test case #2, the results were almost identical to these of method #6(f)). However, for test case #5, the results were excellent (the best of all methods).

The method #6 (apart from test case #3) did not produce any meaningful results. To test this method properly it was necessary to initialize a population by feasible solutions (method #6(f)). This different initialization scheme makes the comparison of these methods even harder. However, an interesting pattern emerged: the method generally gave a quite poor performance. The method was not as stable as other methods on the easiest test case #1 (this was the only method to return a solution of -13.616, far from the optimum), and for the test case #2 only in one run the returned value was below 8000.

No single parameter (number of linear, nonlinear, active constraints, the ratio ρ, type of the function, number of variables) proved its significance as a measure of difficulty of the problem for the evolutionary techniques. For example, all methods approached the optimum quite closely for the test cases #1 and #5 (with $\rho = 0.0111\%$ and $\rho = 0.0003\%$, respectively), whereas most of the methods experienced difficulties for the test case #2 (with $\rho = 0.0010\%$). Two quadratic functions (test cases #1 and #5) with a similar number of constraints (9 and 8, respectively) and an identical number (6) of active constraints at the optimum, gave a different challenge to most of these methods. It seems that other properties (the characteristic of the objective function together with the topology of the feasible region) constitute quite significant measures of the difficulty of the problem. Also, several methods were quite sensitive to the presence of a feasible solution in the initial population.

7.4 Other possibilities

As indicated earlier, several researchers studied heuristics on design of penalty functions. Some hypotheses were formulated in [332]:

- "penalties which are functions of the distance from feasibility are better performers than those which are merely functions of the number of violated constraints,

- for a problem having few constraints, and few full solutions, penalties which are solely functions of the number of violated constraints are not likely to find solutions,

- good penalty functions can be constructed from two quantities, the *maximum completion cost* and the *expected completion cost*,

- penalties should be close to the *expected completion cost*, but should not frequently fall below it. The more accurate the penalty, the better will be the solutions found. When penalty often underestimates the completion cost, then the search may not find a solution."

and in [358]:

- "the genetic algorithm with a variable penalty coefficient outperforms the fixed penalty factor algorithm,"

where variability of the penalty coefficient was determined by a heuristic rule.

This last observation was further investigated by Smith and Tate [360]. In their work they experimented with dynamic penalties, where the penalty measure depends on the number of violated constraints, the best feasible objective function found, and the best objective function value found.

Also, a method of adapting penalties was developed by Bean and Hadj-Alouane [27, 174]. It uses a penalty function, however, one component of the penalty function takes a feedback from the search process. Each individual is evaluated by the formula:

$$eval(\overline{X}) = f(\overline{X}) + \lambda(t) \sum_{j=1}^{m} f_j^2(\overline{X}),$$

where $\lambda(t)$ is updated every generation t in the following way:

$$\lambda(t+1) = \begin{cases} (1/\beta_1) \cdot \lambda(t), & \text{if } \overline{B}(i) \in \mathcal{F} \text{ for all } t-k+1 \leq i \leq t \\ \beta_2 \cdot \lambda(t), & \text{if } \overline{B}(i) \notin \mathcal{F} \text{ for all } t-k+1 \leq i \leq t \\ \lambda(t), & \text{otherwise,} \end{cases}$$

where $\overline{B}(i)$ denotes the best individual, in terms of function *eval*, in generation i, $\beta_1, \beta_2 > 1$ and $\beta_1 \neq \beta_2$ (to avoid cycling). In other words, the method (1) decreases the penalty component $\lambda(t+1)$ for the generation $t+1$, if all best individuals in the last k generations were feasible, and (2) increases penalties, if all best individuals in the last k generations were infeasible. If there are some feasible and infeasible individuals as best individuals in the last k generations, $\lambda(t+1)$ remains without change.

The above approach was applied for integer programming problems. The problem considered in [174] is minimize cx

subject to

$$Ax - b \geq 0, \tag{1}$$
$$\sum_{j=1}^{n_i} x_{ij} = 1, \text{ for } i = 1, \ldots, m, \tag{2}$$
$$x_{ij} \in \{0, 1\}. \tag{3}$$

For each set, $i = 1, \ldots, m$, equations (2) and (3) force exactly one variable in $\{x_{ij}\}_{j=1}^{n_i}$ to be one. Matrix A is $k \times n$ $(n = \sum i = 1^m n_i)$ and b is a constant k-dimensional vector.

As reported in [174], most successful techniques for solving the above problem are branch and bound methods that use either linear programming, Lagrangian relaxation or its variations. Lagrangian relaxation drops some constraints by incorporating a weighted linear penalty for constraint violation. The "correct" weights may result in very good bounds or even optimal solutions to the original problem; a typical Lagrangian relaxation replaces the original problem (the constraints (1)) by the following formulation

minimize $cx - \lambda(Ax - b)$

subject to

$$Ex = e_m, \; x_{ij} \in \{0, 1\}$$

(constraints $Ex = e_m$ are the multiple choice constraints, where e_m is a vector of ones).

The proposed approach replaces the original problem by

minimize $cx + p_\lambda(x)$,

subject to

$$Ex = e_m, \; x_{ij} \in \{0, 1\}$$

where $p_\lambda(x) = \sum_{i=1}^{k} \lambda_i [\min\{0, A_i x - b_i\}]^2$. The function is nonlinear and a genetic algorithm is used to optimize the expression.

There are interesting ideas present in the proposed method. First of all, results of experiments indicated that starting with high values for λ did not lead to efficient algorithm. So the proposed algorithm adjusts the vector λ during the run: a sequence of increasing λ vectors is employed. The experiments [174] indicated that the rate at which this sequence is increased is extremely important (a similar observation was made on the basis of experimental results of GENOCOP II): a slow rate would improve the quality of solutions at the expense of rate of improvement and a fast rate may result in an inefficient genetic evolution. Additionally, the proposed genetic algorithm uses a technique of random keys, where a solution is represented as a vector of random numbers: their sorted order decodes the solution (note the similarity between this method and application of evolution strategies to the traveling salesman problem; see

Chapter 8 on evolution strategies and Chapter 10 on the traveling salesman problem). A solution is represented by a string of length equal to the number of multiple choice sets. Each position i of the string can take any integer in $\{1, \ldots, n_i\}$, where n_i is the number of variables in the multiple choice set.

Other constraint handling methods deserve also some attention. One of them (GENOCOP III, discussed in the next section) is based on repair algorithms: an infeasible solution x is "forced" into the feasible region and its repaired version is evaluated. It is also possible to construct a hybrid algorithms which incorporate some deterministic optimization procedure; for a short survey and a description of one particular method, see [294].

An additional possibility would include the use of the values of objective function f and penalties f_j as elements of a vector and applying multi-objective techniques to minimize all components of the vector. In other words, objective function f and constraint violation measures f_j (for m constraints) constitute a $(m + 1)$-dimensional vector v:

$$v = (f, f_1, \ldots, f_m).$$

Using some multi-objective optimization method, we can attempt to minimize its components: an ideal solution x would have $f_j(x) = 0$ for $1 \le i \le m$ and $f(x) \le f(y)$ for all feasible y (minimization problems). A successful implementation of this approach was presented recently in Surry et al. [376].

Yet another approach was proposed recently by Le Riche et al. [239]. The authors designed a (segregated) genetic algorithm which uses two values of penalty parameters (for each constraint) instead of one; these two values aim at achieving a balance between heavy and moderate penalties by maintaining two subpopulations of individuals. The population is split into two cooperating groups, where individuals in each group are evaluated using either one of the two penalty parameters.

Also, an interesting approach was recently reported by Paredis [308]. The method (described in the context of constraint satisfaction problems) is based on a co-evolutionary model, where a population of potential solutions co-evolves with a population of constraints: fitter solutions satisfy more constraints, whereas fitter constraints are violated by more solutions. This means that individuals from the population of solutions are considered from the whole search space, and that there is no distinction between feasible and infeasible individuals. The evaluation of an individual is determined on the basis of constraint violations measures f_j's; however, better f_j's (e.g., active constraints) would contribute more towards evaluating the solution. It would be interesting to adopt this approach to constrained numerical optimization problems and compare it with the other methods. But the major difficulty to be resolved in such adaptation seems very much the same as in many other methods: how to balance the pressure of feasibility of a solution with the pressure to minimize the objective function.

The research on cultural algorithms [327, 328, 330, 331] was triggered by observations that culture might be another kind of inheritance system. But it is not clear what the appropriate structures and units to represent the adaptation and transmission of cultural information are. Neither it is clear how to describe the interaction between natural evolution and culture. Reynolds developed a few models to investigate the properties of cultural algorithms; in these models, the belief space is used to constrain the combination of traits that individuals can assume. Changes in the belief space represent macroevolutionary change and changes in the population of individuals represent microevolutionary change. Both changes are moderated by the communication link.

The general intuition behind belief spaces is to preserve those beliefs associated with "acceptable" behavior at the trait level (and, consequently, to prune away unacceptable beliefs). The acceptable beliefs serve as constraints that direct the population of traits. It seems that the cultural algorithms may serve as a very interesting tool for numerical optimization problems, where constraints influence the search in a direct way (consequently, the search in constrained spaces may be more efficient than in unconstrained ones!). Very recently Reynolds et al. [329] investigated a possibility of applying cultural algorithms for constrained numerical optimization. The first experiments indicate a great potential behind this approach.

7.5 GENOCOP III

This method incorporates the original GENOCOP system (described in section 7.1, but also extends it by maintaining two separate populations, where a development in one population influences evaluations of individuals in the other population. The first population P_s consists of so-called search points which satisfy linear constraints of the problem (as in the original GENOCOP system). The feasibility (in the sense of linear constraints) of these points is maintained, as before, by specialized operators. The second population P_r consists of so-called reference points; these points are fully feasible, i.e., they satisfy all constraints (if GENOCOP III has difficulties in locating such a reference point for the purpose of initialization, the user is prompted for it. In cases where the ratio ρ between the sizes of feasible and the whole search spaces is very small, it may happen that the initial set of reference points consists of a multiple copies of a single feasible point). Figure 7.3 illustrates these two populations.

Reference points $\overline{R}$, being feasible, are evaluated directly by the objective function (i.e., $eval(\overline{R}) = f(\overline{R})$). On the other hand, infeasible search points are "repaired" for evaluation and the repair process works as follows. Assume, there is a search point $\overline{S}$ is not fully feasible. In such a case the system selects (better reference points have better chances to be selected; a selection method based on nonlinear ranking was used) one of the reference points, say $\overline{R}$, and creates random points $\overline{Z}$ from a segment between $\overline{S}$ and $\overline{R}$ by generating random numbers a from the range $\langle 0, 1 \rangle$: $\overline{Z} = a\overline{S} + (1 - a)\overline{R}$. Figure 7.4 illustrates this repair process.

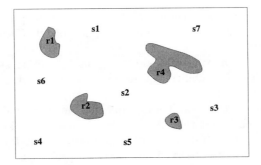

Fig. 7.3. Population $P_s = \{\overline{S_1}, \overline{S_2}, \overline{S_3}, \overline{S_4}, \overline{S_5}\}$ and population $P_r = \{\overline{R_1}, \overline{R_2}, \overline{R_3}, \overline{R_4}\}$

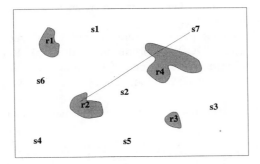

Fig. 7.4. Evaluation of infeasible point $\overline{S}$

Once a feasible $\overline{Z}$ is found, $eval(\overline{S}) = eval(\overline{Z}) = f(\overline{Z})$.[5] Additionally, if $f(\overline{Z})$ is better than $f(\overline{R})$, then the point $\overline{Z}$ replaces $\overline{R}$ as a new reference point. Also, $\overline{Z}$ replaces $\overline{S}$ with some probability of replacement p_r.

GENOCOP III avoids many disadvantages of other systems. It introduces few additional parameters (the population size of reference points, probability of replacement) only. It always returns a feasible solution. A feasible search space is searched by making references from the search points. The neighborhoods of better reference points are explored more often. Some reference points are moved into the population of search-points, where they undergo transformation by specialized operators (which preserve linear constraints).

The preliminary version of GENOCOP III (available from ftp.uncc.edu, directory coe/evol, file genocopIII.tar.Z) was tested on the test cases $G1 - G5$ (given in section 7.3.1). The results of GENOCOP III on $G1$ are identical to these of the original GENOCOP: since there are no nonlinear constraints, there is no need for a population of reference points. Out of the remaining four test

---

[5]Clearly, in different generations the same search point S can evaluate to different values due to the random nature of the repair process.

cases, we experimented with three ($G2$, $G3$, and $G5$); the problem $G4$ contains nonlinear equations NE, and the current version of GENOCOP III does not handle them yet.

GENOCOP III was run for 5,000 iterations (like other systems discussed in section 7.3). The probabilities of all operators were set at 0.08, and both population sizes were 70.

The results were very good. For example, for the problem $G2$ the best result was 7286.650, much better than the best result of the best system from those discussed in section 7.3.1 (for this problem, it was GENOCOP II with 7377.976). Similar performance was observed on two other problems, $G3$ (with 680.640) and $G5$ (with 25.883). Another interesting observation was connected with the stability of the system. GENOCOP III had a very low standard deviation of results. For example, for problem $G3$, all results were between 680.640 and 680.889; on the other hand, other systems produced a variety of results (between 680.642 and 689.660, see [266]).

Of course, all resulting points $\overline{X}$ were feasible, which was not the case with other systems (e.g., GENOCOP II produced a value of 18.917 for the problem $G5$, the systems based on the methods of Homaifar, Lai, and Qi, and Powell and Skolnick gave results of 2282.723 and 2101.367, respectively, for the problem $G2$).

Additional interesting test case emerged recently; the problem [220] is to maximize a function:

$$f(\boldsymbol{x}) = \left| \frac{\sum_{i=1}^{n} \cos^4(x_i) - 2 \prod_{i=1}^{n} \cos^2(x_i)}{\sqrt{\sum_{i=1}^{n} i x_i^2}} \right|,$$

where

$$\prod_{i=1}^{n} x_i > 0.75, \quad \sum_{i=1}^{n} x_i < 7.5n, \text{ and } 0 < x_i < 10 \text{ for } 1 \leq i \leq n.$$

The problem has 2 nonlinear constraints; the function f is nonlinear and its global maximum is unknown.

Keane [220] noted:

"I am currently using a parallel GA with 12-bit binary encoding, crossover, inversion, mutation, niche forming and a modified Fiacco-McCormick constraint penalty function to tackle this. For $n = 20$ I get values like 0.76 after 20,000 evaluations."

GENOCOP III was run for cases of $n = 20$ and $n = 50$. In the former case, the best solution found (in 10,000 generations) was

$$\boldsymbol{x} = (3.16311359, 3.13150430, 3.09515858, 3.06016588, 3.03103566,$$
$$2.99158549, 2.95802593, 2.92285895, 0.48684388, 0.47732279,$$
$$0.48044473, 0.48790911, 0.48450437, 0.44807032, 0.46877760,$$
$$0.45648506, 0.44762608, 0.44913986, 0.44390863, 0.45149332),$$

where $f(\boldsymbol{x}) = 0.80351067$. In the latter case ($n = 50$), the best solution found (in 10,000 generations) was

$$x = (6.28006029, 3.16155291, 3.15453815, 3.14085174, 3.12882447,$$
$$3.11211085, 3.10170507, 3.08703685, 3.07571769, 3.06122732,$$
$$3.05010581, 3.03667951, 3.02333045, 3.00721049, 2.99492717,$$
$$2.97988462, 2.96637058, 2.95589066, 2.94427204, 2.92796040,$$
$$0.40970641, 2.90670991, 0.46131119, 0.48193336, 0.46776962,$$
$$0.43887550, 0.45181099, 0.44652876, 0.43348753, 0.44577143,$$
$$0.42379948, 0.45858049, 0.42931050, 0.42928645, 0.42943302,$$
$$0.43294361, 0.42663351, 0.43437257, 0.42542559, 0.41594154,$$
$$0.43248957, 0.39134723, 0.42628688, 0.42774364, 0.41886297,$$
$$0.42107263, 0.41215360, 0.41809589, 0.41626775, 0.42316407),$$

where $f(x) = 0.83319378.$[6]

Clearly, GENOCOP III is a promising tool for constrained nonlinear optimization problems. However, there are many issues which require further attention and experiments. These include investigation of the significance of the ratio of ρ; note that it is possible to represent some linear constraints as nonlinear constraints; this change in the input file would make the space of reference points smaller and the space of linearly feasible search points larger. However, it is unclear how these changes would affect the performance of the system.

Another group of experiments is connected with a single parameter: probability of replacement p_r. In all experiments reported above, $p_r = 0.15$.

Also, we plan (in a very near future) to extend GENOCOP III to handle nonlinear equations. This would require an additional parameter (ϵ) to define the precision of the system. All nonlinear equations $h_j(\overline{X}) = 0$ (for $j = q + 1, \ldots, m$) would be replaced by a pair of inequalities:

$$-\epsilon \leq h_j(\overline{X}) \leq \epsilon.$$

This new version of GENOCOP III should handle the problem $G4$ directly.

---

[6]This is not the global optimum; Bilchev [43] reported value of the objective function of 0.8348.

8. Evolution Strategies and Other Methods

It were not best that we should think alike;
it is difference of opinion
that makes horse races.

Mark Twain, *Pudd'nhead Wilson*

Evolution strategies (ESs) are algorithms which imitate the principles of natural evolution as a method to solve parameter optimization problems [18], [348]. They were developed in Germany during the 1960s. As stated in [348]:

> "In 1963 two students at the Technical University of Berlin met and were soon collaborating on experiments which used the wind tunnel of the Institute of Flow Engineering. During the search for the optimal shapes of bodies in a flow, which was then a matter of laborious intuitive experimentation, the idea was conceived of proceeding strategically. However, attempts with the coordinate and simple gradient strategies were unsuccessful. Then one of the students, Ingo Rechenberg, now Professor of Bionics and Evolutionary Engineering, hit upon the idea of trying random changes in the parameters defining the shape, following the example of natural mutations. The evolution strategy was born."

(The second student was Hans-Paul Schwefel, now Professor of Computer Science and Chair of System Analysis).

Early evolution strategies may be perceived as evolution programs where a floating point number representation is used, with mutation being the only recombination operator. They have been applied to various optimization problems with continuously changeable parameters. Only recently they were extended for discrete problems [18], [178].

In this chapter we describe the early ESs based on a two-member population and the mutation operator, and various multimembered population ESs (section 8.1). Section 8.2 compares ESs with GAs, whereas section 8.4 presents other strategies proposed recently by various researchers.

olution of evolution strategies

st evolution strategies were based on a population consisting of one individual only. There was also only one genetic operator used in the evolution process: a mutation. However, the interesting idea (not present in GAs) was to represent an individual as a pair of float–valued vectors, i.e., $v = (x, \sigma)$. Here, the first vector x represents a point in the search space; the second vector σ is a vector of standard deviations: mutations are realized by replacing x by

$$x^{t+1} = x^t + N(0, \sigma),$$

where $N(0, \sigma)$ is a vector of independent random Gaussian numbers with a mean of zero and standard deviations σ. (This is in accordance with the biological observation that smaller changes occur more often than larger ones.) The offspring (the mutated individual) is accepted as a new member of the population (it replaces its parent) iff it has better fitness and all constraints (if any) are satisfied. For example, if f is the objective function without constraints to be maximized, an offspring (x^{t+1}, σ) replaces its parent (x^t, σ) iff $f(x^{t+1}) > f(x^t)$. Otherwise, the offspring is eliminated and the population remain unchanged.

Let us illustrate a single step of such an evolution strategy, considering a maximization problem we used as an example (for a simple genetic algorithm) in Chapter 2:

$$f(x_1, x_2) = 21.5 + x_1 \cdot \sin(4\pi x_1) + x_2 \cdot \sin(20\pi x_2),$$

where $-3.0 \leq x_1 \leq 12.1$ and $4.1 \leq x_2 \leq 5.8$.

As explained earlier, a population would consist of a single individual (x, σ), where $x = (x_1, x_2)$ is a point within the search space $(-3.0 \leq x_1 \leq 12.1$ and $4.1 \leq x_2 \leq 5.8)$ and $\sigma = (\sigma_1, \sigma_2)$ represents two standard deviations to be used for the mutation operation. Let us assume that at some time t the single element population consists of the following individual:

$$(x^t, \sigma) = ((5.3, 4.9), (1.0, 1.0)),$$

and that the mutation results in the following change:

$$x_1^{t+1} = x_1^t + N(0, 1.0) = 5.3 + 0.4 = 5.7$$
$$x_2^{t+1} = x_2^t + N(0, 1.0) = 4.9 - 0.3 = 4.6.$$

Since

$$f(x^t) = f(5.3, 4.9) = 18.383705 < 24.849532 = f(5.7, 4.6) = f(x^{t+1}),$$

and both x_1^{t+1} and x_2^{t+1} stay within their ranges, the offspring will replace its parent in the single-element population.

Despite the fact that the population consists of a single individual which undergoes mutation, the evolution strategy discussed above is called a "two-membered evolution strategy". The reason is that the offspring competes with its parent and at the competition stage there are (temporarily) two individuals in the population.

The vector of standard deviations $\boldsymbol{\sigma}$ remains unchanged during the evolution process. If all components of this vector are identical, i.e., $\boldsymbol{\sigma} = (\sigma, \dots, \sigma)$, and the optimization problem is *regular*[1], it is possible to prove the convergence theorem [18]:

Theorem 1 (Convergence Theorem.) *For $\sigma > 0$ and a regular optimization problem with $f_{opt} > -\infty$ (minimalization) or $f_{opt} < \infty$ (maximization),*

$$p\{\lim_{t\to\infty} f(\boldsymbol{x}^t) = f_{opt}\} = 1$$

holds.

The Convergence Theorem states that the global optimum is found with probability one for sufficiently long search time; however, it does not provide any clues for the convergence rate (quotient of the distance covered towards the optimum and the number of elapsed generations needed to cover this distance). To optimize the convergence rate, Rechenberg proposed a "1/5 success rule":

The ratio φ of successful mutations to all mutations should be 1/5. Increase the variance of the mutation operator, if φ is greater than 1/5; otherwise, decrease it.

The 1/5 success rule emerged as a conclusion of the process of optimizing convergence rates of two functions (the so-called corridor model and sphere model; see [18] for details). The rule was applied every k generations (k is another parameter of the method): 1/5 success rule

$$\boldsymbol{\sigma}^{t+1} = \begin{cases} c_d \cdot \boldsymbol{\sigma}^t, & \text{if } \varphi(k) < 1/5, \\ c_i \cdot \boldsymbol{\sigma}^t, & \text{if } \varphi(k) > 1/5, \\ \boldsymbol{\sigma}^t, & \text{if } p_s(k) = 1/5, \end{cases}$$

where $\varphi(k)$ is the success ratio of the mutation operator during the last k generations, and $c_i > 1$, $c_d < 1$ regulate the increase and decrease rates for the variance of the mutation. Schwefel in his experiments [348] used the following values: $c_d = 0.82$, $c_i = 1.22 = 1/0.82$.

The intuitive reason behind the 1/5 success rule is the increased efficiency of the search: if successful, the search would continue in "larger" steps; if not, the steps would be shorter. However, this search may lead to premature convergence

---

[1]An optimization problem is regular if the objective function f is continuous, the domain of the function is a closed set, for all $\epsilon > 0$ the set of all internal points of the domain for which the function differs from the optimal value less than ϵ is non-empty, and for all $\boldsymbol{x}_0$ the set of all points for which the function has values less than or equal to $f(\boldsymbol{x}_0)$ (for minimalization problems; for maximization problems the relationship is opposite) is a closed set.

for some classes of functions — this resulted in a refinement of the method: increased population size.

The multimembered evolution strategy differs from the previous two-membered strategy in the size of the population ($pop\_size > 1$). Additional features of multimembered evolution strategies are:

- all individuals in the populations have the same mating probabilities,

- possibility of introduction of a recombination operator (in the GA community called "uniform crossover"), where two (randomly selected) parents,

$$(\boldsymbol{x}^1, \boldsymbol{\sigma}^1) = ((x_1^1, \ldots, x_n^1), (\sigma_1^1, \ldots, \sigma_n^1)) \text{ and }$$
$$(\boldsymbol{x}^2, \boldsymbol{\sigma}^2) = ((x_1^2, \ldots, x_n^2), (\sigma_1^2, \ldots, \sigma_n^2)),$$

produce an offspring,

$$(\boldsymbol{x}, \boldsymbol{\sigma}) = ((x_1^{q_1}, \ldots, x_n^{q_n}), (\sigma_1^{q_1}, \ldots, \sigma_n^{q_n})),$$

where $q_i = 1$ or $q_i = 2$ with equal probability for all $i = 1, \ldots, n$.

The mutation operator and the adjustment of $\boldsymbol{\sigma}$ remain without changes.

There is still a similarity between two-membered and multimembered evolution strategies: both of them produce a single offspring. In the two-membered strategy, the offspring competes against its parent. In the multimembered strategy the weakest individual (among $pop\_size + 1$ individuals; i.e., original pop_size individuals plus one offspring) is eliminated. A convenient notation, which explains also further refinement of evolution strategies, is:

$(1 + 1)$–ES, for a two membered evolution strategy, and
$(\mu + 1)$–ES, for a multimembered evolution strategy,

where $\mu = pop\_size$.

The multimembered evolution strategies evolved further [348] to mature as

$(\mu + \lambda)$–ESs and (μ, λ)–ESs;

the main idea behind these strategies was to allow control parameters (like mutation variance) to self-adapt rather than changing their values by some deterministic algorithm.

The $(\mu+\lambda)$–ES is a natural extension of a multimembered evolution strategy $(\mu + 1)$–ES, where μ individuals produce λ offspring. The new (temporary) population of $(\mu + \lambda)$ individuals is reduced by a selection process again to μ individuals. On the other hand, in the (μ, λ)–ES, the μ individuals produce λ offspring ($\lambda > \mu$) and the selection process selects a new population of μ individuals from the set of λ offspring only. By doing this, the life of each individual is limited to one generation. This allows the (μ, λ)–ES to perform better on problems with an optimum moving over time, or on problems where the objective function is noisy.

The operators used in the $(\mu + \lambda)$–ESs and (μ, λ)–ESs incorporate two-level learning: their control parameter $\boldsymbol{\sigma}$ is no longer constant, nor it is changed by some deterministic algorithm (like the 1/5 success rule), but it is incorporated in the structure of the individuals and undergoes the evolution process. To produce an offspring, the system acts in several stages:

- select two individuals,

$$(\boldsymbol{x}^1, \boldsymbol{\sigma}^1) = ((x_1^1, \ldots, x_n^1), (\sigma_1^1, \ldots, \sigma_n^1)) \text{ and}$$
$$(\boldsymbol{x}^2, \boldsymbol{\sigma}^2) = ((x_1^2, \ldots, x_n^2), (\sigma_1^2, \ldots, \sigma_n^2)),$$

  and apply a recombination (crossover) operator. There are two types of crossovers:

  - discrete, where the new offspring is

$$(\boldsymbol{x}, \boldsymbol{\sigma}) = ((x_1^{q_1}, \ldots, x_n^{q_n}), (\sigma_1^{q_1}, \ldots, \sigma_n^{q_n})),$$

    where $q_i = 1$ or $q_i = 2$ (so each component comes from the first or second preselected parent),

  - intermediate, where the new offspring is

$$(\boldsymbol{x}, \boldsymbol{\sigma}) = (((x_1^1 + x_1^2)/2, \ldots, (x_n^1 + x_n^2)/2), ((\sigma_1^1 + \sigma_1^2)/2, \ldots, (\sigma_n^1 + \sigma_n^2)/2)).$$

  Each of these operators can be applied also in a global mode, where the new pair of parents is selected for *each* component of the offspring vector.

- apply mutation to the offspring $(\boldsymbol{x}, \boldsymbol{\sigma})$ obtained; the resulting new offspring is $(\boldsymbol{x}', \boldsymbol{\sigma}')$, where

$$\boldsymbol{\sigma}' = \boldsymbol{\sigma} \cdot e^{N(0, \Delta\sigma)} \text{ and}$$
$$\boldsymbol{x}' = \boldsymbol{x} + N(0, \boldsymbol{\sigma}'),$$

  where $\Delta\boldsymbol{\sigma}$ is a parameter of the method.

To improve the convergence rate of ESs, Schwefel [348] introduced an additional control parameter $\boldsymbol{\theta}$. This new control correlates mutations. For the ESs discussed so far (with the dedicated σ_i for each x_i), the preferred direction of the search can be established only along the axes of the coordinate system. Now, each individual in the population is represented as

$$(\boldsymbol{x}, \boldsymbol{\sigma}, \boldsymbol{\theta}).$$

The recombination operators are similar to those discussed in the previous paragraph, and the mutation creates offspring $(\boldsymbol{x}', \boldsymbol{\sigma}', \boldsymbol{\theta}')$ from the $(\boldsymbol{x}, \boldsymbol{\sigma}, \boldsymbol{\theta})$ in the following way:

$$\boldsymbol{\sigma}' = \boldsymbol{\sigma} \cdot e^{N(0, \Delta\sigma)},$$
$$\boldsymbol{\theta}' = \boldsymbol{\theta} + N(0, \Delta\theta), \text{ and}$$
$$\boldsymbol{x}' = \boldsymbol{x} + C(0, \boldsymbol{\sigma}', \boldsymbol{\theta}'),$$

where $\Delta\boldsymbol{\theta}$ is an additional parameter of the method, and $C(0, \boldsymbol{\sigma}', \boldsymbol{\theta}')$ denotes a vector of independent random Gaussian numbers with mean zero and appropriate probability density (for details, see [348] or [18]).

Evolution strategies perform very well in numerical domains, since they were (at least, initially) dedicated to (real) function optimization problems. They are examples of evolution programs which use appropriate data structures (float vectors extended by control strategy parameters) and "genetic" operators for the problem domain.

It is interesting to compare genetic algorithms and evolution strategies, their differences and similarities, their strengths and weaknesses. We discuss these issues in the following section.

8.2 Comparison of evolution strategies and genetic algorithms

The basic difference between evolution strategies and genetic algorithms lies in their domains. Evolution strategies were developed as methods for numerical optimization. They adopt a special hill-climbing procedure with self-adapting step sizes $\boldsymbol{\sigma}$ and inclination angles $\boldsymbol{\theta}$. Only recently have ESs been applied to discrete optimization problems [178]. On the other hand, genetic algorithms were formulated as (general purpose) adaptive search techniques, which allocate exponentially increasing number of trials for above-average schemata. GAs were applied in a variety of domains, and a (real) parameter optimization was just one field of their applications.

For that reason, it is unfair to compare the time and precision performance of ESs and GAs using a numerical function as the basis for the comparison. However, both ESs and GAs are examples of evolution programs and some general discussion of similarities and differences between them is quite natural.

The major similarity between ESs and GAs is that both systems maintain populations of potential solutions and make use of the selection principle of the survival of the fitter individuals. However, there are many differences between these approaches.

The first difference between ESs and classical GAs is in the way they represent the individuals. As mentioned on several occasions, ESs operate on floating point vectors, whereas classical GAs operate on binary vectors.

The second difference between GAs and ESs is hidden in the selection process itself. In a single generation of the ES, μ parents generate intermediate population which consists of λ offspring produced by means of the recombination and mutation operators (for $(\mu + \lambda)$–ES), plus (for (μ, λ)–ES) the original μ parents. Then the selection process reduces the size of this intermediate population back to μ individuals by removing the least fit individuals from the population. This population of μ individuals constitutes the next generation. In a single generation of the GA, a selection procedure selects pop_size individuals from the *pop_size*–sized population. The individuals are selected with repetition, i.e., a strong individual has a good chance to be selected several times to a new population. In the same time, even the weakest individual has a chance of being selected.

In ESs, the selection procedure is deterministic: it selects the best μ out of $\mu + \lambda$ $((\mu + \lambda)$–ES) or λ $((\mu, \lambda)$–ES) individuals (no repetitions). On the other hand, in GAs, the selection procedure is random, selecting *pop_size* out of *pop_size* individuals (with repetition), the chances of selection are proportional to the individual's fitness. Some GAs, in fact, use ranking selection; however, strong individuals can still be selected a few times. In other words, selection in ESs is static, extinctive, and (for (μ, λ)–ES) generational, whereas in GAs selection is dynamic, preservative, and on-the-fly (see Chapter 4).

The relative order of the procedures selection and recombination constitutes the third difference between GAs and ESs: in ESs, the selection process follows application of recombination operators, whereas in GAs these steps occur in the opposite order. In ESs, an offspring is a result of crossover of two parents and a further mutation. When the intermediate population of $\mu + \lambda$ (or λ) individuals is ready, the selection procedure reduces its size back to μ individuals. In GAs, we select an intermediate population first. Then we apply genetic operators (crossover and mutation) to some individuals (selected according to the probabilities of crossover) and some genes (selected according to the probability of mutation).

The next difference between ESs and GAs is that reproduction parameters for GAs (probability of crossover, probability of mutation) remain constant during the evolution process, whereas ESs change them (σ and θ) all the time: they undergo mutation and crossover together with the solution vector x, since an individual is understood as a triplet (x, σ, θ). This is quite important — self-adaptation of control parameters in ESs is responsible for the fine local tuning of the system.

ESs and GAs also handle constraints in a different way. Evolution strategies assume a set of $q \geq 0$ inequalities,

$$g_1(x) \geq 0, \ldots, g_q(x) \geq 0,$$

as part of the optimization problem. If, during some iteration, an offspring does not satisfy all of these constraints, then the offspring is disqualified, i.e., it is not placed in a new population. If the rate of occurrence of such illegal offspring is high, the ESs adjust their control parameters, e.g., by decreasing the components of the vector σ. The major strategy for genetic algorithms (already discussed in Chapter 7) for handling constraints is to impose penalties on individuals that violate them. That reason is that for heavily constrained problems we just cannot ignore illegal offspring (GAs do not adjust their control parameters) — otherwise the algorithm would stay in one place most of the time. At the same time, very often various decoders or repair algorithms are too costly to be considered (the effort to construct a good repair algorithm is similar to the effort to solve the problem). The penalty function technique has many disadvantages, one of which is problem dependence.

The above discussion implies that ESs and GAs are quite different with respect to many details. However, looking closer at the development of ESs and GAs during the last twenty years, one has to admit that the gap between these approaches is getting smaller and smaller.

Let us talk about some issues surrounding ESs and GAs again, this time from a historical perspective.

Quite early, there were signs that genetic algorithms display some difficulties in performing local search for the numerical applications (see Chapter 6). Many researchers experimented with different representations (Gray codes, floating point numbers) and different operators to improve the performance of the GA–based system. Today the first difference between GAs and ESs is not the issue any more: most GA applications for parameter optimization problems use floating point representation [78], adapting operators in appropriate way (see Chapter 4). It seems that the GA community borrowed the idea of vector representation from ESs.

The results of our experiments with our evolution programs provide an interesting observation: neither crossover nor mutation alone is satisfactory in the evolutionary process. Both operators (or rather both families of these operators) are necessary in providing a good performance of the system. The crossover operators are very important in exploring promising areas in the search space and are responsible for earlier (but not premature) convergence; in many systems—in particular those who work on richer data structures (Part III of the book) a decrease in the crossover rates deteriorates their performance. At the same time, the probability of applying mutation operators is quite high: the GENETIC-2 system (Chapter 9) uses a high mutation rate of 0.2.

A similar conclusion was reached by ES community: as a consequence, the crossover operator was introduced into ESs. Note that early ESs were based on a mutation operator only and the crossover operator was incorporated much later [348]. It seems that the score between GAs and ESs is even: the ES community borrowed the idea of crossover operators from GAs.

There are further interesting issues concerning relationships between ESs and GAs. Recently, some other crossover operators were introduced into GAs and ESs simultaneously [269, 270, 352]. Two vectors, x_1 and x_2 may produce two offspring, y_1 and y_2, which are a linear combination of their parents, i.e.,

$$y_1 = a \cdot x_1 + (1 - a) \cdot x_2 \text{ and}$$
$$y_2 = (1 - a) \cdot x_1 + a \cdot x_2.$$

Such a crossover was called

- in GAs: a *guaranteed average crossover* [77] (when $a = 1/2$), or an *arithmetical crossover* [269, 270], and

- in ESs: an *intermediate crossover* [352].

The self-adaptation of control parameters in ESs has its counterpart in GAs research. In general, the idea of adapting a genetic algorithm during its run was expressed some time ago; the ARGOT system [354] adapts the representation of individuals (see section 8.4). The problem of adapting control parameters for genetic algorithms has also been recognized for some time [169, 77, 115]. It was obvious from the start that finding a good settings for GA parameters for a particular problem is not a trivial task. Several approaches were proposed. One approach [169] uses a supervisor genetic algorithm to optimize the parameters of the "proper" genetic algorithm for a class of problems. The parameters considered were population size, crossover rate, mutation rate, generation gap (percentage of the population to be replaced during each generation), and scaling window, and a selection strategy (pure or elitist). The other approach [77] involves adapting the probabilities of genetic operators: the idea is that the probability of applying an operator is altered in proportion to the observed performance of the individuals created by this operator. The intuition is that the operators currently doing "a good job" should be used more frequently. In [115] the author experimented with four strategies for allocating the probability of the mutation operator: (1) constant probability, (2) exponentially decreasing, (3) exponentially increasing, and (4) a combination of (2) and (3).

Also, if we recall non-uniform mutation (described in Chapter 6), we notice that the operator changes its action during the evolution process.

Let us compare briefly the genetic-based evolution program GENOCOP (Chapter 7) with an evolution strategy. Both systems maintain populations of potential solutions and use some selection routine to distinguish between 'good' and 'bad' individuals. Both systems use float number representation. They provide high precision (ES through adaptation of control parameters, GENOCOP through non-uniform mutation). Both systems handle constraints gracefully: GENOCOP takes advantage of the presence of linear constraints, ES works on sets of inequalities. Both systems can easily incorporate the 'constraint handling ideas' from each other. The operators are similar. One employs intermediate crossover, the other arithmetical crossover. Are they really different?

An interesting comparison between ESs and GAs from the perspective of evolution strategies is presented in [187].

A few years ago [117], evolutionary programming (EP) technique (see section 13.1 for a description of the original evolutionary programming technique) were generalized to handle numerical optimization problems. They are quite similar to evolution strategies; they use floating point representation and the mutation is the key operator. The basic differences between evolution strategies and evolutionary programming techniques can be summarized as follows [19]:

- EP do not use any recombination operators,

- EP use a probabilistic selection (tournament selection), whereas ES select the best μ individuals for next generation,

- in EP, fitness values are obtained from objective function values by scaling them and possibly by imposing some random alternation,

- the standard deviation for each individual's mutation is calculated as the square root of a linear transformation of its own fitness value.

For more information on EP for numerical optimization and experimental comparison between ES and EP techniques, the reader is referred to [19, 117, 121].

8.3 Multimodal and multiobjective function optimization

In most chapters of this book we present methods for locating the single, global optimum of a function. However, in many cases either a function may have several optima that we wish to locate (multimodal optimization) or there is more than one criterion for optimization (multiobjective optimization). Clearly, new techniques are necessary to approach these categories of problems; we discuss them in turn.

8.3.1 Multimodal optimization

In many applications it might be important to locate all optima for a given function.[2] A few methods based on evolutionary techniques have been proposed for such multimodal optimization.

The first technique is based on iteration: we just repeat several runs of the algorithm. As discussed in [30], if all optima have an equal likelihood of being found, the number of independent runs should be

$$p \sum_{i=1}^{p} \frac{1}{i} \approx p(\gamma + \log p),$$

where p is the number of optima, and $\gamma \approx 0.577$ is Euler's constant. Unfortunately, in most real-world applications the optima are not equally likely, hence the number of independent runs should be much higher. It is also possible to use a parallel implementation of the iterative method, where several subpopulations evolve (in independent way, i.e., no communication) at the same time.

Goldberg and Richardson [162] described a method based on sharing; the method permits a formation of stable subpopulations (species) of different strings—in this way the algorithm investigate many peaks in parallel (the paper provides also an excellent review of other methods, which incorporated similar ideas on niche methods and species formation). A sharing function determines the degradation of an individual's fitness due to a neighbor at some distance $dist$ [162].[3] A sharing function sh was defined as a function of the distance with the following properties:

---

[2]By 'all' optima we understand all optima of interest, i.e., all optima above a certain threshold.

[3]The distance $dist$ can be defined over on the genotype or phenotype level, i.e., on strings or their interpretation.

- $0 \le sh(dist) \le 1$, for all distances $dist$,

- $sh(0) = 1$, and

- $\lim_{dist \to \infty} = 0$;

there are many sharing functions which satisfy the above conditions. One possibility, as suggested in [162], is

$$sh(dist) = \begin{cases} 1 - (\frac{dist}{\sigma_{sh}})^\alpha & if \ dist < \sigma_{sh} \\ 0 & otherwise \end{cases}$$

where σ_{sh} and α are constants (for a discussion of a significance of these constants, see [162]).

The new (shared) fitness of an individual x is given by:

$$eval'(x) = eval(x)/m(x),$$

where $m(x)$ returns the niche count for a particular individual x:

$$m(x) = \sum_y sh(dist(x, y)).$$

In the above formula the sum over all y in the population includes the string x itself; consequently, if string x is all by itself in its own niche, it fitness value does not decrease $(m(x) = 1)$. Otherwise, the fitness function is decreased proportionally to the number and closeness of neighboring points.

It means, that when many individuals are in the same neighborhood they contribute to one another's share count, thus derating one another's fitness values. As a result this technique limits the uncontrolled growth of particular species within a population [154].

Recently, Beasley, Bull, and Martin [30] described a new (called: sequential niche) technique for multimodal function optimization, which avoids a few disadvantages of the sharing method (e.g., time complexity due to fitness sharing calculations, population size, which should be proportional to the number of optima). The proposed algorithm also uses a distance function $dist$ and a fitness function $eval$, and is based on the following idea: once an optimum is found, the evaluation function can be modified to eliminate this (already found) solution, since there is no interest in re-discovering the same optimum again. In some sense, the subsequent runs of genetic algorithm incorporate the knowledge discovered in the previous runs (as opposed to the simple iterative technique, where each run starts with a randomly generated population). The basic steps of the algorithm are (from [30]):

1. Initialize: equate the modified fitness function with the raw fitness function.

2. Run the GA using the modified fitness function, keeping a record of the best individual found in the run.

3. Update the modified fitness function to give a depression[4] in the region near the best individual, producing a new fitness function.

4. If the raw fitness of the best individual is of interest (i.e., it exceeds the solution threshold), display this as a solution.

5. If not all solutions have been found, return to step 2.

For a detailed discussion of the algorithm and the experimental results, the reader is referred to [30].

Yet another approach was proposed recently by Spears [365]. The proposed algorithm implements the ideas of sharing and restrictive mating. However, the idea of metic distance is dropped and replaced by a concept of labels: each individual in the population has a label (in experiments reported in [365], a label was a n-bit string, consequently, labels could be used to represent 2^n subpopulations). To explain the intuition behind the proposed algorithm, let us cite from [365]:

> "Suppose we have a simple function with two peaks, one peak twice as high as the other, and further suppose we allow one tag bit for each individual. Each tag bit is randomly initialized, so at the beginning of the run we have two subpopulations of roughly equal size. Due to random sampling both subpopulations could eventually settle in on the higher peak, or both could settle in on the lower peak. However, in some cases (again, due to random sampling), each subpopulation will head towards different peaks. If we do not have fitness sharing, the individuals on the higher peak would always get more children than individuals on the lower peak and eventually the subpopulation on the lower peak would vanish. However, with fitness sharing, the higher peak can support only twice as many individuals as can be supported on the lower peak (since it is only twice as high). [...] The fitness sharing mechanism has dynamically adjusted the perceived fitness so that the two peaks have the same perceived height. The result is that both subpopulations can survive in a stable fashion. Furthermore, restricted mating prevents crossover between individuals on the two peaks, which could often result in low fitness individuals."

For more details and experimental results, the reader is referred to [365].

A recent paper by Mahfoud [249] compares several niching methods, which are grouped into to categories: sequential and parallel. Parallel methods form and maintain niches simultaneously within a population; sequential methods locate multiple niches temporally. The results indicate that parallel niching methods outperform sequential ones. For details on this comparison and a complete discussion on advantages of parallel methods, see [249].

---

[4]The description of the algorithm assumes maximization problems.

8.3.2 Multiobjective optimization

For many real-world decision making problems there is a need for simultaneous optimization of multiple objectives. As stated in [182]:

"One possible approach [...] is to use long-run profit maximization as the sole objective. At first glance, this approach appears to have considerable merit. In particular, the objective of long-run profit maximization is specific enough to be used conveniently, and yet it seems to be broad enough to encompass the basic goal of most organizations. In fact, some people tend to feel that all other legitimate objectives can be translated into this one. However, this is such an oversimplification that considerable caution is required! A number of studies have found that, instead of profit maximization, the goal of satisfactory profits combined with other objectives is characteristic of American corporations. In particular, typical objectives might be to maintain stable profits, increase (or maintain) one's share of the market, product diversification, maintain stable prices, improve worker morale, maintain family control of the business, and increase company prestige. These objectives might be compatible with long-run profit maximization, but the relationship is sufficiently obscure that it may not be convenient to incorporate them into this one objective."

Such multiobjective optimization problems require separate techniques, which are very different to the standard optimization techniques for single objective optimization. It is very clear that if there are two objectives to be optimized, it might be possible to find a solution which is the best with respect to the first objective, and another solution, which is the best with respect to the second objective.

It is convenient to classify all potential solutions to the multiobjective optimization problem into *dominated* solutions and *nondominated* (*Pareto-optimal*) solutions. As solution x is dominated if there exits a feasible solution y not worse than x on all coordinates, i.e., for all objectives f_i $(i = 1, \ldots, k)$:

$$f_i(x) \leq f_i(y) \text{ for all } 1 \leq i \leq k.^5$$

If a solution is not dominated by any other feasible solution, we call it nondominated (or Pareto-optimal) solution. All Pareto-optimal solutions might be of some interest; ideally, the system should report back the set of all Pareto-optimal points.

There are some classical methods for multiobjective optimization [369]. These include a method of objective weighting, where multiple objective functions f_i are combined into one overall objective function F:

---

[5] For maximization problems; otherwise the less equal inequality should be replaced by greater equal.

$$F(\boldsymbol{x}) = \sum_{i=1}^{k} w_i f_i(\boldsymbol{x}),$$

where the weights $w_i \in [0..1]$ and $\sum_{i=1}^{k} w_i = 1$. Different weight vectors provide different Pareto-optimal solutions. Another method (method of distance functions) combines multiple objective functions into one on the basis of demand-level vector $\boldsymbol{y}$:

$$F(\boldsymbol{x}) = (\sum_{i=1}^{k} |f_i(\boldsymbol{x}) - y_i|^r)^{\frac{1}{r}},$$

where (usually) $r = 2$ (Euclidean metric).

Multiobjective optimization enjoyed some interest in the GA community. In 1984 Schaffer [339] developed VEGA program (for Vector Evaluated Genetic Algorithm), which was an extension of the GENESIS program [166] to include multicriteria functions. The main idea behind the VEGA system was a division of the population into (equal sized) subpopulations; each subpopulation was "responsible" for a single objective. The selection procedure was performed independently for each objective, but crossover was performed across subpopulation boundaries. Additional heuristics were developed (e.g., wealth redistribution scheme, crossbreeding plan) and studied to decrease a tendency of the system to converge towards individuals which were not the best with respect to any objective.

Recently [369] Srinivas and Deb proposed a new technique, NSGA (for Nondominated Sorting Genetic Algorithm), which is based on several layers of classifications of the individuals. Before the selection is performed, the population is ranked on the basis of nondomination: all nondominated individuals are classify into one category (with a dummy fitness value, which is proportional to the population size, to provide an equal reproductive potential for these individuals). To maintain the diversity of the population, these classified individuals are shared with with their dummy fitness values (see previous subsection). Then this group of classified individuals are ignored and another layer of nondominated individuals is considered. The process continues until all individuals in the population are classified. For the full discussion on the system and first experimental results the reader is referred to [369].

Recently, Fonseca and Fleming [127] published a survey of evolutionary algorithms for multiobjective optimization. They provided an overview of both categories of techniques (these, which combine many criteria into one objective function and return a signle value, and these, which are based on Pareto-optimality and return a set of values), and identified several open research issues.

8.4 Other evolution programs

As we have already discussed earlier, (classical) genetic algorithms are not appropriate tools for local fine tuning. For that reason, GAs give less precise solutions to numerical optimization problems than, for example, ESs, unless the representation of individuals in GAs is changed from binary into floating point and the (evolution) system provides specialized operators (like non–uniform mutation: Chapter 6). However, in the last decade there have been some other attempts to improve (directly or indirectly) this characteristic of GAs.

An interesting modification of GAs, called Delta Coding, was proposed recently by Whitley et al. [400]. The main idea behind this strategy is that it treats individuals in the population not as potential solutions to the problem, but rather as additional (small) values (called: *delta values*), which are added to the current potential solution. The (simplified) Delta Coding algorithm is listed in Figure 8.1.

procedure Delta Coding
begin
    apply GA on level x
    save the best solution ($\boldsymbol{x}$)
    while (**not** termination-condition) **do**
    begin
        apply GA on level δ
        save the best solution ($\boldsymbol{\delta}$)
        modify the best solution (x level):
            $\boldsymbol{x} \leftarrow \boldsymbol{x} + \boldsymbol{\delta}$
    end
end

Fig. 8.1. A (simplified) Delta Coding algorithm

The Delta Coding algorithm applies genetic algorithm techniques on two levels: the level of potential solutions to the problem (level x) and (iterative phase) the level of delta changes (level δ). The best solution found on level x by a single application of a GA is saved ($\boldsymbol{x}$) and kept as a reference point. Then several iterations of the inner (level δ) GA are executed. A termination of a single execution of a GA on this level (i.e., when GA converges) results in the best modification vector δ, which updates the values of $\boldsymbol{x}$. After an update, the next iteration takes place. Each application of GA during the iteration phase reinitializes randomly the population of $\boldsymbol{\delta}$'s. Of course, to evaluate individual $\boldsymbol{\delta}$, we evaluate $\boldsymbol{x} + \boldsymbol{\delta}$.

The original Delta Coding algorithm is more complex, since it operates on bit strings. By doing this, Delta Coding preserves the theoretical foundations of genetic algorithms (since at each iteration there is a single run of GA). The termination conditions for GAs on both levels are expressed by means of the Hamming distance between the best and the worst element in the population (the algorithms terminate if the Hamming distance is not greater than one). Additionally, there is a variable *len* to denote the number of bits representing a single component of the vector δ (actually, only *len*−1 represent the absolute value of the component; the last bit is reserved for the sign of the value). If the best solution from the δ level yields a vector

$$\delta = (0, 0, \ldots, 0),$$

(i.e., no change for the best potential solution x), the variable *len* is increased by one (to increase the precision of the solution), otherwise it is decreased by one. Note also that Delta Coding makes mutations unnecessary, due to reinitialization of populations on level δ for each iteration.

We can simplify the original Delta Coding algorithm (Figure 8.1 provides such a simplified view) and improve its precision and time performance, if we represent both vectors x and δ as sequences of floating point numbers.

Some of the ideas present in Delta Coding algorithm appeared earlier in the literature. For example, Schraudoph and Belew [347] proposed a Dynamic Parameter Encoding (DPE) strategy, where the precision of the encoded individual is dynamically adjusted. In this system, each component of the solution vector is represented by a fixed-length binary string; however, when (in some iteration) a genetic algorithm converges, the most significant bit of the solution is dropped (of course, after saving it!), the remaining bits are shifted one position left, and a new bit is introduced. This new bit in the least significant position increases precision by a finer partitioning of the search space. The process is repeated until some global termination condition is met.

The idea of reinitializing the population was discussed in [153], where Goldberg investigates the properties of systems which use small population size, but reinitialize it every time the genetic algorithm converges (and save the best individuals, of course!). The outline of such a strategy (called serial selection) is given in Figure 8.2.

procedure Serial Selection
begin
    generate a (small) population
    while (**not** termination-condition) **do**
    begin
      apply GA
      save the best solution (x)
      generate a new population by transferring
          the best individuals of the converged
          population and then generating the
          remaining individuals randomly
    end
end

Fig. 8.2. GA based on re-initialization population

Reinitializations of populations introduce diversity among individuals with a positive effect on system performance [153].

Some proposed strategies included a learning component, in a similar way as in evolution strategies. Grefenstette [169] proposed optimizing control parameters of a genetic algorithm (population size, crossover and mutation rates, etc.) by another, supervisor genetic algorithm. Shaefer [354] discussed the AR-GOT strategy (Adaptive Representation Genetic Optimizer Technique), where the system learns the best internal representation for the individuals.

Another interesting evolution program, a selective evolutionary strategy (IRM, for immune recruitment mechanism), was proposed recently [35] as an optimization technique in real spaces (a similar system for optimizing functions in Hamming spaces was called GIRM). The strategy combines some previously seen ideas for directing the search in a desirable direction (e.g., tabu search). As in all evolution programming techniques, an offspring is generated from the current population. In classical genetic algorithms, such offspring replaces its parent. In evolution strategies, the offspring competes with its parent (early ESs), it competes with parents and other offspring ($(\mu + \lambda)$–ES), or it competes with other offspring ((μ, λ)–ES). In IRM systems, an offspring has to pass an additional test of affinity with its neighbors. The test checks whether it displays sufficient similarity with its close neighbors.

In general, a possible candidate k would pass the affinity test, if

$$\sum_i m(k, i) \cdot f_i > T,$$

where i indexes different species already present in the population, f_i is the concentration of the species i, $m(k, i)$ is an affinity function for species k and i, and T is the recruitment threshold.

The IRM strategy directs the search by accepting only individuals which satisfy the affinity test. Similar ideas were formulated by Glover [142], [145] in Scatter and Tabu Search. The Scatter Search techniques, like other evolution programs, maintain a population of potential solutions (vectors $\boldsymbol{x}^i$ are called reference points). This strategy unites preferred subsets of reference points to generate trial points (offspring) by weighted linear combinations, and selects the best members to become the source of new reference points (new population). A new twist here is the use of *multicrossover* (called weighted combination), where several (more than two) parents contribute in producing an offspring. In [145] Glover extended the idea of the Scatter Search by combining it with a Tabu Search — a technique which restricts the selection of new offspring (it requires memory where a historical set of individuals is kept) [143], [144]. The structure of a scatter/tabu search algorithm is shown in Figure 8.3.

After initialization and evaluation, scatter/tabu search algorithm classifies (classify $P(t)$ step) population of solutions $\overline{X}_1, \ldots, \overline{X}_{pop\_size}$ into several sets. These include (1) a set of elite historical generators V consisting of some (fixed) number of best solutions through the whole process, (2) a set of tabu generators $T \subseteq V$ consisting of solutions currently excluded from considerations, (3) a set of selected historical generators V^* consisting of the best elements of $V - T$, and (4) a set of selected current generators S^* consisting of the best elements of S. The classification step (classify $P(t)$) is repeated later in the iteration phase of the algorithm.

During each iteration a set $R(t)$ of trial points is created. The trial points correspond to offspring of the population $P(t)$; they are evaluated and (some of

```
procedure scatter/tabu search
begin
    t = 0
    initialize P(t)
    evaluate P(t)
    classify P(t)
    while (not termination-condition) do
    begin
        t = t + 1
        create R(t)
        evaluate R(t)
        select P(t) from P(t − 1) and R(t)
        classify P(t)
    end
end
```

Fig. 8.3. Scatter/tabu search

them) are incorporated into the new population (select $P(t)$ from $P(t − 1)$ and $R(t)$).

Recently David Fogel applied the ideas of evolutionary programming [126] to real-valued continuous optimization [119]; these extensions include self-adapting independent variances and procedures for optimizing the covariance matrix.

Maniezzo developed the concept of granularity evolution [250], where the algorithm allows a concurrent evolution of objective function samples and of sampling resolution (i.e., granularity). Individuals become variable-length which encoding is interpreted according to a specific resolution level specified within the chromosome.

Also, a concept of a genetic algorithm which processes intervals (called Interval Genetic Algorithm) was investigated by Muselli and Ridella. [293] The interval genetic algorithm combines the ideas of genetic algorithms and simulated annealing; the genetic operators (crossover, which generates a new interval out of two intervals; merging, which generates an offspring from two parents as an intersection of the two intervals; and mutation, which searches the interval (i.e., single parent) for a better point.

It seems that the most promising direction in the search for the "optimal optimizer" lies somewhere among the above ideas. Each strategy provides a new insight which might be useful in developing an evolution program for some class of problems. As stated by Glover [145]:

"The use of structural combinations makes it possible to combine component vectors in a way that is materially different from the results of [classical] crossover operations. Integrating such an approach with genetic algorithms may open the door to new types of search procedures. The fact that weighted and adaptive structured combinations can readily be created to exploit contexts where crossover has no evident meaning (or has difficulty insuring feasibility) suggests that such integrated search procedures may have benefits in settings where genetic algorithms presently have limited applications."

To see this clearly, let us move to the next chapter.

Part III
Evolution Programs

9. The Transportation Problem

Necessity knows no law.

Publilius Syrus, *Moral Sayings*

In Chapter 7 we compared different GA approaches for handling constraints. It seems that for a particular class of problems (like the transportation problem) we can do better: we can use a more appropriate (natural) data structure (for a transportation problem, a matrix) and specialized genetic operators which operate on matrices. Such an evolution program would be much stronger method than GENOCOP: the GENOCOP optimizes any function with linear constraints, whereas the new evolution program optimizes only transportation problems (these problems have precisely $n + k - 1$ equalities, where n and k denote the number of sources and destinations, respectively; see the description of the transportation problem below). However, it would be very interesting to see what can we gain by introducing extra problem-specific knowledge into an evolution program.

Section 9.1 presents an evolution program for the linear transportation problem[1], and section 9.2 presents one for the nonlinear transportation problem.[2]

9.1 The linear transportation problem

The transportation problem (see, for example, [387]) is one of the simplest combinatorial problems involving constraints that has been studied. It seeks the determination of a minimum cost transportation plan for a single commodity from a number of sources to a number of destinations. It requires the specification of the level of supply at each source, the amount of demand at each destination, and the transportation cost from each source to each destination.

Since there is only one commodity, a destination can receive its demand from one or more sources. The objective is to find the amount to be shipped

---

[1]Portions reprinted, with permission, from IEEE Transactions on Systems, Man, and Cybernetics, Vol. 21, No. 2, pp. 445–452, 1991.

[2]Portions reprinted, with permission, from ORSA Journal on Computing, Vol. 3, No. 4, 1991, pp. 307–316, 1991.

from each source to each destination such that the total transportation cost is minimized.

The transportation problem is *linear* if the cost on a route is directly proportional to the amount transported; otherwise, it is *nonlinear*. While linear problems can be solved by OR methods, the nonlinear case lacks a general solving methodology.

Assume there are n sources and k destinations. The amount of supply at source i is $sour(i)$ and the demand at destination j is $dest(j)$. The unit transportation cost between source i and destination j is $cost(i, j)$. If x_{ij} is the amount transported from source i to destination j then the transportation problem is given as:

$$\text{minimize } \sum_{i=1}^{n} \sum_{j=1}^{k} f_{ij}(x_{ij})$$

subject to

$$\sum_{j=1}^{k} x_{ij} \leq sour(i), \text{ for } i = 1, 2, \ldots, n,$$
$$\sum_{i=1}^{n} x_{ij} \geq dest(j), \text{ for } j = 1, 2, \ldots, k,$$
$$x_{ij} \geq 0, \text{ for } i = 1, 2, \ldots, n \text{ and } j = 1, 2, \ldots, k.$$

The first set of constraints stipulates that the sum of the shipments from a source cannot exceed its supply; the second set requires that the sum of the shipments to a destination must satisfy its demand. If $f_{ij}(x_{ij}) = cost_{ij} \cdot x_{ij}$ for all i and j, the problem is linear.

The above problem implies that the total supply $\sum_{i=1}^{k} sour(i)$ must at least equal total demand $\sum_{j=1}^{n} dest(j)$. When the total supply equals the total demand, the resulting formulation is called a *balanced transportation problem*. It differs from the above only in that all the corresponding constraints are equations; that is,

$$\sum_{j=1}^{k} x_{ij} = sour(i), \text{ for } i = 1, 2, \ldots, n,$$
$$\sum_{i=1}^{n} x_{ij} = dest(j), \text{ for } j = 1, 2, \ldots, k.$$

If all $sour(i)$ and $dest(j)$ are integers, any optimal solution to a balanced linear transportation problem is an integer solution, i.e., all x_{ij} ($i = 1, 2, \ldots, n$, $j = 1, 2, \ldots, k$) are integers. Moreover, the number of positive integers among the x_{ij} is at most $k + n - 1$. In this section we assume a balanced linear transportation problem. An example follows. For other information on the transportation problem and balancing, the reader is referred to any elementary text on operations research such as [387].

Example 9.1. Assume 3 sources and 4 destinations. The supply is:

$$sour(1) = 15, \; sour(2) = 25, \text{ and } sour(3) = 5.$$

The demand is:

$$dest(1) = 5, \; dest(2) = 15, \; dest(3) = 15, \text{ and } dest(4) = 10.$$

Note that the total supply and demand equal 45.

The unit transportation cost $cost(i, j)$ $(i = 1, 2, 3,$ and $j = 1, 2, 3, 4)$ is given in the table below.

Cost

10	0	20	11
12	7	9	20
0	14	16	18

The optimal solution is shown below. The total cost is 315. The solution consists of integer values of x_{ij}.

Amount transported

	5	15	15	10
15	0	5	0	10
25	0	10	15	0
5	5	0	0	0

□

9.1.1 Classical genetic algorithms

By a "classical" genetic algorithm we mean, of course, one where the chromosomes (i.e., representations of solutions) are bit strings — lists of 0s and 1s. A straightforward approach in defining a bit vector for a solution in the transportation problem is to create a vector $\langle v_1, v_2, \ldots, v_p \rangle$ ($p = n \cdot k$), such that each component v_i $(i = 1, 2, \ldots, p)$, is a bit vector $\langle w_0^i, \ldots, w_s^i \rangle$ and represents an integer associated with row j and column m in the allocation matrix, where $j = \lfloor (i-1)/k+1 \rfloor$ and $m = (i-1) \bmod k + 1$. The length of the vectors w (parameter s) determines the maximum integer $(2^{s+1} - 1)$ that can be represented.

Let us discuss briefly the consequences of the above representation on constraint satisfaction, the evaluation function and genetic operators.

Constraint satisfaction: It is clear that every solution vector must satisfy the following:

- $v_q \geq 0$ for all $q = 1, 2, \ldots, k \cdot n$,
- $\sum_{i=c \cdot k+1}^{c \cdot k+k} v_i = sour[c + 1]$, for $c = 0, 1, \ldots, n - 1$,
- $\sum_{j=m, step\ k}^{k \cdot n} v_j = dest[m]$, for $m = 1, 2 \ldots, k$.

Note that the first constraint is always satisfied (we interpret a sequence of 0s and 1s as a positive integer). The other two constraints provide the totals for each source and each destination, though these formulas are not symmetrical.

Evaluation function: The natural evaluation function expresses the total cost of transporting items from sources to destinations and is given by the formula:

$$eval(\langle v_1, v_2, \ldots, v_p \rangle) = \sum_{i=1}^{p} v_i \cdot cost[j][m],$$

where $j = \lfloor (i-1)/k + 1 \rfloor$ and $m = (i-1) \bmod k + 1$.

Genetic operators: There is no natural definition of genetic operators for the transportation problem with the above representation. Mutation is usually defined as a change in a single bit in a solution vector. This would correspond to a change of one integer value, v_i. This, in turn, for our problems, would trigger a series of changes in different places (at least three other changes) in order to maintain the constraint equalities. Note also that we always have to remember in which column and row a change was made — despite a vector representation we think and operate in terms of rows and columns (sources and destinations). This is a reason for quite complex formulae; the first sign of this complexity is loss of symmetry in expressing the constraints.

There are some other open questions as well. Mutation is understood as a minimal change in a solution vector, but as we noted earlier, a single change in one integer would trigger at least three other changes in appropriate places. Assume that two random points (v_i and v_m, where $i < m$) are selected such that they do not belong to the same row or column. Let us assume that v_i, v_j, v_k, v_m ($i < j < k < m$) are components of a solution vector (selected for mutation) such that v_i and v_k as well as v_j and v_m belong to a single column, and v_i and v_j as well as v_k and v_m belong to a single row.

That is, in matrix representation:

$$
\begin{array}{ccccccc}
\cdots & \cdot & \cdots & \cdot & \cdots \\
\cdots & \cdot & \cdots & \cdot & \cdots \\
\cdots & v_i & \cdots & v_j & \cdots \\
\cdots & \cdot & \cdots & \cdot & \cdots \\
\cdots & \cdot & \cdots & \cdot & \cdots \\
\cdots & v_k & \cdots & v_m & \cdots \\
\cdots & \cdot & \cdots & \cdot & \cdots \\
\cdots & \cdot & \cdots & \cdot & \cdots \\
\end{array}
$$

Now in trying to determine the smallest change in the solution vector we have a difficulty. Should we increase or decrease v_i? We can choose to change it by 1 (the smallest possible change) or by some random number in the range $\langle 0, 1, \ldots, v_i \rangle$. If we increase the value v_i by a constant C we have to decrease each of the values v_j and v_k by the same amount. What happens if $v_j < C$ or $v_k < C$? We could set $C = \min(v_i, v_j, v_k)$, but then most mutations would result in no change, since the probability of selecting three non-zero elements would be close to zero (less than $1/n$ for vectors of size n^2).

Thus methods involving single bit changes result in inefficient mutation operators with complex expressions for checking the corresponding row or column of the selected element.

The situation is even more complex if we try to repair a chromosome after applying the crossover operator. Breaking a vector at a random point could result in a pair of chromosomes violating numerous constraints. If we try to modify these solutions to obey all constraints, they would lose most similarities with the parents. Moreover, the way to do this is far from obvious: if a vector v is outside the search space, "repairing" it might be as difficult as solving the original problem. Even if we succeeded in building a system based on repair algorithms, such a system would be highly problem specific with little chances for generalizations.

We conclude that the above vector representation is not the most suitable for defining genetic operators in constrained problems of this type.

9.1.2 Incorporating problem-specific knowledge

Can we improve the representation of a solution while preserving the basic structure of this vector representation? We believe so, but we have to incorporate problem-specific knowledge into the representation.

First let us describe a way to create a solution which satisfies all constraints. We will call this procedure an **initialization** — it will be a fundamental component of the mutation operator when we discuss genetic operators for two-dimensional structures. It creates a matrix of at most $k + n - 1$ non-zero elements such that all constraints are satisfied. After sketching the algorithm we explain it using the matrix from Example 9.1.

> **input:** arrays $dest[k]$, $sour[n]$;
> **output:** an array $(v)_{ij}$ such that $v_{ij} \geq 0$ for all i and j,
> $\sum_{j=1}^{k} v_{ij} = dest[i]$ for $i = 1, 2, \ldots, n$, and
> $\sum_{i=1}^{n} v_{ij} = sour[j]$ for $j = 1, 2 \ldots, k$,
> i.e., all constraints are satisfied.

procedure initialization;
begin
    set all numbers from 1 to $k \cdot n$ as unvisited
    repeat
        select an unvisited random number q
            from 1 to $k \cdot n$ and set it as visited
        set (row) $i = \lfloor (q-1)/k + 1 \rfloor$
        set (column) $j = (q-1) \bmod k + 1$
        set $val = \min(sour[i], dest[j])$
        set $v_{ij} = val$
        set $sour[i] = sour[i] - val$
        set $dest[j] = dest[j] - val$
    until all numbers are visited.
end

Example 9.2. With the matrix from Example 9.1, i.e.,

$$sour[1] = 15, \ sour[2] = 25, \text{ and } sour[3] = 5$$
$$dest[1] = 5, \ dest[2] = 15, \ dest[3] = 15, \text{ and } dest[4] = 10.$$

There are altogether $3 \cdot 4 = 12$ numbers, all of them are unvisited at the beginning. Select the first random number, say, 10. This translates into row number $i = 3$ and column number $j = 2$. The $val = \min(sour[3], dest[2]) = 5$, so $v_{32} = 5$. Note also that after the first iteration, $sour[3] = 0$ and $dest[2] = 10$.

We repeat these calculations with the next three random (unvisited) numbers, say 8, 5, and 3 (corresponding to row 2 and column 4, to row 2 and column 1, and to row 1 and column 3, respectively). The resulting matrix v_{ij} (so far) has the following contents:

	0	10	0	0
0		15		
10	5			10
0		5		

Note that the values of $sour[i]$ and $dest[j]$ are those given after 4 iterations.

If the further sequence of random numbers is 1, 11, 4, 12, 7, 6, 9, 2, the final matrix produced (with the assumed sequence of random numbers $\langle$ 10, 8, 5, 3, 1, 11, 4, 12, 7, 6, 9, 2 $\rangle$) is:

	0	0	0	0
0	0	0	15	0
0	5	10	0	10
0	0	5	0	0

Obviously, after 12 iterations all (local copies of) $sour[i]$ and $dest[j] = 0$. Note also, that there are several sequences of numbers for which the procedure

initialization would produce the optimal solution. For example, the optimal solution (given in Example 1) can be achieved for any of the following sequences: $\langle$ 7, 9, 4, 2, 6, *, *, *, *, *, *, * $\rangle$ (where * denotes any unvisited number), as well as for many other sequences.

This technique can generate any feasible solution that contains at most $k + n - 1$ non-zero integer elements. It will not generate other solutions which, though feasible, do not share this characteristic. The initialization procedure would certainly have to be modified when we attempt to solve non-linear versions of the transportation problem.

This knowledge of the problem and its solution characteristics gives us another opportunity to represent a solution to the transportation problem as a vector. A solution vector will be a sequence of $k \cdot n$ distinct integers from the range $\langle 1, k \cdot n \rangle$, which (according to procedure **initialization**) would produce an acceptable solution. In other words, we would view a solution-vector as a permutation of numbers, and we would look for particular permutations which correspond to the optimal solution.

Let us discuss briefly the implications of this representation on constraint satisfaction, evaluation function and genetic operators.

Constraint satisfaction: Any permutation of $k \cdot n$ distinct numbers produces a unique solution which satisfies all constraints. This is guaranteed by procedure **initialization**.

Evaluation function: This is relatively easy: any permutation would correspond to a unique matrix, say, (v_{ij}). The evaluation function is
$\sum_{i=1}^{k} \sum_{j=1}^{n} v_{ij} \cdot cost[i][j]$

Genetic operators: These are also straightforward:

- inversion: any solution vector $\langle x_1, x_2, \ldots, x_q \rangle$ $(q = k \cdot n)$ can be easily inverted into another solution vector $\langle x_q, x_{q-1}, \ldots, x_1 \rangle$

- mutation: any two elements of a solution vector $\langle x_1, x_2, \ldots, x_q \rangle$, say x_i and x_j can be swapped easily resulting in another solution vector.

- crossover: this is little more complex. Note that an arbitrary (blind) crossover operator would result in illegal solutions: applying such a crossover operator to sequences:

$$\langle 1, 2, 3, 4, 5, 6, | 7, 8, 9, 10, 11, 12 \rangle \text{ and }$$
$$\langle 7, 3, 1, 11, 4, 12, | 5, 2, 10, 9, 6, 8 \rangle$$

would result (where the crossover point is after the 6th position) in

$$\langle 1, 2, 3, 4, 5, 6, 5, 2, 10, 9, 6, 8 \rangle \text{ and }$$
$$\langle 7, 3, 1, 11, 4, 12, 7, 8, 9, 10, 11, 12 \rangle$$

neither of which is a legal solution.

Thus we have to use some form of heuristic crossover operator. There is some similarity between these sequences of solution vectors and those for the traveling salesman problem (see [170]). Here we use a heuristic crossover operator (of the PMX family of crossover operators, see [160] and Chapter 10), which, given two parents, creates an offspring by the following procedure:

1. make a copy of the second parent,

2. choose an arbitrary part from the first parent,

3. make minimal changes in the offspring necessary to achieve the chosen pattern.

For example, if the parents are as in the example above, and the chosen part is

$$(4, 5, 6, 7),$$

the resulting offspring is

$$\langle\ 3,\ 1,\ 11,\ 4,\ 5,\ 6,\ 7,\ 12,\ 2,\ 10,\ 9,\ 8\ \rangle.$$

As required, the offspring bears a structural relationship to both parents. The roles of the parents can then be reversed in constructing a second offspring.

A genetic system GENETIC-1 has been built on the above principles. The results of experiments with it are discussed in the next section.

9.1.3 A matrix as a representation structure

Perhaps the most natural representation of a solution for the transportation problem is a two-dimensional structure. After all, this is how the problem is presented and solved by hand. In other words, a matrix $V = (v_{ij})$ $(1 \leq i \leq k,\ 1 \leq j \leq n)$ may represent a solution.

Let us discuss the implications of the matrix representation on constraint satisfaction, evaluation function and genetics operators.

Constraint satisfaction: It is clear that every solution matrix $V = (v_{ij})$ should satisfy the following:

- $v_{ij} \geq 0$ for all $i = 1, \ldots, k$, and $j = 1, \ldots, n$,
- $\sum_{i=1}^{k} v_{ij} = dest[j]$ for $j = 1, \ldots, n$,
- $\sum_{j=1}^{n} v_{ij} = sour[i]$ for $i = 1, \ldots, k$.

This is similar to the set of constraints in the straightforward approach (section 9.1.2), but the constraints are expressed in an easier and more natural way.

Evaluation function: The natural evaluation function expresses is the usual objective function:

$$\text{eval}(v_{ij}) = \sum_{i=1}^{k} \sum_{j=1}^{n} v_{ij} \cdot cost[j][m]$$

Again, the formula is much simpler than in the straightforward approach and faster than in the system GENETICS-1, where each sequence has to be converted (initialized) into a solution matrix before evaluation.

Genetic operators: We define here two genetic operators, mutation and crossover. It is difficult to define a meaningful inversion operator in this case.

- mutation:
  Assume that $\{i_1, i_2, \ldots, i_p\}$ is a subset of $\{1, 2, \ldots, k\}$, and $\{j_1, j_2, \ldots, j_q\}$ is a subset of $\{1, 2, \ldots, n\}$ such that $2 \leq p \leq k$, $2 \leq q \leq n$.
  Let us denote a parent for mutation by $(k \times n)$ matrix $V = (v_{ij})$. Then we can create a $(p \times q)$ submatrix $W = (w_{ij})$ from all elements of the matrix V in the following way: an element $v_{ij} \in V$ is in W if and only if $i \in \{i_1, i_2, \ldots, i_p\}$ and $j \in \{j_1, j_2, \ldots, j_q\}$ (if $i = i_r$ and $j = j_s$, then the element v_{ij} is placed in the r-th row and s-th column of the matrix W).
  Now we can assign new values $sour_W[i]$ and $dest_W[j]$ ($1 \leq i \leq p$, $1 \leq j \leq q$) for matrix W:

  $$sour_W[i] = \sum_{j \in \{j_1, j_2, \ldots, j_q\}} v_{ij}, 1 \leq i \leq p,$$
  $$dest_W[j] = \sum_{i \in \{i_1, i_2, \ldots, i_p\}} v_{ij}, 1 \leq j \leq q.$$

  We can use the procedure **initialization** (section 9.1.3) to assign new values to the matrix W such that all constraints $sour_W[i]$ and $dest_W[j]$ are satisfied. After that, we replace appropriate elements of matrix V by a new elements from the matrix W. In this way all global constraints ($sour[i]$ and $dest[j]$) are preserved.
  The following example will illustrate the mutation operator.

  Example 9.3. Given a problem with 4 sources and 5 destinations and the following constraints:

  $$sour[1] = 8, \ sour[2] = 4, \ sour[3] = 12, \ sour[4] = 6,$$
  $$dest[1] = 3, \ dest[2] = 5, \ dest[3] = 10, \ dest[14] = 7, \ dest[5] = 5.$$

  Assume that the following matrix V is selected as a parent for mutation:

0	0	5	0	3
0	4	0	0	0
0	0	5	7	0
3	1	0	0	2

Select (at random) the two rows $\{2, 4\}$ and three columns $\{2, 3, 5\}$. The corresponding submatrix W is:

4	0	0
1	0	2

Note, that $sour_W[1] = 4$, $sour_W[2] = 3$, $dest_W[1] = 5$, $dest_W[2] = 0$, $dest_W[3] = 2$. After the reinitialization of matrix W, the matrix may get the following values:

2	0	2
3	0	0

So, finally, the offspring of matrix V after mutation is:

0	0	5	0	3
0	2	0	0	2
0	0	5	7	0
3	3	0	0	0

□

- crossover:
  Assume that two matrices $V_1 = (v_{ij}^1)$ and $V_2 = (v_{ij}^2)$ are selected as parents for the crossover operation. Below we describe the skeleton of an algorithm we use to produce the pair of offspring V_3 and V_4.
  Create two temporary matrices: $DIV = (div_{ij})$ and $REM = (rem_{ij})$. These are defined as follows:

$$div_{ij} = \lfloor (v_{ij}^1 + v_{ij}^2)/2 \rfloor$$
$$rem_{ij} = (v_{ij}^1 + v_{ij}^2) \bmod 2$$

Matrix DIV keeps rounded average values from both parents, the matrix REM keeps track of whether any rounding was necessary.

Matrix REM has some interesting properties: the number of 1s in each row and each column is even. In other words, the values of $sour_{REM}[i]$ and $dest_{REM}[j]$ (the marginal sums of rows and columns, respectively, of the matrix REM) are even integers. We use this property to transform the matrix REM into two matrices REM_1 and REM_2 such that

$REM = REM_1 + REM_2$,
$sour_{REM_1}[i] = sour_{REM_2}[i] = sour_{REM}[i]/2$, for $i = 1, \ldots, k$,
$dest_{REM_1}[j] = dest_{REM_2}[j] = dest_{REM}[j]/2$, for $j = 1, \ldots, n$.

Then we produce two offspring of V_1 and V_2:

$$V_3 = DIV + REM_1$$
$$V_4 = DIV + REM_2.$$

The following example will illustrate the case.

Example 9.4. Take the same problem as described in Example 9.1. Let us assume that the following matrices V_1 and V_2 were selected as parents for crossover:

V_1

1	0	0	7	0
0	4	0	0	0
2	1	4	0	5
0	0	6	0	0

V_2

0	0	5	0	3
0	4	0	0	0
0	0	5	7	0
3	1	0	0	2

The matrices DIV and REM are:

DIV

0	0	2	3	1
0	4	0	0	0
1	0	4	3	2
1	0	3	0	1

REM

1	0	1	1	1
0	0	0	0	0
0	1	1	1	1
1	1	0	0	0

The two matrices REM_1 and REM_2 are:

REM_1

0	0	1	0	1
0	0	0	0	0
0	1	0	1	0
1	0	0	0	0

REM_2

1	0	0	1	0
0	0	0	0	0
0	0	1	0	1
0	1	0	0	0

Finally, two offspring V_3 and V_4 are:

V_3

0	0	3	3	2
0	4	0	0	0
1	1	4	4	2
2	0	3	0	1

V_4

1	0	2	4	1
0	4	0	0	0
1	0	5	3	3
1	1	3	0	1

□

An evolution system GENETIC-2 has been built on the above principles. We have carried out experiments in first tuning and then comparing the the modified classical (vector based) GENETIC-1 and the alternative (matrix based) GENETIC-2 versions of the algorithm to solve the standard linear transportation problem.

Our purpose is not, of course, to compare the genetic methods with the standard optimization algorithm, for on every efficiency measure we choose, genetic methods will be unable to compete. This situation is different when we apply the methods to the nonlinear case. Rather, by using a range of problems of different sizes with known solutions, we aim to investigate the effects of problem representation (GENETIC-1 versus GENETIC-2).

Some randomly generated artificial problems and some published examples comprise the test problems. The artificial problems had randomly generated unit costs, supply and demand values, though the problems remained balanced. We felt that published examples would contain more typical cost structures than artificial problems. For example, a production-inventory problem has a recognizable pattern of costs when represented as a transportation problem.

In every case the problem was first solved using a standard transportation algorithm so that the optimum value was known for use as a stopping criterion and for later comparison of the techniques.

The problems were limited in size by the computers we were using (Macintosh SE/30, Macintosh II, AT&T 3B2 and Sun 3/60 machines, the latter two running under versions of the Unix operating system). The problems are referenced in Table 9.1.

Problem Name	size	Reference
prob01 to prob15	4 by 4 to 10 by 10	problems generated randomly
sas218	5 by 5	[338], p.218
taha170	3 by 4	[387], p.170
taha197	5 by 4	[387], p.197
win268	9 by 5	[405], p.268

Table 9.1. The problems used

In comparing optimization algorithms one first has to decide on the criteria to be used. One obvious criterion is the number of generations required to reach an optimum value, perhaps combined with the time or the number of operations required to complete each generation. A serious problem is that in some cases the genetic algorithms take many generations to reach an optimum. Moreover with some settings of the parameters no optimum is reached before the run has to be stopped. The number of generations needed also varies markedly with the random number starting seed used as well as the problem being solved. We felt that this measure, though natural, could not easily be used for these particular experiments.

Alternative, and more practical, criteria are based on the closeness to the optimum value reached in a fixed number of generations. We chose the percentage above the known optimum value reached in 100 generations. Observations were also made for 1000 generations but in most cases, for the problems studied here, the solution reached by then is at or very close to the optimum for both algorithms and comparisons are inconclusive.

Other workers have found (see, for example, [169]) that changing the parameters of the genetic algorithm can make a difference to its performance. We first deal with the results from experiments in tuning the parameters for the two methods. We kept the population size fixed at 40 and the number of the solutions chosen for reproduction in each generation fixed at 10 (25% of the population). This latter number is also the number of solutions removed each generation. We were then able to adjust the values of the mutation parameters, *cross*, *inv*, and *mut*, i.e., the number of parents chosen to reproduce by *crossover*, *inversion*, and *mutation*, respectively, keeping their sum at 10. The number subject to crossover, *cross*, has to be even since crossovers occur between pairs of parents. Inversion is only possible in the vector-based version, GENETIC-1. We also fixed the probability distribution parameter *sprob* that controls the geometric distribution used for choosing parents and those to be removed.

Over 1000 runs were carried out during the tuning process. Runs were made for five different random number seeds for each chosen combination of parameters. The average objective value of the five runs was converted into a percentage above the known optimum.

Figure 9.1 shows an example of the effect of varying the number of crossover pairs, *cross*, in the two programs for one particular published problem, sas218. Similar, though not identical, results were found for other published problems and for the artificial problems. Because we fixed the total number of parents, more crossovers implies fewer mutations and, for the vector model, fewer inversions. Each point in the figure is the average of five runs with different starting seeds. The percentage above the known optimum reached in 100 generations is graphed.

In general the matrix based GENETIC-2 version gave smoother curves as a function of the number of crossover pairs. Usually, the fewer the crossovers the better, with the best results occurring at zero crossover pairs. But results are affected by the choice of problem. Problem sas218 is different from most others in that the best results are obtained with 2 to 4 crossover pairs. The GENETIC-1 results show generally increasing percentage values with crossovers, though, again, this is not universal. In both cases, as the figure demonstrates, the situation deteriorates when all pairs are used in a crossover mode, leaving none for random mutation. The GENETIC-2 model is particularly sensitive to this effect and in this case is much worse than GENETIC-1 for the runs demonstrated here.

For the class of problems studied we find that, though results differ over the problems, a small proportion of crossovers works best for both models, and, for the vector model, zero inversions are best.

Once optimum tuning parameters had been obtained, we carried out comparisons between the models running on the collection of problems. Once again the results quoted are in every case the average of five runs with different starting seeds.

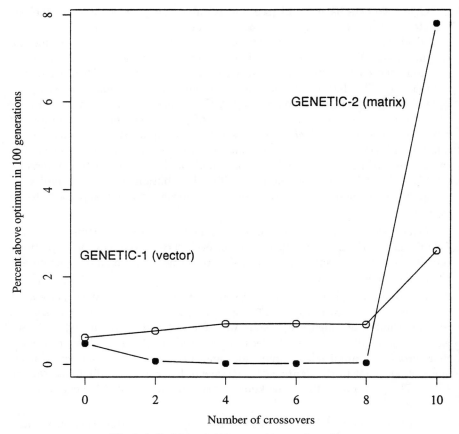

Fig. 9.1. Problem sas218 using the two algorithms

Figure 9.2 shows the results of runs on the whole set of problems for 2 crossover pairs graphed against problem size, $(n * k)$. In every case the matrix-based GENETIC-2 performs better (i.e., gets closer to the optimum in 100 generations) than the vector-based GENETIC-1. Never did the vector version outperform the matrix version on the same problem. GENETIC-1 was also much more unreliable than GENETIC-2. This effect is particularly striking with some problems, as can be seen by the outliers in the figure.

9.1.4 Conclusions

We conclude that in these experiments the matrix based algorithm (GENETIC-2) performs better than the vector based version (GENETIC-1) using our criterion. Note that this comparison is between a matrix based version and a *specially developed* vector based model. While the matrix based algorithm in-

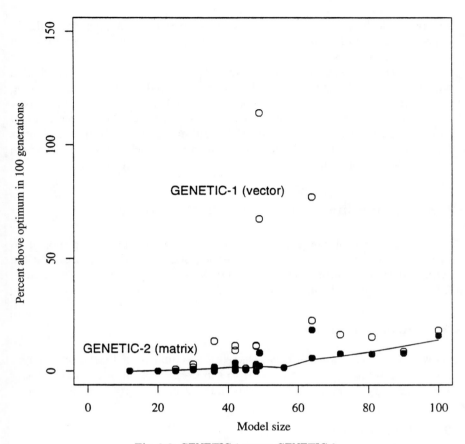

Fig. 9.2. GENETIC-1 versus GENETIC-2

cludes problem-specific knowledge in a natural manner, the vector based model also must rely on additional assumptions in order to proceed: a special **initialization** routine, based on a detailed analysis of the problem, had to be devised in order to make the vector based version work at all. For that reason the special vector based version (GENETIC-1) cannot easily be generalized but the matrix approach (GENETIC-2) is potentially very fruitful for generalization, including to the more complicated nonlinear versions of the transportation problem. Here, GENETIC-1 cannot work properly: procedure **initialization**, which serves as a basis of this system, depends heavily on knowledge of the solution form in the linear case. When the optimal solution need not be a matrix of integers and the number of non-zeros can be much larger than $k + n - 1$, the GENETIC-1 system must fail. Even if we change **initialization** (for example, for each selected i-th row and j-th column, variable val is assigned a random number from the range $\langle 0, \ min(sour[i], \ dest[j]) \rangle$), a vector representing the sequence of initialization points should be extended to record all selected random numbers as well. We conclude that all of these difficulties are due to the artificial representation of a solution as a vector.

The genetic algorithm is very slow and can in no way be compared with the special optimizing techniques based on the standard linear programming algorithm. The latter solves the problem in a number of iterations of the order of the problem size $(n * k)$, whereas each generation of the genetic method involves constructing a set of potential solutions to the problem. However it holds promise of being useful for non-linear and fixed-charge problems where the standard transportation methods cannot be used (next section).

9.2 The nonlinear transportation problem

We discuss our evolution program for the balanced nonlinear transportation problem in terms of the five components for genetic algorithms: representation, initialization, evaluation, operators, and parameters. The algorithm was named GENETIC-2 (as in the linear case).

9.2.1 Representation

As with the linear case, we have selected a two-dimensional structure for representing a solution (a chromosome) to the transportation problem: a matrix $V = (x_{ij})$ $(1 \leq i \leq k,\ 1 \leq j \leq n)$. This time each x_{ij} is a real number.

9.2.2 Initialization

The initialization procedure is identical to the one from the linear case (section 9.1.3). As in the linear case, it creates a matrix of at most $k + n - 1$ non-zero elements such that all constraints are satisfied. Although other initialization procedures are feasible, this method will generate a solution that is at a vertex of the simplex which describes the convex boundary of the constrained solution space.

9.2.3 Evaluation

In this case we have to minimize cost, a nonlinear function of the matrix entries. A number of functions were selected (section 9.2.6) and the results of experiments are presented in section 9.2.7.

9.2.4 Operators

We define two genetic operators, *mutation* and *arithmetical crossover*.

- **mutation:**

  Two types of mutation operators are defined. The first, **mutation-1**, is identical to that used in the linear case and introduces as many zero entries into the matrix as possible. The second, **mutation-2**, is modified to avoid choosing zero entries by selecting values from a range. The **mutation-2** operator is identical to **mutation-1** except that in recalculating the contents of the chosen sub-matrix a modified version of the **initialization** routine is used.

  It is changed from that described in section 9.1.3 as follows: The line

  set $val = \min(sour[i], dest[j])$

  is replaced by:

  set $val_1 = \min(sour[i], dest[j])$
  if (i is the last available row) **or**
      (j is the last available column)
      then $val = val_1$
      else set val = random (real) number from $\langle 0, val_1 \rangle$

  This change provides real numbers instead of integers and zeros but the procedure must be further modified as it currently produces a matrix which may violate the constraints.

  For example, using the matrix from Example 9.1, suppose that the sequence of selected numbers is $\langle 3, 6, 12, 8, 10, 1, 2, 4, 9, 11, 7, 5 \rangle$ and that the first real number generated for number 3 (first row, third column) is 7.3 (which is within the range $\langle 0.0, \min(sour[1], dest[3]) \rangle = \langle 0.0, 15.0 \rangle$). The second random real number for 6 (second row, second column) is 12.1, and the rest of the real numbers generated by the new initialization algorithm are: $3.3, 5.0, 1.0, 3.0, 1.9, 1.7, 0.4, 0.3, 7.4, 0.5$. The resulting matrix is:

	5.0	**15.0**	**15.0**	**10.0**
15.0	3.0	1.9	7.3	1.7
25.0	0.5	12.1	7.4	5.0
5.0	0.4	1.0	0.3	3.3

  Only by adding 1.1 to the element x_{11} can we satisfy the constraints. So we need to add a final line to the **mutation-2** algorithm:

  make necessary additions

  This completes the modification of the **initialization** procedure.

- **Crossover**

  Starting with two parents (matrices U and V) the crossover operator will produce two children X and Y, where

$$X = c_1 \cdot U + c_2 \cdot V \text{ and } Y = c_1 \cdot V + c_2 \cdot U,$$

(where $c_1, c_2 \geq 0$ and $c_1 + c_2 = 1$). Since the constraint set is convex this operation ensures that both children are feasible if both parents are. This is a significant simplification of the linear case where there was an additional requirement to maintain all components of the matrix as integers.

9.2.5 Parameters

In addition to the set of control parameters used for the linear case (population size, mutation and crossover rates, random number starting seed, etc.) a few more are needed. These are the crossover proportions, c_1 and c_2, and m_1, a parameter to determine the proportion of mutation-1 in the mutations applied.

9.2.6 Test cases

To get some indication of the usefulness of the proposed approach, we have selected a single example of a 7×7 transportation problem (Table 9.2) and experimented with various objective functions.

	20	20	20	23	26	25	26
27	x_1	x_2	x_3	x_4	x_5	x_6	x_7
28	x_8	x_9	x_{10}	x_{11}	x_{12}	x_{13}	x_{14}
25	x_{15}	x_{16}	x_{17}	x_{18}	x_{19}	x_{20}	x_{21}
20	x_{22}	x_{23}	x_{24}	x_{25}	x_{26}	x_{27}	x_{28}
20	x_{29}	x_{30}	x_{31}	x_{32}	x_{33}	x_{34}	x_{35}
20	x_{36}	x_{37}	x_{38}	x_{39}	x_{40}	x_{41}	x_{42}
20	x_{43}	x_{44}	x_{45}	x_{46}	x_{47}	x_{48}	x_{49}

Table 9.2. The 7×7 transportation problem

The problem is to minimize a function

$$f(\boldsymbol{x}) = f(x_1, \ldots, x_{49}),$$

subject to fourteen (thirteen independent) equations:

$$x_1 + x_2 + x_3 + x_4 + x_5 + x_6 + x_7 = 27$$
$$x_8 + x_9 + x_{10} + x_{11} + x_{12} + x_{13} + x_{14} = 28$$
$$x_{15} + x_{16} + x_{17} + x_{18} + x_{19} + x_{20} + x_{21} = 25$$
$$x_{22} + x_{23} + x_{24} + x_{25} + x_{26} + x_{27} + x_{28} = 20$$
$$x_{29} + x_{30} + x_{31} + x_{32} + x_{33} + x_{34} + x_{35} = 20$$
$$x_{36} + x_{37} + x_{38} + x_{39} + x_{40} + x_{41} + x_{42} = 20$$
$$x_{43} + x_{44} + x_{45} + x_{46} + x_{47} + x_{48} + x_{49} = 20$$

$$x_1 + x_8 + x_{15} + x_{22} + x_{29} + x_{36} + x_{43} = 20$$
$$x_2 + x_9 + x_{16} + x_{23} + x_{30} + x_{37} + x_{44} = 20$$
$$x_3 + x_{10} + x_{17} + x_{24} + x_{31} + x_{38} + x_{45} = 20$$
$$x_4 + x_{11} + x_{18} + x_{25} + x_{32} + x_{39} + x_{46} = 23$$
$$x_5 + x_{12} + x_{19} + x_{26} + x_{33} + x_{40} + x_{47} = 26$$
$$x_6 + x_{13} + x_{20} + x_{27} + x_{34} + x_{41} + x_{48} = 25$$
$$x_7 + x_{14} + x_{21} + x_{28} + x_{35} + x_{42} + x_{49} = 26$$

The behavior of nonlinear optimization algorithms depends markedly on the form of the objective function. It is clear that different solution techniques may respond quite differently.

For purposes of testing, we have arbitrarily classified potential objective functions into those that might conceivably be seen in practical OR problems (practical), those that are mainly seen in textbooks on optimization (reasonable) and those that are more often seen as difficult test cases for optimization techniques (other). In brief, these may be described as follows:

- practical functions
  Typically piece-wise linear cost functions, these appear often in practice either because of data limitations or because the operation of the facility has domains where different costs apply. Often they are not smooth and certainly the derivatives can be discontinuous. They will often cause difficulties for gradient methods though approximations to turn them into differentiable functions are possible. Examples: $A(x)$ and $B(x)$.

- reasonable functions
  These functions are smooth and often simple powers of the flows. They can be further classified into convex and concave functions. Examples: $C(x)$ and $D(x)$.

- other functions
  These typically have multiple valleys (or peaks) with sub-optima that will cause difficulties for any gradient method. They are invented as severe tests of optimization algorithms and, we conjecture, infrequently appear in practice. Examples: $E(x)$ and $F(x)$.

Listed below are the two examples from each group of objective functions used in the tests. They are all separable functions of the components of the solution vector with no cross terms. The continuous versions of their graphs (already modified for the GAMS system) are presented in Figure 9.3.

- function A

$$A(x) = \begin{cases} 0, & \text{if } 0 < x \leq S \\ c_{ij}, & \text{if } S < x \leq 2S \\ 2c_{ij}, & \text{if } 2S < x \leq 3S \\ 3c_{ij}, & \text{if } 3S < x \leq 4S \\ 4c_{ij}, & \text{if } 4S < x \leq 5S \\ 5c_{ij}, & \text{if } 5S < x \end{cases}$$

where S is less than a typical x value.

- function B

$$B(x) = \begin{cases} c_{ij}\frac{x}{S}, & \text{if } 0 \leq x \leq S \\ c_{ij}, & \text{if } S < x \leq 2S \\ c_{ij}(1 + \frac{x-2S}{S}), & \text{if } 2S < x \end{cases}$$

where S is of the order of a typical x value.

- function C

$$C(x) = c_{ij}x^2$$

- function D

$$D(x) = c_{ij}\sqrt{x}$$

- function E

$$E(x) = c_{ij}\left(\frac{1}{1 + (x - 2S)^2} + \frac{1}{1 + (x - \frac{9}{4}S)^2} + \frac{1}{1 + (x - \frac{7}{4}S)^2}\right)$$

where S is of the order of a typical x value.

- function F

$$F(x) = c_{ij}x(\sin(x\frac{5\pi}{4S}) + 1)$$

where S is of the order of a typical x value.

The objective function for the transportation problem is of the form

$$\sum_{ij} f(x_{ij})$$

where $f(x)$ is one of the functions above, the parameters c_{ij} are obtained from the parameter matrix (see Figure 9.4), and S is obtained from the attributes of the problem to be tested.

To derive S, it is necessary to estimate the value of a typical x value; this was done by the way of preliminary runs to estimate the number and magnitudes of non-zero x_{ij}'s. In this way the average flow on each arc was estimated and a value for S found. For function A we used $S = 2$, while for B, E, and F, we used $S = 5$.

Note that the objective function is identical on each arc, so a cost-matrix was used to provide a variation between arcs. The matrix provides the c_{ij}'s which act to scale the basic function shape, thus providing 'one degree' of variability.

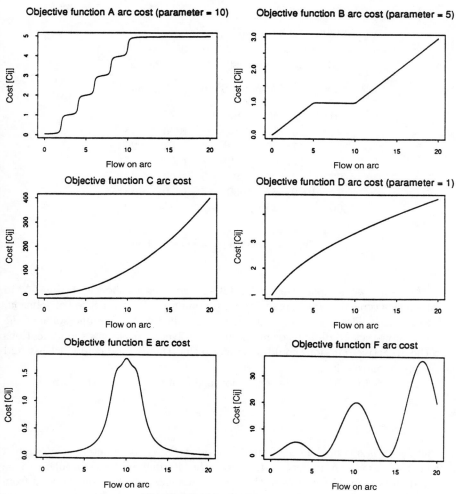

Fig. 9.3. Six test functions A – F

9.2.7 Experiments and results

In testing the GENETIC-2 algorithm on the linear transportation problem (section 9.1) we can compare its solution with the known optimum found using the standard algorithm. Hence we can determine how efficient the genetic algorithm is in absolute terms. Once we move to nonlinear objective functions, the optimum may not be known. Testing is reduced to comparing the results with those of other nonlinear solution methods that may themselves have converged to a local optimum.

Number of Sources: 7
Number of Destinations: 7

Source Flows: 27 28 25 20 20 20 20
Destination Flows: 20 20 20 23 26 25 26

Arc Parameter Matrix (Source by Destination):

$$
\begin{array}{ccccccc}
0 & 21 & 50 & 62 & 93 & 77 & 1000 \\
21 & 0 & 17 & 54 & 67 & 1000 & 48 \\
50 & 17 & 0 & 60 & 98 & 67 & 25 \\
62 & 54 & 60 & 0 & 27 & 1000 & 38 \\
93 & 67 & 98 & 27 & 0 & 47 & 42 \\
77 & 1000 & 67 & 1000 & 47 & 0 & 35 \\
1000 & 48 & 25 & 38 & 42 & 35 & 0
\end{array}
$$

Fig. 9.4. Example problem description

As usual, we compare the GENETIC-2 algorithm method with the GAMS system as a typical example of an industry-standard efficient method of solution. This system, being essentially a gradient-controlled method, found some of the problems we set up difficult or impossible to solve. In these cases modifications to the objective functions could be made so that the method could at least find an approximate solution.

The objective for the transportation problem was then of the form

$$\sum_{ij} f(x_{ij})$$

where $f(x)$ is one of the six selected functions, the c_{ij} parameters are obtained from the parameter matrix and S from the attributes of the problem to be tested. S is approximated from the average non-zero arc flow determined from a number of preliminary runs to make sure the flows occurred in the interesting part of the objective function.

In some sense it is desirable to use completely randomly structured objective functions on each arc. Given that our objective is to demonstrate how the algorithm performs on a variety of problems the question reduces to asking how much variation between arcs is required for a particular function form. When the function is identical on each arc the problem may have many solutions with the same cost, reducing the information obtained when analyzing the algorithm.

In our experiments a cost-matrix was used to provide variation between arcs. The matrix provides the c_{ij}'s which act to scale the basic function shape, thus providing 'one degree' of variability. More matrices (providing more degrees of variability) were not required.

For functions C, E, and F, the GAMS application was straightforward: using built–in nonlinear functions. Due to the requirement for gradient estimation of objective functions, GAMS could not handle functions A, B and D directly. In the case of A and B, the expression could not be formulated in GAMS while in the case of D (the square root function) difficulty was encountered in measuring gradients near zero. Therefore, we made the following modifications to the problem for the GAMS runs:

- function A
  Separate arc-tangent functions are used to approximate each of the five steps. A parameter, P_A, was used to control the 'tightness' of the fit. The cost on arc [i,j] is:

$$c_{ij} \cdot \begin{pmatrix} \arctan(P_A(x_{ij} - S))/\pi + \frac{1}{2} + \\ \arctan(P_A(x_{ij} - 2S))/\pi + \frac{1}{2} + \\ \arctan(P_A(x_{ij} - 3S))/\pi + \frac{1}{2} + \\ \arctan(P_A(x_{ij} - 4S))/\pi + \frac{1}{2} + \\ \arctan(P_A(x_{ij} - 5S))/\pi + \frac{1}{2} \end{pmatrix}$$

- function B
  The arc-tangent function was again used, this time to approximate each of the three gradients. A parameter, P_B, was used to control the tightness of the fit. The cost on arc [i,j] is:

$$c_{ij} \cdot \begin{pmatrix} (\frac{x_{ij}}{S}) \cdot (\arctan(P_B x_{ij})/\pi + \frac{1}{2}) + \\ (1 - \frac{x_{ij}}{S}) \cdot (\arctan(P_B(x_{ij} - S))/\pi + \frac{1}{2}) + \\ (\frac{x_{ij}}{S} - 2) \cdot (\arctan(P_B(x_{ij} - 2S))/\pi + \frac{1}{2}) \end{pmatrix}$$

- function D
  In order to avoid gradient problems at or near zero, the function D was changed to:

$$D'(x) = D(x + \epsilon) .$$

As for the linear case, for each problem, multiple GAMS runs were made under different values of the modification parameter and the best result chosen. The best values for the three parameters were found to be: P_A between 1 and 20, P_B very large (e.g., 1000), and ϵ (for function D) between 1 and 7. The final result values were always calculated after the optimization using the unmodified function, instead of the modified function.

For the main set of experiments, five 10×10 transportation matrices were used with each function. They were constructed from a set of independent uniformly distributed c_{ij} values and randomly chosen source and destination vectors with a total flow of 100 units. Each function–matrix combination was given 5 runs using different random number starting seeds for the genetic algorithm. Problems were run for 10,000 generations.

For function A, S was set to 2, while for functions B, E, and F a value of 5 was used.

The 10×10 node problems reach the limit of the student version of GAMS (where allowable problem size is restricted). From a listing of some example problems tested on the GAMS system, it appears that with the full version (where problem size is limited by available memory and internal limits) on a 640k memory AT computer, a 25×25 node problem should be possible. Note that an $N \times N$ node problem would be formulated by GAMS as having N^2 variables, 2N constraints and a nonlinear objective function. Clearly, larger problems could be formulated on bigger systems (especially a mainframe) or with specialized solvers.

However, using much larger problems to compare the genetic system with nonlinear programming type solvers may be of limited value. Results of the 10×10 runs demonstrate the tendency for GAMS (and presumably, similar systems) to fall into local (non-global) optima. Ignoring the time spent evaluating the objective function and using the number of solutions tested as the measure of time it is clear that standard nonlinear programming techniques will always 'finish' faster than genetic systems. This is because they typically explore only a particular path within the current local optimum zone. They will do well only if the local optimum is a relatively good one.

A set of parameters were chosen for GENETIC-2 after experience with the linear problems and on the basis of tuning runs with the nonlinear problems. The population size was fixed at 40. The mutation rate was $p_m = 20\%$ with the proportion of mutation-1 being 50%, and the crossover rate was $p_c = 5\%$. The crossover proportions were $c_1 = .35$ and $c_2 = .65$.

It may appear that the chosen mutation rate is too high and the crossover rate too low in comparison with classical genetic algorithms. However, our operators are different from the classical ones, because (1) we select parents for mutations and crossovers, i.e., the *whole* structure (as opposed to single bits) undergoes mutation, and (2) mutation-1 creates an offspring 'pushing' the parent towards the surface of the solution space, whereas crossover and mutation-2 'push' the offspring towards the center of the solution space.

The use of high mutation rates may also suggest that the algorithm is nearly a random search. However, the random search algorithm (crossover rate 0%) performs quite poorly in comparison to the tuned algorithm used here. To demonstrate this, we tabulate below (Table 9.3) some typical results for different values of parameters p_m and p_c using functions A and F and for a particular 7×7 transportation problem. The values given are the average minimum cost achieved for 5 runs with different seeds in 10,000 generations.

Function	$p_c = 0\%$ $p_m = 25\%$	$p_c = 25\%$ $p_m = 0\%$	$p_c = 5\%$ $p_m = 20\%$
A	45.8	181.0	0.0
F	178.7	189.6	110.9

Table 9.3. Results for different values of p_c and p_m

The 7×7 transportation problem used is given in Figure 9.4; the solutions found by the algorithm GENETIC-2 for different values of parameters p_c and p_m are given in Figure 9.5.

$p_c = 0\%$, $p_m = 25\%$, function A, Cost $= 45.8$

20.00	0.00	0.00	1.00	2.00	2.00	2.00
0.00	20.00	1.00	2.00	2.00	2.00	1.00
0.00	0.00	19.00	0.00	2.00	1.00	3.00
0.00	0.00	0.00	20.00	0.00	0.00	0.00
0.00	0.00	0.00	0.00	20.00	0.00	0.00
0.00	0.00	0.00	0.00	0.00	20.00	0.00
0.00	0.00	0.00	0.00	0.00	0.00	20.00

$p_c = 25\%$, $p_m = 0\%$, function A, Cost $= 181.2$

18.25	0.00	0.11	1.81	3.30	3.53	0.00
0.00	18.21	3.94	3.05	0.00	1.97	0.82
1.75	1.79	13.24	1.46	1.46	1.48	3.83
0.00	0.00	1.91	14.87	1.18	0.35	1.69
0.00	0.00	0.72	0.97	18.10	0.21	0.00
0.00	0.00	0.02	0.71	1.96	17.30	0.00
0.00	0.00	0.05	0.14	0.00	0.16	19.65

$p_c = 5\%$, $p_m = 20\%$, function A, Cost $= 0.0$

19.87	0.00	0.68	1.80	1.33	1.80	1.51
0.08	20.00	1.00	1.90	1.48	1.61	1.92
0.00	0.00	18.32	1.89	1.08	1.78	1.93
0.05	0.00	0.00	17.09	1.91	0.96	0.00
0.00	0.00	0.00	0.00	19.92	0.00	0.08
0.00	0.00	0.00	0.00	0.00	18.60	1.40
0.00	0.00	0.00	0.31	0.28	0.25	19.16

$p_c = 0\%$, $p_m = 25\%$, function F, Cost $= 178.7$

15.00	6.00	0.00	6.00	0.00	0.00	0.00
0.00	14.00	0.00	14.00	0.00	0.00	0.00
0.00	0.00	20.00	0.00	0.00	0.00	5.00
5.00	0.00	0.00	3.00	6.00	0.00	6.00
0.00	0.00	0.00	0.00	20.00	0.00	0.00
0.00	0.00	0.00	0.00	0.00	20.00	0.00
0.00	0.00	0.00	0.00	0.00	5.00	15.00

$p_c = 25\%$, $p_m = 0\%$, function F, Cost $= 189.6$

20.00	0.25	0.00	0.75	0.00	6.00	0.00
0.00	5.75	0.00	22.25	0.00	0.00	0.00
0.00	0.00	20.00	0.00	0.00	0.00	5.00
0.00	14.00	0.00	0.00	6.00	0.00	0.00
0.00	0.00	0.00	0.00	20.00	0.00	0.00
0.00	0.00	0.00	0.00	0.00	14.00	6.00
0.00	0.00	0.00	0.00	0.00	5.00	15.00

$p_c = 5\%$, $p_m = 20\%$, function F, Cost $= 110.9$

14.31	6.31	6.39	0.00	0.00	0.00	0.00
0.00	13.69	0.31	14.00	0.00	0.00	0.00
0.00	0.00	13.31	0.00	0.00	0.00	5.69
5.69	0.00	0.00	3.00	6.00	0.00	5.31
0.00	0.00	0.00	0.00	20.00	0.00	0.00
0.00	0.00	0.00	0.00	0.00	19.31	0.69
0.00	0.00	0.00	0.00	0.00	5.69	14.31

Fig. 9.5. Solutions found by GENETIC-2 for different values of p_c and p_m

The GENETIC-2 system was run on SUN SPARCstation 1 computers while GAMS was run on an Olivetti 386. Although speed comparisons between the two machines are difficult it should be noted that in general GAMS finished each run well before the genetic system. An exception is case A (in which GAMS evaluates numerous arc-tangent functions) where the genetic algorithm took no more than 15 minutes to complete while GAMS averaged about twice that. For cases A,B, and D, where the extra GAMS modification parameter meant that multiple runs had to be performed to find its best solution, the genetic system overall was much faster.

A typical comparison of the optima between GENETIC-2 (averaged over 5 seeds) and GAMS is shown in the Table 9.4 for a single 10×10 problem; its description is given in Figure 9.6.

Figure 9.7 displays the results for all five considered problems. For the class of 'practical' problems, A and B, GENETIC-2 is, on average, better than GAMS by 24.5% in case A and by 11.5% in case B. For the 'reasonable' functions the results were different. In case C (the square function), the genetic system performed worse by 7.5% while in case D (the square-root function), the genetic system was better by just 2.0%, on average. For the 'other' functions, E and F,

Function	GAMS	GENETIC-2	% difference
A	281.0	202.0	−28.1%
B	180.8	163.0	−9.8%
C	4402.0	4556.2	+3.5%
D	408.4	391.1	−4.2%
E	145.1	79.2	−45.4%
F	1200.8	201.9	−83.2%

Table 9.4. Comparison between GAMS and GENETIC-2

```
Number of Sources:        10
Number of Destinations:   10
Source Flows:        8   8   2  26  12   1   6  18  18   1
Destination Flows:  19   2  33   5  11  11   2  14   2   1
```

Arc Parameter Matrix (Source by Destination):

```
15   3  23   1  19  14   6  16  41  33
13  17  30  36  20  17  26  19   3  33
37  17  30   5  48  27   8  25  36  21
13  13  31   7  35  11  20  41  34   3
31  24   8  30  28  33   2   8   1   8
32  36  12   9  18   1  44  49  11  11
49   6  17   0  42  45  22   9  10  47
 2  21  18  40  47  27  27  40  19  42
13  16  25  21  19   0  32  20  32  35
23  42   2   0   9  30   5  29  31  29
```

Fig. 9.6. Example problem description

the genetic system dominates; it resulted in improvements of 33.0% and 54.5% over GAMS, averaging over the five problems.

9.2.8 Conclusions

Our objective was to investigate the behavior of a type of genetic algorithm on problems with multiple constraints and nonlinear objective functions. The transportation problem was chosen for study as it provided a relatively simple convex feasible set. This was to make it is easier to maintain feasibility in the solutions. We were then able to examine the influence of the objective function alone on the algorithm's behavior.

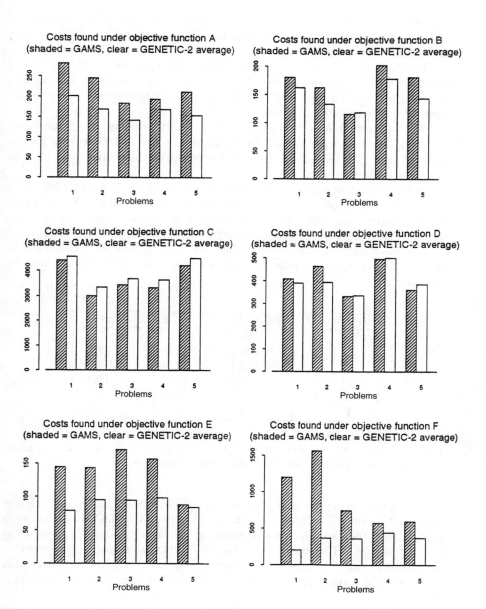

Fig. 9.7. Results

The results demonstrate the efficiency of the genetic method in finding the global optimum in difficult problems, though GAMS did well on the smooth monotonic, 'reasonable', functions. The gradient controlled techniques are most suited to these situations. For function C, GAMS found better solutions much faster than GENETIC-2.

For the 'practical' problems, the gradient techniques have difficulty 'seeing around the corner' to new zones of better costs. The genetic type of algorithm, taking a more global approach, is able to move to new zones readily, hence generating much better solutions.

The 'other' problems, although they are both smooth, have significant structural features that were admittedly designed to cause real difficulties for the gradient methods. GENETIC-2 excelled over GAMS here even more than in the 'practical' cases.

It is also interesting to compare GENOCOP (Chapter 7) with GENETIC-2 (see Table 9.5). In general, their results are very similar. However, note again that the matrix approach was tailored to the specific (transportation) problem, whereas GENOCOP is problem independent and works without any hard-coded domain knowledge. In other words, while one might expect the GENOCOP to perform similarly well for other constrained problems, GENETIC-2 cannot be used at all.

Function	GENETIC-2	GENOCOP	% difference
A	00.00	24.15	
B	203.81	205.60	0.87%
C	2564.23	2571.04	0.26%
D	480.16	480.16	0.00%
E	204.73	204.82	0.04%
F	110.94	119.61	7.24%

Table 9.5. GENETIC-2 versus GENOCOP: the results for the 7×7 problem, with transportation cost functions A–F and cost matrix given in Figure 7.3

While comparing all three systems (GAMS, GENOCOP, GENETIC-2), it is important to underline that two of them, GAMS and GENOCOP, are problem independent: they are capable of optimizing any function subject to any set of linear constraints. The third system, GENETIC-2, was designed for transportation problems only: the particular constraints are incorporated into matrix data structures and special "genetic" operators (for further comparisons, see Chapter 14).

GENETIC-2 was specifically tailored to transportation problems but an important characteristic is that it handles any type of cost function (which need not even be continuous). It is also possible to modify it to handle many similar operations research problems including allocation and some scheduling problems. This seems to be a promising research direction which may result in a generic technique for solving matrix based constrained optimization problems.

10. The Traveling Salesman Problem

In the next chapter, we present several examples of evolution programs tailored to specific applications (graph drawing, partitioning, scheduling). The traveling salesman problem (TSP) is just one of such applications; however, we treat it as a special problem — the mother of all problems — and discuss it in a separate chapter. What are the reasons?

Well, there are many. First of all, the TSP is conceptually very simple: the traveling salesman must visit every city in his territory exactly once and then return to the starting point. Given the cost of travel between all cities, how should he plan his itinerary for minimum total cost of the entire tour? The search space for the TSP is a set of permutations of n cities. Any single permutation of n cities yields a solution (which is a complete tour of n cities). The optimal solution is a permutation which yields the minimum cost of the tour. The size of the search space is $n!$.

The TSP is a relatively old problem: it was documented as early as 1759 by Euler (though not by that name), whose interest was in solving the knights' tour problem. A correct solution would have a knight visit each of the 64 squares of a chessboard exactly once in its tour.

The term 'traveling salesman' was first used in a 1932 a German book *The traveling salesman, how and what he should do to get commissions and be successful in his business*, written by a veteran traveling salesman (see [236]). Though not the main topic of the book, the TSP and scheduling are discussed in the last chapter.

The TSP was introduced by the RAND Corporation in 1948. The Corporation's reputation helped to make the TSP a well known and popular problem. The TSP also became popular at that time due to the new subject of linear programming and attempts to solve combinatorial problems.

The Traveling Salesman Problem was proved to be NP-hard [134]. It arises in numerous applications and the number of cities might be quite significant — as stated in [207]:

"Circuit board drilling applications with up to 17,000 cities are mentioned in [246], X-ray crystallography instances with up to 14,000 cities are mentioned in [44], and instances arising in VLSI fabrication have been reported with as many as 1.2 million cities [227]. Moreover, 5 hours on a multi-million dollar computer for an optimal solution may not be cost-effective if one can get within a few percent in seconds on a PC. Thus there remains a need for heuristics."

During the last decades, several algorithms emerged to approximate the optimal solution: nearest neighbor, greedy algorithm, nearest insertion, farthest insertion, double minimum spanning tree, strip, space-filling curve, algorithms by Karp, Litke, Christofides, etc. [207] (some of these algorithms assume that the cities correspond to points in the plane under some standard metric). Another group of algorithms (2-opt, 3-opt, Lin-Kernighan) aims at a local optimization: an improvement of a tour by local perturbations. The TSP also became a target for the GA community: several genetic-based algorithms were reported [131], [160], [168], [170], [206], [241], [288], [299], [353], [370], [375], [389], [402]. These algorithms aim at producing near-optimal solutions by maintaining a population of potential solutions which undergoes some unary and binary transformations ('mutations' and 'crossovers') under a selection scheme biased towards fit individuals. It is interesting to compare these approaches, paying particular attention to the representation and genetic operators used — this is what we intend to do in this chapter. In other words, we shall trace the evolution of evolution programs for TSP.

To underline some important characteristics of the TSP, let us consider the CNF-satisfiability problem first. A logical expression in conjunctive normal form (CNF) is a sequence of clauses separated by the Boolean operator $\wedge$; a clause is a sequence of literals separated by the Boolean operator $\vee$; a literal is a logical variable or its negation; a logical variable is a variable that may be assigned values TRUE or FALSE (1 or 0).

For example, the following logical expression is in CNF:

$$(a \vee \bar{b} \vee c) \wedge (b \vee c \vee d \vee \bar{e}) \wedge (\bar{a} \vee c) \wedge (a \vee \bar{c} \vee \bar{e}),$$

where a, b, c, d, and e are logical variables; $\bar{a}$ denotes the negation of variable a ($\bar{a}$ has the value TRUE if and only if a has the value FALSE).

The problem is to determine whether there exists a truth assignment for the variables in the expression, so that the whole expression evaluates to TRUE. For example, the above CNF logical expression has several truth assignments, for which the whole expression evaluates to TRUE, e.g., any assignment with a = TRUE and c = TRUE.

If we try to apply a genetic algorithm to the CNF-satisfiability problem, we notice that it is hard to imagine a problem with better suited representation: a binary vector of fixed length (the length of the vector corresponds to the number of variables) should do the job. Moreover, there are no dependencies between bits: any change would result in a legal (meaningful) vector. Thus we can apply mutations and crossovers without any need for decoders or repair algorithms. However, the choice of the evaluation function is the hardest task. Note that all logical expressions evaluate to TRUE or FALSE, and if a specific truth assignment evaluates the whole expression to TRUE, then the solution to the problem is found. The point is that during the search for a solution, all chromosomes (vectors) in a population would evaluate to FALSE (unless a solution is found), so it is impossible to distinguish between 'good' and 'bad' chromosomes. In short, the CNF-satisfiability problem has natural representation and operators, without any natural evaluation function. For a further discussion on the problems related to a selection of an appropriate evaluation function, the reader is referred to [90].

On the other hand, the TSP has an extremely easy (and natural) evaluation function: for any potential solution (a permutation of cities), we can refer to the table with distances between all cities and (after $n - 1$ addition operations) we get the total length of the tour. Thus, in a population of tours, we can easily compare any two of them. However, the choice of the representation of a tour and the choice of operators to be used are far from clear.

An additional reason for treating the TSP in a separate chapter is that similar techniques were used for variety of other sequencing problems, like scheduling and partitioning. Some of these problems are discussed in the next chapter.

There is an agreement in the GA community that the binary representation of tours is not well suited for the TSP. It is not hard to see why: after all, we are interested in the best permutation of cities, i.e.,

$$(i_1, i_2, \ldots, i_n),$$

where $(i_1, i_2, \ldots, i_n)$ is a permutation of $\{1, 2, \ldots, n\}$. The binary code of these cities will not provide any advantage. Just the opposite is true: the binary representation would require special repair algorithms, since a change of a single bit may result in a illegal tour. As observed in [402]:

> "Unfortunately, there is no practical way to encode a TSP as a binary string that does not have ordering dependencies or to which operators can be applied in a meaningful fashion. Simply crossing strings of cities produces duplicates and omissions. Thus, to solve this problem some variation on standard genetic crossover must be used. The ideal recombination operator should recombine critical information from the parent structures in a non-destructive, meaningful manner."

It is interesting to note that a recent paper by Lidd [241] describes a GA approach for the TSP with a binary representation and classical operators (crossover and mutation). The illegal tours are evaluated on the basis of complete (not necessarily legal) tours created by a greedy algorithm. The reported results are of surprisingly high quality, however, the largest considered test case consisted of 100 cities only.

During the last few years there have been three vector representations considered in connection with the TSP: *adjacency, ordinal,* and *path* representations. Each of these representations has its own "genetic" operators — we shall discuss them in turn. Since it is relatively easy to come up with some sort of mutation operator which would introduce a small change into a tour, we shall concentrate on crossover operators. In all three representations, a tour is described as a list of cities. In the following discussions we use a common example of 9 cities numbered from 1 to 9.

Adjacency Representation:

The adjacency representation represents a tour as a list of n cities. The city j is listed in the position i if and only if the tour leads from city i to city j. For example, the vector

(2 4 8 3 9 7 1 5 6)

represents the following tour:

$$1 - 2 - 4 - 3 - 8 - 5 - 9 - 6 - 7$$

Each tour has only one adjacency list representation; however, some adjacency lists can represent illegal tours, e.g.,

(2 4 8 1 9 3 5 7 6),

which leads to

$$1 - 2 - 4 - 1,$$

i.e., the (partial) tour with a (premature) cycle.

The adjacency representation does not support the classical crossover operator. A repair algorithm might be necessary. Three crossover operators were defined and investigated for the adjacency representation: *alternating edges, subtour chunks,* and *heuristic* crossovers [168].

- alternating-edges crossover builds an offspring by choosing (at random) an edge from the first parent, then selects an appropriate edge from the second parent, etc. — the operator extends the tour by choosing edges from alternating parents. If the new edge (from one of the parents) introduces a cycle into the current (still partial) tour, the operator selects instead a (random) edge from the remaining edges which does not introduce cycles. For example, the first offspring from the two parents

$$p_1 = (2\ 3\ 8\ 7\ 9\ 1\ 4\ 5\ 6) \text{ and}$$
$$p_2 = (7\ 5\ 1\ 6\ 9\ 2\ 8\ 4\ 3)$$

might be

$$o_1 = (2\ 5\ 8\ 7\ 9\ 1\ 6\ 4\ 3),$$

where the process started from the edge (1,2) from the parent p_1, and the only random edge introduced during the process of alternating edges was (7,6) instead of (7,8), which would have introduced a premature cycle.

- subtour-chunks crossover constructs an offspring by choosing a (random length) subtour from one of the parents, then choosing a (random length) subtour from another parent, etc. — the operator extends the tour by choosing edges from alternating parents. Again, if some edge (from one of the parents) introduces a cycle into the current (still partial) tour, the operator selects instead a (random) edge from the remaining edges which does not introduce cycles.

- heuristic crossover builds an offspring by choosing a random city as the starting point for the offspring's tour. Then it compares the two edges (from both parents) leaving this city and selects the better (shorter) edge. The city on the other end of the selected edge serves as a starting point in selecting the shorter of the two edges leaving this city, etc. If, at some stage, a new edge would introduce a cycle into the partial tour, then the tour is extended by a random edge from the remaining edges which does not introduce cycles.

In [206], the authors modified the above heuristic crossover by changing two rules: (1) if the shorter edge (from a parent) introduces a cycle in the offspring tour, check the other (longer) edge. If the longer edge does not introduce a cycle, accept it; otherwise (2) select the shortest edge from a pool of q randomly selected edges (q is a parameter of the method).

The effect of this operator is to glue together short subpaths of the parent tours. However, it may leave undesirable crossings of edges — it is why the heuristic crossover is not appropriate for fine local tuning of the tours. Suh and Gucht [375] introduced an additional heuristic operator (based on 2-opt algorithm [245]) appropriate for local tuning. The operator randomly selects two edges, $(i\ j)$ and $(k\ m)$, and checks whether

$$dist(i,j) + dist(k,m) > dist(i,m) + dist(k,j),$$

where $dist(a,b)$ is a given distance between cities a and b. If this is the case, the edges $(i\ j)$ and $(k\ m)$ in the tour are replaced by edges $(i\ m)$ and $(k\ j)$.

An advantage of adjacency representation is that it allows schemata analysis similar to one discussed in Chapter 3, where binary strings were considered. Schemata correspond to natural building blocks, i.e., edges; for example, the schema

$$(* * * 3 * 7 * * *)$$

denotes the set of all tours with edges (4 3) and (6 7). However, the main disadvantage of this representation is relatively poor results for all operators. The alternating-edges crossover often disrupts good tours due to its own operation by alternating edges from two parents. The subtour-chunk crossover performs better than alternating-edges crossover, since the disruption rate is lower. However, the performance is still quite low. The heuristic crossover, of course, is the best operator here. The reason is that the first two crossovers are blind, i.e., they do not take into account the actual lengths of the edges. On the other hand, heuristic crossover selects the better edgne out of two possible edges — this is the reason it performs much better than the other two. However, the performance of the heuristic crossover is not outstanding: in three experiments reported [168] on 50, 100, and 200 cities, the system found tours within 25%, 16%, and 27% of the optimum, in approximately 15000, 20000, and 25000 generations, respectively.

Ordinal Representation:

The ordinal representation represents a tour as a list of n cities; the i-th element of the list is a number in the range from 1 to $n - i + 1$. The idea behind the ordinal representation is as follows. There is some ordered list of cities C, which serves as a reference point for lists in ordinal representations. Assume, for example, that such an ordered list (reference point) is simply

$$C = (1\ 2\ 3\ 4\ 5\ 6\ 7\ 8\ 9).$$

A tour

$$1 - 2 - 4 - 3 - 8 - 5 - 9 - 6 - 7$$

is represented as a list l of references,

$$l = (1\ 1\ 2\ 1\ 4\ 1\ 3\ 1\ 1),$$

and should be interpreted as follows:

- the first number on the list l is 1, so take the first city from the list C as the first city of the tour (city number 1), and remove it from C. The partial tour is

  1

- the next number on the list l is also 1, so take the first city from the current list C as the next city of the tour (city number 2), and remove it from C. The partial tour is

  $1 - 2$

- the next number on the list l is 2, so take the second city from the current list C as the next city of the tour (city number 4), and remove it from C. The partial tour is

  $1 - 2 - 4$

- the next number on the list l is 1, so take the first city from the current list C as the next city of the tour (city number 3), and remove it from C. The partial tour is

    $$1 - 2 - 4 - 3$$

- the next number on the list l is 4, so take the fourth city from the current list C as the next city of the tour (city number 8), and remove it from C. The partial tour is

    $$1 - 2 - 4 - 3 - 8$$

- the next number on the list l is again 1, so take the first city from the current list C as the next city of the tour (city number 5), and remove it from C. The partial tour is

    $$1 - 2 - 4 - 3 - 8 - 5$$

- the next number on the list l is 3, so take the third city from the current list C as the next city of the tour (city number 9), and remove it from C. The partial tour is

    $$1 - 2 - 4 - 3 - 8 - 5 - 9$$

- the next number on the list l is 1, so take the first city from the current list C as the next city of the tour (city number 6), and remove it from C. The partial tour is

    $$1 - 2 - 4 - 3 - 8 - 5 - 9 - 6$$

- the last number on the list l is 1, so take the first city from the current list C as the next city of the tour (city number 7, the last available city), and remove it from C. The final tour is

    $$1 - 2 - 4 - 3 - 8 - 5 - 9 - 6 - 7$$

The main advantage of the ordinal representation is that the classical crossover works! Any two tours in the ordinal representation, cut after some position and crossed together, would produce two offspring, each of them being a legal tour. For example, the two parents

$$p_1 = (1\ 1\ 2\ 1\ |\ 4\ 1\ 3\ 1\ 1) \text{ and}$$
$$p_2 = (5\ 1\ 5\ 5\ |\ 5\ 3\ 3\ 2\ 1),$$

which correspond to the tours

$$1 - 2 - 4 - 3 - 8 - 5 - 9 - 6 - 7 \text{ and}$$
$$5 - 1 - 7 - 8 - 9 - 4 - 6 - 3 - 2,$$

with the crossover point marked by '|', would produce the following offspring:

$$o_1 = (1\ 1\ 2\ 1\ 5\ 3\ 3\ 2\ 1) \text{ and}$$
$$o_2 = (5\ 1\ 5\ 5\ 4\ 1\ 3\ 1\ 1);$$

these offspring correspond to

$$1 - 2 - 4 - 3 - 9 - 7 - 8 - 6 - 5 \text{ and}$$
$$5 - 1 - 7 - 8 - 6 - 2 - 9 - 3 - 4.$$

It is easy to see that partial tours to the left of the crossover point do not change, whereas partial tours to the right of the crossover point are disrupted in a quite random way. Poor experimental results indicate [168] that this representation together with classical crossover is not appropriate for the TSP.

Path Representation:
The path representation is perhaps the most natural representation of a tour. For example, a tour

$$5 - 1 - 7 - 8 - 9 - 4 - 6 - 2 - 3$$

is represented simply as

$$(5\ 1\ 7\ 8\ 9\ 4\ 6\ 2\ 3).$$

Until recently, three crossovers were defined for the path representation: *partially -mapped* (PMX), *order* (OX), and *cycle* (CX) crossovers. We will now discuss them in turn.

- PMX — proposed by Goldberg and Lingle [160] — builds an offspring by choosing a subsequence of a tour from one parent and preserving the order and position of as many cities as possible from the other parent. A subsequence of a tour is selected by choosing two random cut points, which serve as boundaries for swapping operations. For example, the two parents (with two cut points marked by '|')

$$p_1 = (1\ 2\ 3\ |\ 4\ 5\ 6\ 7\ |\ 8\ 9) \text{ and}$$
$$p_2 = (4\ 5\ 2\ |\ 1\ 8\ 7\ 6\ |\ 9\ 3)$$

would produce offspring in the following way. First, the segments between cut points are swapped (the symbol 'x' can be interpreted as 'at present unknown'):

$$o_1 = (x\ x\ x\ |\ 1\ 8\ 7\ 6\ |\ x\ x) \text{ and}$$
$$o_2 = (x\ x\ x\ |\ 4\ 5\ 6\ 7\ |\ x\ x).$$

This swap defines also a series of mappings:

$$1 \leftrightarrow 4,\ 8 \leftrightarrow 5,\ 7 \leftrightarrow 6, \text{ and } 6 \leftrightarrow 7.$$

Then we can fill further cities (from the original parents), for which there is no conflict:

$o_1 = (\text{x } 2 \text{ } 3 \mid 1 \text{ } 8 \text{ } 7 \text{ } 6 \mid \text{x } 9)$ and
$o_2 = (\text{x x } 2 \mid 4 \text{ } 5 \text{ } 6 \text{ } 7 \mid 9 \text{ } 3)$.

Finally, the first x in the offspring o_1 (which should be 1, but there was a conflict) is replaced by 4, because of the mapping $1 \leftrightarrow 4$. Similarly, the second x in the offspring o_1 is replaced by 5, and the x and x in the offspring o_2 are 1 and 8. The offspring are

$o_1 = (4 \text{ } 2 \text{ } 3 \mid 1 \text{ } 8 \text{ } 7 \text{ } 6 \mid 5 \text{ } 9)$ and
$o_2 = (1 \text{ } 8 \text{ } 2 \mid 4 \text{ } 5 \text{ } 6 \text{ } 7 \mid 9 \text{ } 3)$.

The PMX crossover exploits important similarities in the value and ordering simultaneously when used with an appropriate reproductive plan [160].

- OX — proposed by Davis [71] — builds offspring by choosing a subsequence of a tour from one parent and preserving the relative order of cities from the other parent. For example, two parents (with two cut points marked by '|')

$p_1 = (1 \text{ } 2 \text{ } 3 \mid 4 \text{ } 5 \text{ } 6 \text{ } 7 \mid 8 \text{ } 9)$ and
$p_2 = (4 \text{ } 5 \text{ } 2 \mid 1 \text{ } 8 \text{ } 7 \text{ } 6 \mid 9 \text{ } 3)$

would produce the offspring in the following way. First, the segments between cut points are copied into offspring:

$o_1 = (\text{x x x} \mid 4 \text{ } 5 \text{ } 6 \text{ } 7 \mid \text{x x})$ and
$o_2 = (\text{x x x} \mid 1 \text{ } 8 \text{ } 7 \text{ } 6 \mid \text{x x})$.

Next, starting from the second cut point of one parent, the cities from the other parent are copied in the same order, omitting symbols already present. Reaching the end of the string, we continue from the first place of the string. The sequence of the cities in the second parent (from the second cut point) is

$$9 - 3 - 4 - 5 - 2 - 1 - 8 - 7 - 6;$$

after removal of cities 4, 5, 6, and 7, which are already in the first offspring, we get

$$9 - 3 - 2 - 1 - 8.$$

This sequence is placed in the first offspring (starting from the second cut point):

$o_1 = (2 \text{ } 1 \text{ } 8 \mid 4 \text{ } 5 \text{ } 6 \text{ } 7 \mid 9 \text{ } 3)$.

Similarly we get the other offspring:

$o_2 = (3 \text{ } 4 \text{ } 5 \mid 1 \text{ } 8 \text{ } 7 \text{ } 6 \mid 9 \text{ } 2)$.

The OX crossover exploits a property of the path representation, that the order of cities (not their positions) are important, i.e., the two tours

$$9 - 3 - 4 - 5 - 2 - 1 - 8 - 7 - 6 \text{ and}$$
$$4 - 5 - 2 - 1 - 8 - 7 - 6 - 9 - 3$$

are in fact identical.

- CX — proposed by Oliver [299] — builds offspring in such a way that each city (and its position) comes from one of the parents. We explain the mechanism of the cycle crossover using the following example. Two parents

$$p_1 = (1\ 2\ 3\ 4\ 5\ 6\ 7\ 8\ 9) \text{ and}$$
$$p_2 = (4\ 1\ 2\ 8\ 7\ 6\ 9\ 3\ 5)$$

would produce the first offspring by taking the first city from the first parent:

$$o_1 = (1\ x\ x\ x\ x\ x\ x\ x\ x).$$

Since every city in the offspring should be taken from one of its parents (from the same position), we do not have any choice now: the next city to be considered must be city 4, as the city from the parent p_2 just "below" the selected city 1. In p_1 this city is at position '4', thus

$$o_1 = (1\ x\ x\ 4\ x\ x\ x\ x\ x).$$

This, in turn, implies city 8, as the city from the parent p_2 just "below" the selected city 4. Thus

$$o_1 = (1\ x\ x\ 4\ x\ x\ x\ 8\ x);$$

Following this rule, the next cities to be included in the first offspring are 3 and 2. Note, however, that the selection of city 2 requires selection of city 1, which is already on the list — thus we have completed a cycle

$$o_1 = (1\ 2\ 3\ 4\ x\ x\ x\ 8\ x).$$

The remaining cities are filled from the other parent:

$$o_1 = (1\ 2\ 3\ 4\ 7\ 6\ 9\ 8\ 5).$$

Similarly,

$$o_2 = (4\ 1\ 2\ 8\ 5\ 6\ 7\ 3\ 9).$$

The CX preserves the absolute position of the elements in the parent sequence.

It is possible to define other operators for the path representation. For example, Syswerda [383] defined two modified versions of the order crossover operator. (However, this work was in connection with the scheduling problem and we will discuss this problem in the next chapter). The first modification (called order-based crossover) selects (randomly) several positions in a vector, and the order of cities in the selected positions in one parent is imposed on the corresponding cities in the other parent. For example, consider two parents

$$p_1 = (1\ 2\ 3\ 4\ 5\ 6\ 7\ 8\ 9) \text{ and}$$
$$p_2 = (4\ 1\ 2\ 8\ 7\ 6\ 9\ 3\ 5).$$

Assume that the selected positions are 3rd, 4th, 6th, and 9th; the ordering of the cities in these positions from parent p_2 will be imposed on parent p_1. The cities at these positions (in the given order) in p_2 are 2, 8, 6, and 5. In parent p_1 these cities are present at positions 2, 5, 6, and 8. In the offspring the elements on these positions are reordered to match the order of the same elements from p_2 (the order is $2 - 8 - 6 - 5$). The first offspring is a copy of p_1 on all positions except positions 2, 5, 6, and 8:

$$o_1 = (1\ \text{x}\ 3\ 4\ \text{x}\ \text{x}\ 7\ \text{x}\ 9).$$

All other elements are filled in the order given in parent p_2, i.e., 2, 8, 6, 5, so finally,

$$o_1 = (1\ 2\ 3\ 4\ 8\ 6\ 7\ 5\ 9).$$

Similarly, we can construct the second offspring:

$$o_2 = (3\ 1\ 2\ 8\ 7\ 4\ 6\ 9\ 5).$$

The second modification (called position-based crossover) is more similar to the original order crossover. The only difference is that in position-based crossover, instead of selecting one subsequence of cities to be copied, several cities are (randomly) selected for that purpose.

It is interesting to note that these two operators (order-based crossover and position based crossover) are, in some sense, equivalent to each other. An order-based crossover with some number of positions selected as crossover points, and a position-based crossover with compliment positions as its crossover points will always produce the same result. This means that if the average number of crossover points is $m/2$ (m is the total number of cities), these two operators should give the same performance. However, if the average number of crossover points is, say, $m/10$, then the two operators display different characteristics. For more information on these operators and some theoretical and empirical results comparing some of them, the reader is referred to [154], [131], [299], [370], and [382].

In surveying different reordering operators which have emerged during the last few years, we should mention the inversion operator as well. Simple inversion [188] selects two points along the length of the chromosome, which is cut at these points, and the substring between these points is reversed. For example, a chromosome:

(1 2 | 3 4 5 6 | 7 8 9)

with two cut points marked by '|', is changed into

(1 2 | 6 5 4 3 | 7 8 9).

Such simple inversion guarantees that the resulting offspring is a legal tour; some theoretical investigations [188] indicate that the operator should be useful in finding good string orderings. It is reported [402] that in a 50-city TSP, a system with inversion outperformed a system with a "cross and correct" operator. However, an increase in the number of cut points decreases the performance of the system. Also, inversion (like a mutation) is a unary operator, which can only supplement recombination operators — the operator is unable to recombine information by itself. Several versions of the inversion operator have been investigated [154]. Holland [188] provides a modification of a schema theorem to include its effect.

At this point we should also mention recent attempts to solve the TSP using evolution strategies [178], [352]. One of the attempts [178] experimented with four different mutations operators (mutation is still the basic operator in evolution strategies — see Chapter 8):

- *inversion* — as described above;

- *insertion* — selects a city and inserts it in a random place;

- *displacement* — selects a subtour and inserts it in a random place;

- *reciprocal exchange* — swaps two cities.

Also, a version of the heuristic crossover operator was used. In this modification, several parents contribute in producing offspring. After selecting the first city of the offspring tour (randomly), all left and right neighbors of that city (from all parents) are examined. The city which yields the shortest distance is selected. The process continues until the tour is completed.

Another application of evolution strategy [352] generates a (float) vector on n numbers (n corresponds to the number of cities). The evolution strategy is applied as for any continuous problem. The trick is in coding. Components of the vector are sorted and their order determines the tour. For example, the vector

$$v = (2.34, -1.09, 1.91, 0.87, -0.12, 0.99, 2.13, 1.23, 0.55)$$

corresponds to the tour

$$2 - 5 - 9 - 4 - 6 - 8 - 3 - 7 - 1,$$

since the smallest number, -1.09 is the second component of the vector v, the second smallest number, -0.12 is the fifth component of the vector v, etc.

Most of the operators discussed so far take into account cities (i.e., their positions and order) as opposed to edges — links between cities. What might be important is not the particular position of a city in a tour, but rather the linkage of this city with other cities. As observed by Homaifar and Guan [194]:

"Considering the problem carefully, we can argue that the basic building blocks for TSP are edges as opposed to the position representation of cities. A city or short path in a given position without adjacent or surrounding information has little meaning for constructing a good tour. However, it is hard to argue that injecting city a in position 2 is better than injecting it in position 5. Although this is the extreme case, the underlying assumption is that a good operator should extract edge information from parents as much as possible. This assumption can be partially explained from the experimental results in Oliver's paper [299] that OX does 11% better than PMX, and 15% better than the cycle crossover."

Grefenstette [170] developed a class of heuristic operators that emphasizes edges. They work along the following lines:

1. randomly select a city to be the current city c of the offspring,

2. select four edges (two from each parent) incident to the current city c,

3. define a probability distribution over selected edges based on their cost. The probability for the edge associated with a previously visited city is 0,

4. select an edge. If at least one edge has non-zero probability, selection is based on the above distribution; otherwise, selection is random (from unvisited cities),

5. the city on 'the other end' of the selected edge becomes the current city c,

6. if the tour is complete, stop; otherwise, go to step 2.

However, as reported in [170], such operators transfer around 60% of the edges from parents — which means that 40% of edges are selected randomly.

Whitley, Starweather, and Fuquay [402] have developed a new crossover operator: the *edge recombination* crossover (ER), which transfers more than 95% of the edges from the parents to the single offspring. The ER operator explores the information on edges in a tour, e.g., for the tour

$$(3\ 1\ 2\ 8\ 7\ 4\ 6\ 9\ 5),$$

the edges are (3 1), (1 2), (2 8), (8 7), (7 4), (4 6), (6 9), (9 5), and (5 3). After all, edges — not cities — carry values (distances) in the TSP. The objective function to be minimized is the total of edges which constitute a legal tour. The position of a city in a tour is not important: tours are circular. Also, the direction of an edge is not important: both edges (3 1) and (1 3) signal only that cities 1 and 3 are directly connected.

The general idea behind the ER crossover is that an offspring should be built exclusively from the edges present in both parents. This is done with help of the edge list created from both parent tours. The edge list provides, for each city

c, all other cities connected to city c in at least one of the parents. Obviously, for each city c there are at least two and at most four cities on the list. For example, for the two parents

$p_1 = (1\ 2\ 3\ 4\ 5\ 6\ 7\ 8\ 9)$ and
$p_2 = (4\ 1\ 2\ 8\ 7\ 6\ 9\ 3\ 5)$,

the edge list is

city 1: edges to other cities: 9 2 4
city 2: edges to other cities: 1 3 8
city 3: edges to other cities: 2 4 9 5
city 4: edges to other cities: 3 5 1
city 5: edges to other cities: 4 6 3
city 6: edges to other cities: 5 7 9
city 7: edges to other cities: 6 8
city 8: edges to other cities: 7 9 2
city 9: edges to other cities: 8 1 6 3.

The construction of the offspring starts with a selection of an initial city from one of the parents. In [402] the authors selected one of the initial cities (e.g., 1 or 4 in the example above). The city with the smallest number of edges in the edge list is selected. If these numbers are equal, a random choice is made. Such selection increases the chance that we complete a tour with all edges selected from the parents. With a random selection, the chance of having edge failure, i.e., being left with a city without a continuing edge, would be much higher. Assume we have selected city 1. This city is directly connected with three other cities: 9, 2, and 4. The next city is selected from these three. In our example, cities 4 and 2 have three edges, and city 9 has four. A random choice is made between cities 4 and 2; assume city 4 was selected. Again, the candidates for the next city in the constructed tour are 3 and 5, since they are directly connected to the last city, 4. Again, city 5 is selected, since it has only three edges as opposed to the four edges of city 3. So far, the offspring has the following shape:

$(1\ 4\ 5\ x\ x\ x\ x\ x\ x)$.

Continuing this procedure we finish with the offspring

$(1\ 4\ 5\ 6\ 7\ 8\ 2\ 3\ 9)$,

which is composed entirely of edges taken from the two parents. From a series of experiments [402], edge failure occurred at a very low rate (1% – 1.5%).

The ER operator was tested [402] on three TSPs with 30, 50, and 75 cities — in all cases it returned a solution better than the previously "best known" sequence.

Two years later, the edge recombination crossover was further enhanced [370]. The idea was that the 'common subsequences' were not preserved in the ER crossover. For example, if the edge list contains the row with three edges city 4: edges to other cities: 3 5 1,

one of these edges repeats itself. Referring to the previous example, it is the edge (4 5). This edge is present in both parents. However, it is listed as other edges, e.g., (4 3) and (4 1), which are present in one parent only. The proposed solution [370] modifies the edge list by storing 'flagged' cities:

city 4: edges to other cities: 3 -5 1;

the character '-' means simply that the flagged city 5 should be listed twice. In the previous example of two parents

$p_1 = (1\ 2\ 3\ 4\ 5\ 6\ 7\ 8\ 9)$ and
$p_2 = (4\ 1\ 2\ 8\ 7\ 6\ 9\ 3\ 5),$

the (enhanced) edge list is:

city 1: edges to other cities: 9 -2 4
city 2: edges to other cities: -1 3 8
city 3: edges to other cities: 2 4 9 5
city 4: edges to other cities: 3 -5 1
city 5: edges to other cities: -4 6 3
city 6: edges to other cities: 5 -7 9
city 7: edges to other cities: -6 -8
city 8: edges to other cities: -7 9 2
city 9: edges to other cities: 8 1 6 3.

The algorithm for constructing a new offspring gives priority to flagged entries: this is important only in the cases where three edges are listed — in two other cases either there are no flagged cities, or both cities are flagged. This enhancement (plus a modification for making better choices when random edge selection is necessary) further improved the performance of the system [370].

The edge recombination operators indicate clearly that the path representation might be too poor to represent important properties of a tour — this is why it was complemented by the edge list. Are there other representations more suitable for the traveling salesman problem? Well, we cannot give a positive 'yes' for the answer. However, it is worthwhile to experiment with other, possibly non-vector, representations.

During the last two years, there were at least three independent attempts to construct an evolution program using matrix representation for chromosomes. These were by Fox and McMahon [131], Seniw [353], and Homaifar and Guan [194], We discuss them briefly, in turn.

Fox and McMahon [131] represented a tour as a precedence binary matrix M. Matrix element m_{ij} in row i and column j contains a 1 if and only if the city i occurs before city j in the tour. For example, a tour

$(3\ 1\ 2\ 8\ 7\ 4\ 6\ 9\ 5)$

	1	2	3	4	5	6	7	8	9
1	0	1	0	1	1	1	1	1	1
2	0	0	0	1	1	1	1	1	1
3	1	1	0	1	1	1	1	1	1
4	0	0	0	0	1	1	0	0	1
5	0	0	0	0	0	0	0	0	0
6	0	0	0	0	1	0	0	0	1
7	0	0	0	1	1	1	0	0	1
8	0	0	0	1	1	1	1	0	1
9	0	0	0	0	1	0	0	0	0

Fig. 10.1. Matrix representation of a tour

is represented in matrix form in Figure 10.1.

In this representation, the $n \times n$ matrix M representing a tour (total order of cities) has the following properties:

1. the number of 1s is exactly $\frac{n(n-1)}{2}$,

2. $m_{ii} = 0$ for all $1 \leq i \leq n$, and

3. if $m_{ij} = 1$ and $m_{jk} = 1$ then $m_{ik} = 1$.

If the number of 1s in the matrix is less than $\frac{n(n-1)}{2}$, and the two other requirements are satisfied, then the cities are partially ordered. This means that we can complete such a matrix (in at least one way) to get a legal tour (total order of cities). As stated in [131]:

> "The Boolean matrix representation of a sequence encapsulates all of the information about the sequence, including both the micro-topology of individual city-to-city connections and the macro-topology of predecessors and successors. The Boolean matrix representation can be used to understand existing operators and to develop new operators that can be applied to sequences to produce desired effects while preserving the necessary properties of the sequence."

The two new operators developed in [131] were *intersection* and *union*. Both are binary operators (crossover-like operators). As for other evolution programs (e.g., GENETIC-2 for the transportation problem; Chapter 9), such operators should combine the features of both parents and preserve constraints (requirements) at the same time.

The intersection operator is based on the observation that the intersection of bits from both matrices results in a matrix where (1) the number of 1s is not greater than $\frac{n(n-1)}{2}$, and (2) the two other requirements are satisfied. Thus, we can complete such a matrix to get a legal tour (total order of cities).

For example, two parents

$$p_1 = (1\ 2\ 3\ 4\ 5\ 6\ 7\ 8\ 9)\ \text{and}\ p_2 = (4\ 1\ 2\ 8\ 7\ 6\ 9\ 3\ 5)$$

are represented by two matrices (Figure 10.2).

The intersection of these two matrices gives the matrix displayed in Figure 10.3.

	1	2	3	4	5	6	7	8	9
1	0	1	1	1	1	1	1	1	1
2	0	0	1	1	1	1	1	1	1
3	0	0	0	1	1	1	1	1	1
4	0	0	0	0	1	1	1	1	1
5	0	0	0	0	0	1	1	1	1
6	0	0	0	0	0	0	1	1	1
7	0	0	0	0	0	0	0	1	1
8	0	0	0	0	0	0	0	0	1
9	0	0	0	0	0	0	0	0	0

	1	2	3	4	5	6	7	8	9
1	0	1	1	0	1	1	1	1	1
2	0	0	1	0	1	1	1	1	1
3	0	0	0	0	1	0	0	0	0
4	1	1	1	0	1	1	1	1	1
5	0	0	0	0	0	0	0	0	0
6	0	0	1	0	1	0	0	0	1
7	0	0	1	0	1	1	0	0	1
8	0	0	1	0	1	1	1	0	1
9	0	0	1	0	1	0	0	0	0

Fig. 10.2. Two parents

	1	2	3	4	5	6	7	8	9
1	0	1	1	0	1	1	1	1	1
2	0	0	1	0	1	1	1	1	1
3	0	0	0	0	1	0	0	0	0
4	0	0	0	0	1	1	1	1	1
5	0	0	0	0	0	0	0	0	0
6	0	0	0	0	0	0	0	0	1
7	0	0	0	0	0	0	0	0	1
8	0	0	0	0	0	0	0	0	1
9	0	0	0	0	0	0	0	0	0

Fig. 10.3. First phase of the intersection operator

The partial order imposed by the result of intersection requires that city 1 precedes cities 2, 3, 5, 6, 7, 8, and 9; city 2 precedes cities 3, 5, 6, 7, 8, and 9; city 3 precedes city 5; city 4 precedes cities 5, 6, 7, 8, and 9; and cities 6, 7, and 8 precede city 9.

During the next stage of the intersection operator, one of the parents is selected; some 1s (that are unique to this parent) are 'added', and the matrix is completed into a sequence through an analysis of the sums of the rows and

	1	2	3	4	5	6	7	8	9
1	0	1	1	1	1	1	1	1	1
2	0	0	1	1	1	1	1	1	1
3	0	0	0	0	1	0	0	0	1
4	0	0	1	0	1	1	1	1	1
5	0	0	0	0	0	0	0	0	1
6	0	0	1	0	1	0	0	0	1
7	0	0	1	0	1	1	0	0	1
8	0	0	1	0	1	1	1	0	1
9	0	0	0	0	0	0	0	0	0

Fig. 10.4. Final result of the intersection

columns. For example, the matrix from Figure 10.4 is a possible result after completing the second stage; it represents a tour (1 2 4 8 7 6 3 5 9).

The union operator is based on the observation that the subset of bits from one matrix can be safely combined with a subset of bits from the other matrix, provided that these two subsets have empty intersection. The operator partitions the set of cities into two disjoint groups (in [131] a special method was used to make this partition). For the first group of cities, it copies the bits from the first matrix; and for the second group of cities, it copies the bits from the second matrix. Finally, it completes the matrix into a sequence through an analysis of the sums of the rows and columns (as for intersection operator).

For example, the two parents p_1 and p_2 and the partition of cities into {1, 2, 3, 4} and {5, 6, 7, 8, 9} produce the matrix shown in Figure 10.5, which is completed as for the intersection operator.

	1	2	3	4	5	6	7	8	9
1	0	1	1	1	x	x	x	x	x
2	0	0	1	1	x	x	x	x	x
3	0	0	0	1	x	x	x	x	x
4	0	0	0	0	x	x	x	x	x
5	x	x	x	x	0	0	0	0	0
6	x	x	x	x	1	0	0	0	1
7	x	x	x	x	1	1	0	0	1
8	x	x	x	x	1	1	1	0	1
9	x	x	x	x	1	0	0	0	0

Fig. 10.5. First phase of the union operator

The experimental results on different topologies of the cities (random, clusters, concentric circles) reveal an interesting characteristic of the union and intersection operators, which makes progress even when the elitism (preserving the best) option was not used. This was not the case for either ER or PMX operators. A solid comparison of several binary and unary (swap, slice, and invert) operators in terms of performance, complexity, and execution time is provided in [131].

The second approach in using matrix representation was described by one of my Master students, David Seniw [353]. Matrix element m_{ij} in the row i and column j contains a 1 if and only if the tour goes from city i directly to city j. This means that there is only one nonzero entry for each row and each column in the matrix (for each city i there is exactly one city visited prior to i, and exactly one city visited next to i). For example, a chromosome in Figure 10.6(a) represents a tour that visits the cities (1, 2, 4, 3, 8, 6, 5, 7, 9) in this order. Note also that this representation avoids the problem of specifying the starting city, i.e., Figure 10.6(a) also represents the tours (2, 4, 3, 8, 6, 5, 7, 9, 1), (4, 3, 8, 6, 5, 7, 9, 1, 2), etc.

	1	2	3	4	5	6	7	8	9
1	0	1	0	0	0	0	0	0	0
2	0	0	0	1	0	0	0	0	0
3	0	0	0	0	0	0	0	1	0
4	0	0	1	0	0	0	0	0	0
5	0	0	0	0	0	0	1	0	0
6	0	0	0	0	1	0	0	0	0
7	0	0	0	0	0	0	0	0	1
8	0	0	0	0	0	1	0	0	0
9	1	0	0	0	0	0	0	0	0

(a)

	1	2	3	4	5	6	7	8	9
1	0	1	0	0	0	0	0	0	0
2	0	0	0	1	0	0	0	0	0
3	0	0	0	0	0	0	0	1	0
4	0	0	0	0	1	0	0	0	0
5	0	0	0	0	0	0	1	0	0
6	0	0	0	0	0	0	0	0	1
7	1	0	0	0	0	0	0	0	0
8	0	0	0	0	0	1	0	0	0
9	0	0	1	0	0	0	0	0	0

(b)

Fig. 10.6. Binary matrix chromosomes

It is interesting to note that each complete tour is represented as a binary matrix with only one bit in each row and one bit in each column set to one. However, not every matrix with these properties would represent a single tour. Binary matrix chromosomes may represent multiple subtours. Each subtour will eventually loop back onto itself, without connecting to any other subtour in the chromosome. For example, a chromosome from Figure 10.6(b) represents two subtours (1, 2, 4, 5, 7) and (3, 8, 6, 9).

The subtours were allowed in the hope that natural clustering would take place. After the evolution program terminated, the best chromosome is reduced to a single tour by successively combining pairs of subtours using a deterministic algorithm. Subtours of one city (a tour leaving a city to travel right back to itself), having a distance cost of zero, were not allowed. A lower limit of $q = 3$

cities in a subtour was set in an attempt to prevent the GA from reducing a TSP problem to a large number of subtours each with very few cities (q is a parameter of the method).

Figure 10.7(a) depicts the subtours resulting from a sample run of the algorithm on a number of cities intentionally placed in clusters. As expected, the algorithm developed isolated subtours. Figure 10.7(b) depicts the tour after the subtours have been combined.

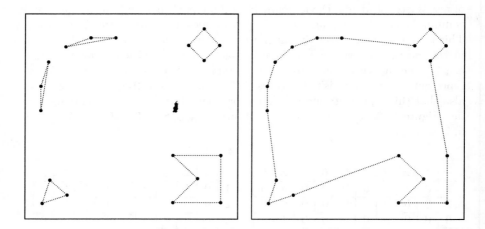

Fig. 10.7. Separate subtours and the final tour

Two genetics operators were defined: mutation and crossover. The mutation operator takes a chromosome, randomly selects several rows and columns in that chromosome, removes the set bits in the intersections of those rows and columns, and randomly replaces them in possibly a different configuration.

For example, let us consider the tour from Figure 10.6(a), representing the tour:

$$(1, 2, 4, 3, 8, 6, 5, 7, 9).$$

Assume that rows 4, 6, 7, 9 and columns 1, 3, 5, 8, and 9 are randomly selected to participate in a mutation. The marginal sums for these rows and columns are calculated. The bits at the intersections of these rows and columns are removed and replaced randomly, though they must agree with the marginal sums. In other words, the submatrix corresponding to rows 4, 6, 7, and 9, and columns 1, 3, 5, 8, and 9 from the original matrix (Figure 10.8(a)), is replaced by another submatrix (Figure 10.8(b)).

The resulting chromosome represents a chromosome with two subtours:

$$(1, 2, 4, 5, 7) \text{ and } (3, 8, 6, 9)$$

0	1	0	0	0
0	0	1	0	0
0	0	0	0	1
1	0	0	0	0

(a)

0	0	1	0	0
0	0	0	0	1
1	0	0	0	0
0	1	0	0	0

(b)

Fig. 10.8. Part of chromosome before (a) and after (b) mutation

and is represented in Figure 10.6(b).

The crossover operator begins with a child chromosome that has all bits reset to zero. The operator first examines the two parent chromosomes, and when it discovers the same bit (identical row and column) set (i.e., 1) in both parents, it sets a corresponding bit in the child (phase 1). The operator then alternately copies one set bit from each parent, until no bits exist in either parent which may be copied without violating the basic restrictions of chromosome construction (phase 2). Finally, if any rows in the child chromosome still do not contain a set bit, the chromosome will be filled in randomly (final phase). As the crossover traditionally produces two child chromosomes, the operator is executed a second time with the parent chromosomes transposed.

The following example of crossover starts with the first parent chromosome in Figure 10.9(a) representing two subtours:

(1, 5, 3, 7, 8) and (2, 4, 9, 6).

and the second parent chromosome (Figure 10.9(b)) representing a single tour:

(1, 5, 6, 2, 7, 8, 3, 4, 9).

	1	2	3	4	5	6	7	8	9
1	0	0	0	0	1	0	0	0	0
2	0	0	0	1	0	0	0	0	0
3	0	0	0	0	0	0	1	0	0
4	0	0	0	0	0	0	0	0	1
5	0	0	1	0	0	0	0	0	0
6	0	1	0	0	0	0	0	0	0
7	0	0	0	0	0	0	0	1	0
8	1	0	0	0	0	0	0	0	0
9	0	0	0	0	0	1	0	0	0

(a)

	1	2	3	4	5	6	7	8	9
1	0	0	0	0	1	0	0	0	0
2	0	0	0	0	0	0	1	0	0
3	0	0	0	1	0	0	0	0	0
4	0	0	0	0	0	0	0	0	1
5	0	0	0	0	0	1	0	0	0
6	0	1	0	0	0	0	0	0	0
7	0	0	0	0	0	0	0	1	0
8	0	0	1	0	0	0	0	0	0
9	1	0	0	0	0	0	0	0	0

(b)

Fig. 10.9. First (a) and second (b) parents

The first two phases of building the first offspring are displayed in Figure 10.10.

The first offspring for crossover, after the final phase, is displayed in Figure 10.11. It represents a subtour:

	1	2	3	4	5	6	7	8	9
1	0	0	0	0	1	0	0	0	0
2	0	0	0	0	0	0	0	0	0
3	0	0	0	0	0	0	0	0	0
4	0	0	0	0	0	0	0	0	1
5	0	0	0	0	0	0	0	0	0
6	0	1	0	0	0	0	0	0	0
7	0	0	0	0	0	0	0	1	0
8	0	0	0	0	0	0	0	0	0
9	0	0	0	0	0	0	0	0	0

(a)

	1	2	3	4	5	6	7	8	9
1	0	0	0	0	1	0	0	0	0
2	0	0	1	0	0	0	0	0	0
3	0	0	0	0	0	0	0	0	0
4	0	0	0	0	0	0	0	0	1
5	0	0	0	0	0	1	0	0	0
6	0	1	0	0	0	0	0	0	0
7	0	0	0	0	0	0	0	1	0
8	1	0	0	0	0	0	0	0	0
9	0	0	0	0	0	0	0	0	0

(b)

Fig. 10.10. Offspring for crossover, after (a) phase 1 and (b) phase 2

	1	2	3	4	5	6	7	8	9
1	0	0	0	0	1	0	0	0	0
2	0	0	1	0	0	0	0	0	0
3	0	0	0	1	0	0	0	0	0
4	0	0	0	0	0	0	0	0	1
5	0	0	0	0	0	1	0	0	0
6	0	1	0	0	0	0	0	0	0
7	0	0	0	0	0	0	0	1	0
8	1	0	0	0	0	0	0	0	0
9	0	0	0	0	0	0	1	0	0

Fig. 10.11. Offspring for crossover, after the final phase

(1, 5, 6, 2, 3, 4, 9, 7, 8).

The second offspring represents

(1, 5, 3, 4, 9) and (2, 7, 8, 6).

Note that there are common segments of the parent chromosomes in both offspring.

This evolution program gave a reasonable performance on several test cases from 30 cities to 512 cities. However, it is not clear what is the influence of the parameter q (minimum number of cities in a subtour) on the quality of the final solution. Also, the algorithms for combining several subtours into a single tour are far from obvious. On the other hand, the method has some similarities with Litke's recursive clustering algorithm [246], which recursively replaces clusters of size B by single representative cities until less than B cities remain. Then, the smaller problem is solved optimally. All clusters are expanded one by one and the algorithm sequences the expanded set between the two neighbors in the current tour. Also, the approach might be useful for solving the multiple traveling salesman problem, where several salesman should complete their separate (non-overlapping) tours.

The third approach based on matrix representation was recently proposed by Homaifar and Guan [194]. As in the previous approach, the m_{ij} element of the binary matrix M is set to 1 if and only if there is an edge from city i to city j. However, they used different crossover operators and heuristic inversion — we discuss them in turn.

Two matrix crossover (MX) operators were defined [194]. These operators exchange all entries of the two parent matrices either after a single crossover point (1-point crossover) or between two crossover points (2-point crossover). An additional "repair algorithm" is run to (1) remove duplications, i.e., to ensure that each row and each column has precisely one 1, and (2) cut and connect cycles (if any) to produce a legal tour.

A 2-point crossover is illustrated by the following example. Two parent matrices are given in Figure 10.12; they represent two legal tours:

(1 2 4 3 8 6 5 7 9) and (1 4 3 6 5 7 2 8 9)

Two crossovers points were selected; these are points between columns 2 and 3 (first point), and between columns 6 and 7 (second point). The crossover points cut the matrices vertically: for each matrix, the first two columns constitute the first part of the division, columns 3, 4, 5, and 6 the middle part, and the last three columns the third part. After the first step of the 2-point MX operator, entries of both matrices are exchanged between the crossover points (i.e., entries in columns 3, 4, 5, and 6). The intermediate result is given in Figure 10.13.

Both offspring, (a) and (b) are illegal; however, the total number of 1s in each intermediate matrix is correct (i.e., 9). The first step of the "repair algorithm" moves some 1s in matrices in such a way that each row and each column has precisely one 1. For example, in the offspring from Figure 10.13(a) the duplicate 1s occur in rows 1 and 3. The algorithm may move the entry $m_{14} = 1$ into m_{84}, and the entry $m_{38} = 1$ into m_{28}. Similarly, in the other offspring (Figure 10.13(b)) the duplicate 1s occur in rows 2 and 8. The algorithm may move the entry $m_{24} = 1$ into m_{34}, and the entry $m_{86} = 1$ into m_{16}. After the completion of the first step of the repair algorithm, the first offspring represents a (legal) tour,

(1 2 8 4 3 6 5 7 9),

and the second offspring represents a tour which consists of two subtours,

(1 6 5 7 2 8 9) and (3 4).

The second step of the repair algorithm should be applied to the second offspring only. During this stage, the algorithm cuts and connects subtours to

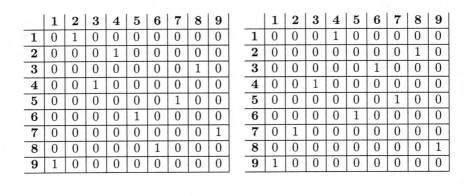

	1	2	3	4	5	6	7	8	9
1	0	1	0	0	0	0	0	0	0
2	0	0	0	1	0	0	0	0	0
3	0	0	0	0	0	0	0	1	0
4	0	0	1	0	0	0	0	0	0
5	0	0	0	0	0	0	1	0	0
6	0	0	0	0	1	0	0	0	0
7	0	0	0	0	0	0	0	0	1
8	0	0	0	0	0	1	0	0	0
9	1	0	0	0	0	0	0	0	0

	1	2	3	4	5	6	7	8	9
1	0	0	0	1	0	0	0	0	0
2	0	0	0	0	0	0	0	1	0
3	0	0	0	0	0	1	0	0	0
4	0	0	1	0	0	0	0	0	0
5	0	0	0	0	0	0	1	0	0
6	0	0	0	0	1	0	0	0	0
7	0	1	0	0	0	0	0	0	0
8	0	0	0	0	0	0	0	0	1
9	1	0	0	0	0	0	0	0	0

(a)   (b)

Fig. 10.12. Binary matrix chromosomes with crossover points marked

	1	2	3	4	5	6	7	8	9
1	0	1	0	1	0	0	0	0	0
2	0	0	0	0	0	0	0	0	0
3	0	0	0	0	0	1	0	1	0
4	0	0	1	0	0	0	0	0	0
5	0	0	0	0	0	0	1	0	0
6	0	0	0	0	1	0	0	0	0
7	0	0	0	0	0	0	0	0	1
8	0	0	0	0	0	0	0	0	0
9	1	0	0	0	0	0	0	0	0

	1	2	3	4	5	6	7	8	9
1	0	0	0	0	0	0	0	0	0
2	0	0	0	1	0	0	0	1	0
3	0	0	0	0	0	0	0	0	0
4	0	0	1	0	0	0	0	0	0
5	0	0	0	0	0	0	1	0	0
6	0	0	0	0	1	0	0	0	0
7	0	1	0	0	0	0	0	0	0
8	0	0	0	0	0	1	0	0	1
9	1	0	0	0	0	0	0	0	0

(a)   (b)

Fig. 10.13. Two intermediate offspring after the first step of MX operator

produce a legal tour. The cut and connect phase takes into account the existing edges in the original parents. For example, the edge (2 4) is selected to connect these two subtours, since this edge is present in one of the parents. Thus the complete tour (a legal second offspring) is

(1 6 5 7 2 4 3 8 9).

The second operator used by Homaifar and Guan [194] to complement MX crossover was heuristic inversion. The operator reverses the order of cities between two cut points (just as simple inversion did — we discussed it earlier in this chapter). If the distance between the two cut points is large (high order inversion), the operator explores connections between 'good' paths, otherwise (low order inversion), the operator performs local search. However, there are two differences between classical and proposed inversion operators. The first is that the resulting offspring is accepted only if the new tour is better than the original. The second difference is that the inversion procedure selects a single city in a tour and checks for improvement for inversions of the (lowest possible) order 2. The first inversion which results in an improvement is accepted and the inversion procedure terminates. Otherwise, inversions of the order 3 are considered, and so on.

The reported results [194] indicate that evolution program with 2-point MX and inversion operators performed successfully on 30–100-city TSP problems. In the most recent experiment, the result of this algorithm for a 318-city problem was only 0.6% away from the optimal solution.

In this chapter we did not provide the exact results of various experiments for different data structures and 'genetic' operators. Rather, we made a general overview of numerous attempts in building a successful evolution program for the TSP. One of the reasons is that most of the quoted performance results heavily depend on many details (population size, number of generations, size of the problem, etc.). Moreover, many results were related to relatively small sizes of the TSP (up to 100 cities); as observed in [207]:

> "It does appear that instances as small as 100 cities must now be considered to be well within the state of the global optimization art, and instances must be considerably larger than this for us to be sure that heuristic approaches are really called for."

However, most of the papers cited in this chapter compare the proposed approach with other approaches. For these comparisons, two families of test cases were used:

- random collection of cities. Here, an empirical formula for the expected length of L^* of a minimal TSP tour is useful:

$$L^* = k\sqrt{n \cdot R},$$

  where n is the number of cities, R is the area of the square box within which the cities were randomly placed, and k is an empirical constant of approximately 0.765 [1] (see [372]).

---

[1] According to David Johnson [209], such experimental comparison should be done very carefully. First, currently he leans toward a slightly lower estimate of the asymptotic constant, more like 0.7128 (± 0.0005). Second, the ratio $L^*/\sqrt{n}$ (for $R = 1$) converges very slowly — even for $n = 100000$ it is 0.7134, and for $n = 1000$ it is something like 0.7306 (both estimates are from David Johnson [209]). Third, the variations between instances are quite large for the smaller sizes. For $n = 1000$ the standard deviation is 0.0064 or so, and for $n = 100$ it is 0.0224. The conclusion is that one should compare tour lengths to the Held-Karp bound for the given instance, rather than an estimate of the expected optimal. The Held-Karp lower bound [179, 180] requires an iterative process involving evaluations of several spanning trees followed by Lagrangian relaxations.

- publicly available collection of cities (partly with optimal solutions), ftp softlib.rice.edu, directory /pub/tsplib. There are two files, tsplib.sh and tsplib.tar; these are 6Mb and 2Mb respectively.

To get a complete picture of the application of genetic algorithm techniques to the TSP, we should report the work of other researchers, e.g., [288] and [389], who used GAs for local optimization of the TSP. Local optimization algorithms (2-opt, 3-opt, Lin-Kernighan) are quite efficient. As stated in [207]:

> "Given that the problem is NP-hard, and hence polynomial-time algorithms for finding optimal tours are unlikely to exist, much attention has been addressed to the question of efficient approximation algorithms, fast algorithms that attempt only to find near-optimal tours. To date, the best such algorithms in practice have been based on (or derived from) a general technique known as local optimization, in which a given solution is iteratively improved by making local changes."

Local optimization algorithms, for a given (current) tour, specify a set of neighboring tours and replace the current tour by a (possibly) better neighbor. This step is applied until a local optimum is reached. For example, the 2-opt algorithm defines neighboring tours as tours where one can be obtained from the other by modifying two edges only.

Local search algorithms served as the basis for the development of the genetic local search algorithm [389], which

- uses a local search algorithm to replace each tour in the current population (of size μ) by a (local optimum) tour,

- extends the population by additional λ tours — offspring of the recombination operator applied to some tours in the current population,

- uses (again) a local search algorithm to replace each of the λ offspring in the extended population by a (local optimum) tour,

- reduces the extended population to its original size, μ, according to some selection rules (survival of the fittest),

- repeats the last three steps until some stopping condition is met (evolution process).

Note that there are some similarities between the genetic local search algorithm and the $(\mu + \lambda)$ evolution strategy (Chapter 8). As in the $(\mu + \lambda)$-ES, μ individuals produce λ offspring and the new (extended) population of $(\mu + \lambda)$ individuals is reduced by a selection process again to μ individuals.

The above genetic local search algorithm [389] is similar to a genetic algorithm for the TSP proposed earlier by Mühlenbein, Gorges-Schleuter, and Krämer [288], which encourages "intelligent evolution" of individuals. The algorithm

- uses a local search algorithm (opt-2) to replace each tour in the current population by a (local optimum) tour,

- selects partners for mating (above-average individuals get more offspring),

- reproduces (crossover and mutation),

- searches for the minimum by each individual (reduction, problem solver, expansion),

- repeats the last three steps until some stopping condition is met (evolution process).

The crossover used in this algorithm is a version of the order crossover (OX). Here, two parents (with two cut points marked by '|'),

$$p_1 = (1\ 2\ 3\ |\ 4\ 5\ 6\ 7\ |\ 8\ 9)\text{ and}$$
$$p_2 = (4\ 5\ 2\ |\ 1\ 8\ 7\ 6\ |\ 9\ 3),$$

would produce the offspring in the following way. First, the segments between cut points are copied into offspring:

$$o_1 = (\text{x x x}\ |\ 4\ 5\ 6\ 7\ |\ \text{x x})\text{ and}$$
$$o_2 = (\text{x x x}\ |\ 1\ 8\ 7\ 6\ |\ \text{x x}).$$

Next, (instead of starting from the second cut point of one parent as was the case for OX), the cities from the other parent are copied in the same order from the beginning of the string, omitting symbols already present:

$$o_1 = (2\ 1\ 8\ |\ 4\ 5\ 6\ 7\ |\ 9\ 3)\text{ and}$$
$$o_2 = (2\ 3\ 4\ |\ 1\ 8\ 7\ 6\ |\ 5\ 9).$$

The experimental results were at least encouraging. The algorithm found a tour for the 532 city problem and the length was found to be 27702, which is within 0.06% of the optimal solution (27686, found by Padberg and Rinaldi [302]).

Two additional approaches were reported recently. The first approach, by Craighurst and Martin [68] concentrated on exploring a connection between incest prevention (see also Chapter 4) and the performance of genetic algorithm for the TSP. The authors used a GA with the following features for their experiments: population size of 128, generation based selection based on ranking, where 128 offspring compete with 128 parents, selection for MPX crossover is rank based, local hill climbing (2-opt) is performed on offspring at creation, mutation rate is 0.005, and the termination condition is 500 consecutive generations without improvement. The approach to incest was family-oriented: the authors were concerned only with ancestors of two individuals selected for crossover, and they introduced several incest laws. The k-th incest law prohibits mating of an individual with $k - 1$ ancestors (i.e., for $k = 0$, there are no restrictions, for $k = 1$ an individual can not mate with itself, for $k = 2$, it can not mate with itself, with its parents, with children, nor with its siblings, and so on). Several experiments were made (on six test problems (from the softlib.rice.edu library mentioned earlier in the chapter) with the number of cities from the range 48–101). The results indicated a strong and interesting interdependence between incest prohibition laws and the mutation rate: for low mutation rates incest prevention improves the results, however, if mutation rate grows, the significance of incest prevention mechanism decreases until (for high mutation rates ≈ 0.1) it impairs the results of the system. Also, different laws of incest prevention did not influence the diversity of the population (where a similarity between two individuals is measured as a ratio of a difference between the total number of edges in a solution and the number of common edges shared between solutions, and the total number of edges in a solution) in a significant way. The final conclusion was a negative answer for the following question: "is more prohibition better?". For more discussion of the results, see [68].

Valenzuela and Jones [390] proposed an interesting approach for applying evolutionary algorithms to hard combinatorial problems; their method was based on the idea of the divide and conquer technique of Karp–Steele algorithms for the TSP [219, 379]. Their Evolutionary Divide and Conquer (EDAC) algorithm can be applied to any problem in which some knowledge of good solutions of subproblems is useful in constructing a global solution, however, they applied this technique to the geometric TSP. Several bisection methods can be considered; these methods cut a rectangle with n cities into two smaller rectangles (e.g., one of the methods partition the problem by exactly bisecting the area of the rectangle parallel to the shorter side; other method intersects the $int(n/2)$ closest city to the shorter edge of the rectangle, thus providing a "shared" city between two sub-rectangles). Final subproblems are quite small (typically between 5 and 8 cities), which are relatively easy to solve (2-opt was chosen as the method for this case because of its speed and simplicity). The patching algorithm replaces some edges in two separate tours to get one larger tour. Now, the major role of a genetic algorithm is to determine the direction of bisection (horizontal or vertical) used at each stage. It is interesting to note that the data structure used for chromosomal representation of individuals was a $p \times p^2$ binary array M, which was correlated with the geometric regions of the TSP square. If, at some stage of the divide and conquer algorithm, a rectangle is to be bisected, a bit is selected from the matrix M which most closely corresponds to the center of the rectangle; the value of this bit determines a direction of the current cut (horizontal versus vertical). The genetic operators used in [390] were straightforward: crossover swaps binary elements between two arrays, which are cut (in two points) along x or y axis[3] Mutation flips all bits of array with constant probability (0.1 was used, i.e., 10% of bits in the array are mutated). For experimental results and a discussion on the overall contribution of genetic algorithm in this approach, see [390].

---

[2]The authors considered a square region, hence the matrix has equal dimensions.

[3]The authors used the additional restriction that the distance between these two cutting points was between one third and two thirds along the axis to ensure a reasonable proportion of genetic material from each parent.

It seems that a good evolution program for the TSP should incorporate local improvement operators (mutation group), based on algorithms for local optimization, together with carefully designed binary operator(s) (crossover group), which would incorporate heuristic information about the problem. We conclude this chapter by a simple observation: the quest for an evolution program for the TSP, which would include 'the best' representation and 'genetic' operators to be performed on them, is still going on!

11. Evolution Programs for Various Discrete Problems

As stated in the Introduction, it seems that most researchers modified their implementations of genetic algorithms either by using non-standard chromosome representation and/or by designing problem-specific genetic operators (e.g., [141], [385], [65], [76], etc.) to accommodate the problem to be solved, thus building efficient evolution programs. Such modifications were discussed in detail in the previous two chapters (Chapters 9 and 10) for the transportation problem and the traveling salesman problem, respectively. In this chapter, we have made a somewhat arbitrary selection of a few other evolution programs developed by the author and other researchers, which are based on non-standard chromosome representation and/or problem-specific knowledge operators. We discuss some systems for scheduling problems (section 11.1), the timetable problem (section 11.2), partitioning problems (section 11.3), and the path planning problem in mobile robot environment (section 11.4). The chapter concludes with an additional section 11.5, which provides some brief remarks on a few other, interesting problems.

The described systems and the results of their applications provide an additional argument to support the evolution programming approach, which promotes creation of data structures together with operators for a particular class of problems.

11.1 Scheduling

A job shop is a process–organized manufacturing facility; its main characteristic is a great diversity of jobs to be performed [182]. A job shop produces goods (parts); these parts have one or more alternative process plans. Each process plan consists of a sequence of operations; these operations require resources and have certain (predefined) durations on machines. A job shop processes orders, where each order is for some number of the same part. The task of planning, scheduling, and controlling the work is very complex, and only limited analytical procedures are available to assist in these tasks [182], [132].

The job shop scheduling problem is to select a sequence of operations together with an assignment of start/end times and resources for each operation. The main considerations to be taken into account are the cost of having idle machine and labor capacity, the cost of carrying in-process inventory, and the need to meet certain order completion due dates. As stated in [182]:

> "Unfortunately, these considerations tend to conflict with each other. One can have a low cost of idle machine and labor capacity by providing only a minimum amount of machinery and manpower. However, this would result in considerable work waiting to be done and, therefore, large in-process inventories and difficulty in meeting completion due dates. On the other hand, one can essentially guarantee meeting completion due dates by providing so much machine and labor capacity that orders usually would not have to wait to be processed. However, this would result in excessive costs for idle machine and labor capacity. Therefore, it is necessary to strive for an economic compromise between these considerations."

There are various versions of the job shop scheduling problem, each characterized by some additional constraints (e.g., maintenance, machine down and setup times, etc.).

Let us consider a simple example of a job shop problem to illustrate the above description.

Example 11.1. Assume there are three orders, o_1, o_2, and o_3. For each order, the parts and the number of units to be produced are:

o_1 : 30× part a;
o_2 : 45× part b;
o_3 : 50× part a.

Each part has one or more alternative process plans:

a : plan # 1_a (opr_2, opr_7, opr_9);
a : plan # 2_a (opr_1, opr_3, opr_7, opr_8);
a : plan # 3_a (opr_5, opr_6);
b : plan # 1_b (opr_2, opr_6, opr_7);
b : plan # 2_b (opr_1, opr_9);

where terms opr_i denote the required operations to be performed. Each operation requires some times on one or more machines; these are:

opr_1 : $(m_1\ 10)$ $(m_3\ 20)$;
opr_2 : $(m_2\ 20)$;
opr_3 : $(m_2\ 20)$ $(m_3\ 30)$;
opr_4 : $(m_1\ 10)$ $(m_2\ 30)(m_3\ 20)$;
opr_5 : $(m_1\ 10)$ $(m_3\ 30)$;
opr_6 : $(m_1\ 40)$;
opr_7 : $(m_3\ 20)$;
opr_8 : $(m_1\ 50)$ $(m_2\ 30)$ $(m_3\ 10)$;
opr_9 : $(m_2\ 20)$ $(m_3\ 40)$.

Finally, each machine has its setup time necessary for changes in operation:

m_1 : 3;
m_2 : 5;
m_3 : 7.

□

The job shop problem enjoyed some interest in GA community. One of the first attempts to approach this problem was reported by Davis [72]. The main idea of his approach was to encode the representation of a schedule in such a way, that (1) the genetic operators would operate in a meaningful way, and (2) a decoder would always produce a legal solution to the problem. This strategy, to encode solutions for operations and to decode them for evaluation, is quite general and might be applied to a variety of constrained problems — the same idea was used by Jones [211] to approach the partitioning problem (see section 11.5).

In general, we would like to represent information on schedules, e.g., "machine m_2 performs operation o_1 on part a from time t_1 to time t_2". However, most operators (mutations, crossovers) applied to such a message would result in illegal schedules — this is why Davis [72] used an encoding/decoding strategy.

Let us see how the encoding strategy was applied to the job shop problem. The system developed by Davis [72] maintained a list of preferences for each machine; these preferences were linked to times. An initial member of a list is a time at which the list went into effect, the remaining part of the list is made up of some permutation of the orders, plus two additional elements: 'wait' and 'idle'. The decoding procedure simulated the job's operations in such way that whenever a machine was ready to make a choice, the first allowable operation from its preference list was taken. So if the preference list for the machine m_1 was

m_1 : (40 o_3 o_1 o_2 'wait' 'idle'),

then the decoding procedure at time 40 would search for a part from the order o_3 for the machine m_1 to work on. If unsuccessful, the decoding procedure would search for a part from the orders o_1 and o_2 (i.e., first from o_1; in the case of failure, from o_2). This representation guarantees a legal schedule.

The operators were problem specific (they were derived from deterministic methods):

run-idle: this operator is applied only to preference lists of the machines that have been waiting for more than an hour. It inserts the 'idle' as the second member of the preference list and reset the first member (time) of the preference list to 60 (minutes);

scramble: the operator "scrambles" the members of a preference list;

crossover: the operator exchanges preference lists for selected machines.

The probabilities of these operators varied, from 5% and 40% for scramble and crossover, respectively, at the beginning of a run, down to 1% and 5%. The probability of run-idle was set to the percentage of the time the machine spent waiting, divided by the total time of the simulation.

However, the experiments were made on a small example of two orders, six machines, and three operations [72], thus it is difficult to evaluate the usefulness of this approach.

Another group of researchers approached the job shop problem from the TSP point of view [62], [383], [384], [402]. The motivation was that most operators developed for the TSP were 'blind', i.e., they did not use any information about the actual distances between cities (see Chapter 10). This means that these operators might be useful in other sequencing problems, where there is no distance between two points (cities, orders, jobs, etc.). However, it need not be the case. Although both problems, the TSP and the scheduling problem, are sequencing problems, they display different (problem-specific) characteristics. For the TSP the important information is adjacency information on cities, whereas in the scheduling problem the relative order of items is the main concern. The adjacency information is useless for the scheduling problem, whereas the relative order is not important for the TSP due to the cyclic nature of the tours: tours (1 2 3 4 5 6 7 8) and (4 5 6 7 8 1 2 3) are, in fact, identical. This is why we need different operators for different applications. As observed in [370]:

"Gil Syswerda [383] conducted a study in which 'edge recombination' (a genetic operator specifically designed for the TSP) performed poorly relative to other operators on a job sequence scheduling task. While the population size used by Syswerda was small (30 strings) and good results were obtained on this problem using mutation alone (no recombination), Syswerda's discussion of the relative importance of position, order, and adjacency for different sequencing tasks raises an issue that has not been adequately addressed. Researchers, including ourselves [402], [403], seem to tacitly assume that all sequencing tasks are similar and that one genetic operator should suffice for all types of sequencing problems."

A similar observation was made one year earlier by Fox and McMahon [131]:[1]

---

[1]1991, Reprinted with permission from Rawlins, G., *Foundations of Genetic Algorithms.*

"An important concern is the applicability of each genetic operator to a variety of sequencing problems. For example, in the TSP, the value of a sequence is equivalent to the value of that sequence in reverse order. This trait is not true of all sequencing problems. In scheduling problems, this is a gross error."

In [370] six sequencing operators (order crossover, partially mapped crossover, cycle crossover, enhanced edge recombination, order-based crossover, and position-based crossover — all these operators were discussed in the previous chapter) were compared on two different sequencing tasks: a 30-city (blind) TSP and a 195-element sequencing task for a scheduling application. As expected, the results of schedule optimization (as far as the 'goodness' of the six operators is concern) were almost the opposite of the results from the TSP. In the case of schedule optimization, the enhanced edge recombination operator was the best, followed closely by order crossover, order-based crossover, and position-based crossover, with PMX and cycle crossovers being the worst. On the other hand, in the case of the TSP, the best were position-based and order-based crossovers, followed by the cycle crossover and PMX, with order crossover and enhanced edge recombination being the worst. These differences can be explained by examining how these operators preserve adjacency (for the TSP) and order (for the scheduling problem) information.

Similar observations can be made for other sequencing (ordering) problems. In [78] Davis describes an order-based genetic algorithm for the following graph coloring problem:

Given a graph with weighted nodes and n colors, achieve the highest score by assigning colors to nodes, such that no pair of connected (by a direct link) nodes can have the same color; the score is the total of weights of the colored nodes.

A simple greedy algorithm would sort the set of nodes in order of decreasing weights and process nodes (i.e., assigning the first legal color to the node from the list of colors) in this order. Clearly, this is a sequencing problem — at least one permutation of nodes would return the maximum profit, so we search for the optimal sequence of nodes. It is also clear that the simple greedy algorithm does not guarantee the optimum solution: some other techniques should be used. Again, on the surface, the problem is similar to the TSP, where we were after the best order of cities to be visited by a salesman. However, the 'nature' of the problem is very different: for example, in the graph coloring problem there are weights for nodes, whereas in the TSP the weights are distributed between nodes (as distances). In [78] Davis represented an ordering as a list of nodes (e.g., (2 4 7 1 4 8 3 5 9), as the path representation for the TSP) and used two operators: the order-based crossover (discussed in the previous chapter as an operator used for the TSP) and scramble sublist mutation. Even mutation, which should carry out a local modification of a chromosome, seems to be problem dependent [78]:

> "It is tempting to think of mutations as the swapping of the values of two fields on the chromosome. I have tried this on several different problems, however, and it doesn't work as well for me as an operator I call scramble sublist mutation."

The scramble sublist mutation selects a sublist of nodes from a parent and scrambles it in the offspring, i.e., the parent (with the beginning and the end of the selected sublist marked by |):

$$p = (2\ 4\ |\ 7\ 1\ 4\ 8\ |\ 3\ 5\ 9)$$

may produce the offspring

$$o = (2\ 4\ |\ 4\ 8\ 1\ 7\ |\ 3\ 5\ 9).$$

However, it remains to be seen how this operator performs for other ordering or scheduling problems. Again, let us cite from [78]:

> "Many other types of mutations can be employed on order-based problems. Scramble sublist mutation is the most general one I have used. To date nothing has been published on these types of operators, although this is a promising topic for future work."

Let us return to the scheduling problems. As mentioned earlier, Syswerda [383] developed an evolution program for scheduling problems. However, a simple chromosome representation was chosen:

> "In choosing a chromosome representation for the [...] scheduler, we have two basic elements to choose from. The first is the list of tasks to be scheduled. This list is very much like the list of cities to be visited in the TSP problem. [...] An alternative to using a sequence of tasks is directly to use a schedule as a chromosome. This may seem like an overly cumbersome representation, necessitating complicated operators to work on them, but is has a decided advantage when dealing with complicated real-world problems like scheduling. [...] In our case, the appeal of a clean and simple chromosome representation won over the more complicated one. The chromosome syntax we use for the scheduling problem is what was described above for the TSP, but instead of cities we will use orderings of tasks."

The chromosome was interpreted by a schedule builder — a piece of software which 'understands' the details of the scheduling task. This representation was supported with specialized operators. Three mutations were considered: position-based mutation (two tasks are selected at random, and the second task is placed before the first), order-based mutation (two tasks selected at random are swapped), and scramble mutation (same as Davis' scramble sublist mutation described in the previous paragraph). All three performed much better than random search, with order-based mutation being the clear winner. As mentioned earlier, the best crossover operators for the scheduling problem were order-based and position-based crossovers.

It seems, however, that the choice of a simple representation was not the best. Judging from other (unrelated) experiments, e.g., transportation problem (Chapter 9), we feel that the chromosome representation used should be much closer to the scheduling problem. It is true that in a such cases a significant effort must be placed in designing problem-specific 'genetic' operators; however, this effort would pay off in increased speed and improved performance of the system. Moreover, some operators might not be quite so simple [383]:

> "A simple greedy algorithm running over the schedule could find a place for the high-priority task by removing a low-priority task or two and replacing them with the high-priority task."

We believe that in general, and for the scheduling problems in particular, this is the direction to follow: to incorporate the problem-specific knowledge not only in operators (as was done for a simple chromosome representation), but in the chromosome structures as well. The first attempts to apply such an approach have already emerged. In their study, Husbands, Mill, and Warrington [198] represented a chromosome as a sequence

$$(opr_1 \ m_1 \ s_1) \ (opr_2 \ m_2 \ s_2) \ (opr_3 \ m_3 \ s_3) \ ... \ ,$$

where opr_i, m_i, and s_i denote the i-th operation, machine, and setup, respectively.

In [21] the authors compared three representations, from the simplest (representation–1):

$$(o_1) \ (o_2) \ (o_3) \ ... \ ,$$

through intermediate (representation–2):

$$(o_1 \ \text{plan} \ \# \ 1_a) \ (o_2 \ \text{plan} \ \# \ 2_b) \ (o_3 \ \text{plan} \ \# \ 2_a) \ ... \ ,$$

to the most complex (representation–3):

$$(o_1 \ \langle opr_2 : m_2, \ opr_7 : m_3, \ opr_9 : m_2 \rangle) \ (o_2 \ \langle opr_1 : m_3, \ opr_9 : m_2 \rangle)$$
$$(o_3 \ \langle opr_1 : m_1, \ opr_3 : m_2, \ opr_7 : m_3, \ opr_8 : m_1 \rangle) \ ...$$

The results for representation–3 were significantly better than for the two other representations. In the concluding discussion, the authors observed [21]:

> "The operators themselves must be adjusted to suit the domain requirements. The chromosome representation should contain all the information that pertain to the optimization problem."

In summary, it is possible to classify all GA-based approaches as various scheduling problems on the basis of chromosome representations. These fall into two categories [21]:

- **indirect representations**, where a transformation from a chromosome representation to a legal production schedule has to be performed by a special decoder (schedule builder); only then an individual solution can be evaluated. Further, these representations can be divided [53] into domain independent and problem-specific representations; we have seen both these cases in earlier paragraphs.

- **direct representations**, where the production schedule itself is used as a chromosome (e.g., [198]). Usually this representation requires some problem-specific operators [53].

For a complete survey of evolutionary algorithms for scheduling problems, see, for example, [54].

11.2 The timetable problem

One of the most interesting problems in Operations Research is the timetable problem. The timetable problem has important practical applications: it has been intensively studied and it is known to be NP-hard [106].

The timetable problem incorporates many nontrivial constraints of various kinds — this is probably why it was only recently that the first (as far as the author is aware) attempt was made [63] to apply genetic algorithm techniques to approach this problem. There are many versions of the timetable problem; one of them can be described by

- a list of teachers $\{T_1, \ldots, T_m\}$,

- a list of time intervals (hours) $\{H_1, \ldots, H_n\}$,

- a list of classes $\{C_1, \ldots, C_k\}$.

The problem is to find the optimal timetable (teachers – times – classes); the objective function aims to satisfy some goals (soft constraints). These include didactic goals (e.g., spreading some classes over the whole week), personal goals (e.g., keeping afternoons free for some part-time teachers), and organizational goals (e.g., each hour has an additional teacher available for temporary teaching post).

The constraints include:

- there is a predefined number of hours for every teacher and every class; a legal timetable must "agree" with these numbers,

- there is only one teacher in a class at a time,

- a teacher cannot teach two classes at a time,

- for each class scheduled at some time slot, there is a teacher.

It seems that the most natural chromosome representation of a potential solution of the timetable problem is a matrix representation: a matrix $(R)_{ij}$ ($1 \leq i \leq m$, and $1 \leq j \leq n$), where each row corresponds to a teacher, and each column to an hour; the elements of the matrix R are classes ($r_{ij} \in \{C_1, \ldots, C_k\}$).[2]

In [63] the constraints were managed mainly by genetic operators (the authors used also a repair algorithm to eliminate cases where more than one teacher is present in the same class at the same time). The following genetic operators were used:

mutation of order k: the operator selects two contiguous sequences of k elements from the same row in the matrix R, and swaps them,

day mutation: this operator is a special case of the previous one: it selects two groups of columns (hours) of the matrix R which correspond to different days, and swaps them,

crossover: given two matrices R_1 and R_2, the operator sorts the rows of the first matrix in order of decreasing values of a so-called local fitness function (a part of the fitness function due only to characteristics specific to each teacher) and the best b rows (b is a parameter determined by the system on the basis of the local fitness function and both parents) are taken as a building block; the remaining $m - b$ rows are taken from the matrix R_2.

The resulting evolution program was successfully tested on data for a large school in Milan, Italy [63].

Timetable problems have been also studied by Paechter, Luchian, and Petriuc [303], who compared two evolutionary methods (time-space permutation method and the place and seek method) for a large, real world timetable problem. Recently, Burke et al. [57] described a hybrid genetic algorithm for highly constrained timetabling problems. This approach combines a direct representation of a timetable with a few heuristic crossover operators and heuristic mutation operator. Thus the algorithm maintains the feasibility of solutions by specialized data structures and operators.

11.3 Partitioning objects and graphs

There is an interesting class of partitioning[3] problems, which require partitioning n objects into k categories. This class contain many well-known problems, like bin packing problem (assigning items to bins), graph coloring problem (assigning nodes of a graph to specific colors), etc. Many different systems have been developed for various types of partitioning problems; in this section we discuss some of them.

---

[2]Actually, in [63] the elements of the matrix R were classes with three possible subscripts to include the concepts of sections, temporary teaching posts, etc.

[3]Sometimes these problems are called grouping problems [109].

One of the categories of evolution programs was based on representing all objects (e.g., items in the bin packing problem, or nodes in the graph coloring problem) as a permutation list; special operators can be applied, and a decoder makes the decisions on the assignments. For example, for the graph coloring problem, Davis [78] represents a permutation of nodes in a chromosome and applies specialized operators (uniform order-based crossover, order-based mutation) to this structure. In the same time, a greedy decoder is used for interpretation of a structure, which works as follows: consider a particular color and paint (if possible) all nodes (in the order given in the chromosome) using this color. When no more coloring is possible, switch to the next color. Davis provides [78] experimental results for a 100-node graph coloring problem.

An interesting approach to the partitioning problem[4] is presented in [235]. Von Laszewski encodes partitions using group-number encoding, i.e., partitions are represented as n-strings of integer numbers,

$$(i_1, \ldots, i_n),$$

where the j-th integer $i_j \in \{1, \ldots, k\}$ indicates the group number assigned to object j. However, this representation is supported by "intelligent structural operators": structural crossover and structural mutation. We discuss them in turn.

structural crossover: the mechanism of the structural crossover is explained by the following example. Assume, there are two selected parents (12-strings)

$$p_1 = (112311232233) \text{ and } p_2 = (112123122333).$$

These strings decode into the following partitions:

$$p_1 : \{1, 2, 5, 6\}, \{3, 7, 9, 10\}, \{4, 8, 11, 12\}, \text{ and}$$
$$p_2 : \{1, 2, 4, 7\}, \{3, 5, 8, 9\}, \{6, 10, 11, 12\}.$$

First, a random partition is chosen: say, partition #2. This partition is copied from p_1 into p_2:

$$p_2' = (112123222233).$$

The copying process (as seen in the above example) usually destroys the requirement of equal partition sizes, hence we apply a repair algorithm. Note that in the original p_2 there were elements assigned to partition #2, which were not elements of the copied partition: these were elements 5 and 8. These elements are erased,

$$p_2'' = (1121 * 32 * 2233),$$

---

[4]The additional requirement in this version of the partitioning problem was that the sizes of the partitions are equal or nearly equal.

and replaced (randomly) by numbers of other partitions, which were over-written in the copying step. Thus the final offspring might be

$$p_2''' = (112133212233).$$

As mentioned earlier, in the group-number coding, two identical partitions may be represented by different strings due to different numbering of partitions. To take care of this, before the crossover is executed, the codings are adapted to minimize the difference between the two parents.

structural mutation: typically, a mutation would replace a single component of a string by some random number; however, this would destroy the requirement of the equal sizes of partitions. Structural mutation was defined as a swap of two numbers in the string. Thus a parent

$$p = (112133212233)$$

may produce the following offspring (the numbers on positions 4 and 6 are swapped):

$$p' = (112331212233).$$

The algorithm was implemented as a parallel genetic algorithm enhanced with additional strategies (such as a parent replacement strategy); on random graphs of 900 nodes with maximum node degree of 4, this evolution program significantly outperformed other heuristic algorithms [235]. A similar approach was tried by Mühlenbein [287], who experimented with the same group-number encoding, and who also used "intelligent" crossover, which transmits whole partitions rather than separate objects.

Several evolution programs were constructed by Jones and Beltramo [211] for this class. These programs used different representations and several operators to manipulate them. It is interesting to observe the influence of the incorporation of the problem-specific knowledge on the performance of the developed evolution programs. Two test problems were selected:

- to divide n numbers into k groups to minimize the differences among the group sums, and

- to partition the 48 states of the continental U.S. into 4 color groups to minimize the number of bordering state pairs that are in the same group.

The first group of evolution programs encoded partitions as n-strings of integer numbers

$$(i_1, \ldots, i_n),$$

where the j-th integer $i_j \in \{1, \ldots, k\}$ indicates the group number assigned to object j; this is a *group-number* encoding.

The group-number encoding creates a possibility of applying standard operators. A mutation would replace a single (randomly selected) gene i_j by a (random) number from $\{1, \ldots, k\}$. Crossovers (single-point or uniform) would always produce a legitimate offspring. However, as pointed out in [211], an offspring (after mutation or crossover) may contain less than k groups; moreover, an offspring of two parents, both representing the same partition, may represent a totally different partition, due to different numbering of groups. Special repair algorithms (rejection method, renumbering the parents) were used to eliminate these problems. Also, we can consider applying edge-based crossover (defined in the previous chapter). Here we assume that two objects are connected by an edge, if and only if they are in the same group. Edge-based crossover constructs an offspring by combining edges from the parents.

It is interesting to note that several experiments on the two test problems indicate the superiority of the edge-based operator [211]; however, the representation used does not support this operator. As reported, it took 2–5 times more computation time per iteration than the other crossover methods. This is due to inappropriate representation: for example, two parents

$$p_1 = (11222233) \text{ and}$$
$$p_2 = (12222333)$$

represent the following edges:

edges for p_1: (12), (34), (35), (36), (45), (46), (56), (78),
edges for p_2: (23), (24), (25), (34), (35), (45), (67), (68), (78).

An offspring should contain edges present in at least one parent, e.g.,

(11222333)

represents the following edges:

(12), (34), (35), (45), (67), (68), (78).

However, the process of selection of edges is not straightforward: the selection of (56) and (67) — where both these edges are represented above — implies the presence of the edge (57), which is not there.

It seems that some other representations might be more suitable for the problem. The second group of evolution programs encoded partitions as $n+k-1$-strings of distinct integer numbers

$$(i_1, \ldots, i_{n+k-1});$$

integers from the range $\{1, \ldots, n\}$ represent the objects, and integers from the range $\{n+1, \ldots, n+k-1\}$ represent separators; this is a *permutation with separators* encoding. For example, the 7-string

(1122233)

is represented as a 9-string

(128345967),

where 8 and 9 are separators.

Of course, all $k-1$ separators must be used; also, they cannot appear at the first or last position, and they cannot appear together, one next to the other (otherwise, a string would decode into less than k groups).

As usual, care should be taken in designing operators. These would be similar to some operators used for solving the traveling salesman problem, where the TSP is represented as a permutation of cities. A mutation would swap two objects (separators are excluded). Two crossovers were considered: order crossover (OX) and partially matched crossover (PMX) — these were discussed in Chapter 10. A crossover would repeat its operation until the offspring[5] decodes into a partition with k groups.

Generally, the results of the evolution programs based on permutation with separators encoding were better than for the programs based on group-number encoding. However, neither coding method makes significant use of the problem-specific knowledge. One can build a third family of evolution programs, incorporating knowledge in the objective function. This very thing was done in [211] in the following way.

The representation used was the simplest one: each n-string represents n objects:

$$(i_1, \ldots, i_n),$$

where the $i_j \in \{1, \ldots, n\}$ denotes the object number — hence $i_j \neq i_p$ for $j \neq p$. The interpretation of this representation uses a greedy heuristic: the first k objects in the string are used to initialize k groups, i.e., each of the first k objects is placed in a separate group. The remaining objects are added on first-come first-go basis, i.e., they are added in the order they appear in the string; they are placed in a group which yields the best objective value.

This greedy heuristic also simplifies operators: every permutation encodes a valid partition, so we can use the same operators as for the traveling salesman problem. Needless to say, the "greedy decoding" approach significantly outperforms evolution programs based on other decodings: group-number and permutation with separators [211].

Recently, Falkenauer [109] proposed so-called Grouping Genetic Algorithm (GGA) to deal with a variety of grouping (partitioning) problems; his efforts aimed at designing appropriate chromosomal representation to capture the structure of the problem. In this approach a chromosome consists of two parts: an *object part* and a *group part*. The object part uses the group-number encoding (discussed earlier in this section): it consist of a n-string of integer numbers

$$(i_1, \ldots, i_n),$$

---

[5]Jones and Beltramo [211] produced only one offspring per crossover.

where the j-th integer $i_j \in \{1, \ldots, k\}$ indicates the group number assigned to object j. The group part of the chromosome is of variable length and represents only groups. For example, the following chromosome:

$$(1\ 2\ 1\ 3\ 3\ 3\ 1\ 1\ 2 : 1\ 2\ 3)$$

is interpreted in the following way. The second part of the chromosome indicates that there are 3 groups (1, 2, and 3 — as specified). The first part of the chromosome allows for interpretation of allocations: group number 1 includes objects $\{1, 3, 7, 8\}$; group number 2 has objects $\{2, 9\}$; and group number 3 — objects $\{4, 5, 6\}$. Note that we can replace digit '3' by digit '5' in the above representation (in both parts, of course), and the meaning of the allocations remains unchanged.

The key concept behind such a representation is that the genetic operators work with the group part (i.e., second part) of the chromosomes, whereas the first part of the chromosomes is just used for identification of allocations.

For example, for the bin packing problem (i.e., pack n objects into a minimum number of bins of constant capacity), the proposed Bin Packing Crossover Operator BPCX works as follows. Assume that the two parent chromosomes are:

$$(1\ 2\ 1\ 3\ 3\ 3\ 1\ 1\ 2 : 1\ 2\ 3),\ \text{and}$$
$$(2\ 3\ 3\ 5\ 1\ 4\ 2\ 2\ 6 : 2\ 3\ 4\ 5\ 6).$$

Two crossing sites are selected (at random) for each group part of the chromosomes; say:

$$(1\ 2\ 1\ 3\ 3\ 3\ 1\ 1\ 2 : 1\ |\ 2\ 3\ |),\ \text{and}$$
$$(2\ 3\ 3\ 5\ 1\ 4\ 2\ 2\ 6 : 2\ |\ 3\ 4\ |\ 5\ 6).$$

Then the contents of the crossing section of the first parent are inserted at the first crossing of the second parent (for another offspring, the roles of first and second parent are switched):

$$(\ldots : 2_2\ 2_1\ 3_1\ 3_2\ 4_2\ 5_2\ 6_2)$$

(subscripts denote the parent). Now we can eliminate conflicts; note that the contents of bins listed above are as follows:

bin 2_2 — objects 1, 7, and 8
bin 2_1 — objects 2 and 9
bin 3_1 — objects 4, 5, and 6
bin 3_2 — objects 2 and 3
bin 4_2 — object 6
bin 5_2 — object 4
bin 6_2 — object 9.

Since there are common objects in "new" bins (from the first parent) and "old" bins — from the second parent, we remove these "old" bins (responsible for conflicts) from the second part of the chromosome, which is now:

$(... : 2_2\ 2_1\ 3_1)$

Note, that we removed bin 6_2 (because of object 9 already in 2_1), bin 5_2 (because of object 4 already in 3_1), bin 4_2 (because of object 6 already in 3_1), and bin 3_2 (because of object 2 already in 2_1). At this stage the offspring chromosome has the following shape:

$(2_2\ 2_1\ ?\ 3_1\ 3_1\ 3_1\ 2_2\ 2_2\ 2_1 : 2_2\ 2_1\ 3_1)$.

After renaming bin numbers 2_2, 2_1, and 3_1 as 1, 2, and 3, respectively, the chromosome is

$(1\ 2\ ?\ 3\ 3\ 3\ 1\ 1\ 2 : 1\ 2\ 3)$.

Note that due to the above method of conflict resolution, the third object lost its allocation (this is marked by a question mark '?' in the first part of the chromosome). Thus we may use some heuristic repair algorithm to complete a feasible individual; Falkenauer [109] used the first fit descending heuristic to insert missing objects. If there is no room for the object number 3 in any of the three bins in the above chromosome, it is necessary to create an additional bin:

$(1\ 2\ 4\ 3\ 3\ 3\ 1\ 1\ 2 : 1\ 2\ 3\ 4)$.

Note that such a crossover is responsible for transmitting as much as possible of the meaningful information from both parents.

In the above approach, the mutation operator was very simple and useful. It selects and eliminates a few bins (at random). Objects without allocation are re-inserted into bins by the first fit heuristic (in random order).

The experimental results were very good; Falkenauer concluded his paper by the following remark [109]:

"We also hope to have made a convincing case for the importance of adequate encoding (and, consequently, genetic operators) for a successful application of the GA paradigm."

It is hard not to agree.

11.4 Path planning in a mobile robot environment

Navigation is the science (or art) of directing the course of a mobile robot as it traverses the environment. Inherent in any navigation scheme is the desire to reach a destination without getting lost or crashing into any objects.

Often, a path is planned off-line for the robot to follow, which can lead the robot to its destination assuming that the environment is perfectly known and stationary and the robot can track perfectly. Early path planners were such off-line planners or were only suitable for such off-line planning (e.g., [248, 218, 205]). However, the limitations of off-line planning led researchers to study on-line planning, which relies on knowledge acquired from sensing the local environment [222] to handle unknown obstacles as the robot traverses the environment.

The evolution program that we describe here, i.e., the Evolutionary Navigator (EN), unifies off-line and on-line planning with a simple map of high fidelity and an efficient planning algorithm [243, 244]. The first part of the algorithm (off-line planner) searches for the optimal global path from the start to the destination, whereas the second part (on-line planner) is responsible for handling possible collisions or previously unknown objects by replacing a part of the original global path by the optimal subtour. It is important to point out that both parts of the EN use the same evolutionary algorithm, just with different values of various parameters.

During the last five years other researchers have been experimenting with evolutionary computation techniques for the path planning problem. Davidor [69] used dynamic structures of chromosomes and a modified crossover operator to optimize some real world processes (including robot paths applications). In [355] a genetic algorithm to the path planning problem is described, and in [356] a genetic algorithm for the development of real-time multi-heuristic search strategies is presented. Both approaches assume a predefined map consisting of knot points. Other researchers used classifier systems [414] or genetic programming paradigm [176] to approach the path planning problem. Our approach is unique in the sense that the Evolutionary Navigator (1) operates in the entire free space and does not make any a priori assumptions about feasible knot points of a path, and (2) it combines together off-line and on-line planning algorithms.

Before we explain the algorithm in detail, let us first explain the map structure. In order to support path search in the entire, continuous free space, vertex graphs are used to represent objects in the environment. Currently, we restrict the environment to be two-dimensional with polygonal objects only and motions of the robot to be translational only. Therefore, the robot can be shrunk to a point while the objects in the environment "grow" accordingly [248]. A mobile robot equipped with ultrasonic sensors (e.g., a Denning robot) is assumed for the EN. A known object is represented by the ordered list (clockwise fashion) of its vertices. On-line encountered unknown obstacles are modeled by pieces of "wall", where each piece of "wall" is a straight-line and represented by the list of its two end points. This representation is consistent with the representation of known objects, while it also accommodates the fact that only partial information about an unknown obstacle can be obtained from sensing at a particular location. Finally, the entire environment is defined as a rectangular area.

Now it is important to define paths that the EN generates. A path consists of one or more straight-line segments, with the starting location, the goal location, and (possibly) the intersection locations of two adjacent segments defining the *nodes*. A feasible path consists of feasible nodes; an infeasible path contains at least one infeasible node. Assume there is a path $p = \langle m_1, m_2, \ldots, m_n \rangle$ $(n \geq 2)$, where m_1 and m_n denote start and goal nodes, respectively. A node m_i $(i = 1, \ldots, n-1)$ is infeasible if it is either not connectable to the next node m_{i+1} due to obstacles, or it is located inside (or too close to) some obstacle. We assume that the start and goal nodes are located outside the obstacles, and not too close to them. Note, however, that the start node need not be feasible (it may be not connectable to the next node), whereas the goal node is always feasible. Note also that different paths may have different numbers of nodes.

Now we are ready to go through the EN procedure (Figure 11.1).

```
procedure Evolutionary Navigator
begin
    begin (off-line planner)
        get map
        obtain the task
        perform planning:
            current path := FEG(start, goal)
    end (off-line planner)
    if current path is feasible then
    begin (on-line planner)
    repeat
        move along the current path while
        sensing the environment
        if too close to any object then
        begin
            local_start := current location
            local_goal := next node on the current path
            if the object is new
                then update the object map
                else virtually grow the object
                    at the closest spot
            perform planning:
                local_path := NEG(local_start, local_goal)
            update current path
        end
    until (at goal) or (failure condition)
    end (on-line planner)
end
```

Fig. 11.1. The structure of the Evolutionary Navigator

The EN first reads the map and obtain the start and goal locations of the task. Then the oFf-line Evolutionary alGorithm (FEG) generates a near-optimal global path, a piece-wise straight-line path consisting of feasible knot points or nodes. Figure 11.2 shows such a global path generated by FEG. (The filled circle simulates the robot).

As the robot starts to follow the path to move towards the goal, it senses the environment for its proximity to nearby objects, and the oN-line Evolutionary alGorithm (NEG) is used to generate local paths to deal with unexpected

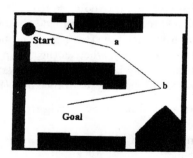

Fig. 11.2. An environment and a global path

collisions and objects. To simulate the effect of unknown objects in the environment, additional data files were created to represent such obstacles (like pieces of "wall" as explained earlier). We experimented with five different sets of unknown objects; Figure 11.3a–d presents the actions of the robot on one of these sets.

When the robot moved too close to the lower left corner of the nearby object 'A', the NEG virtually "grew" 'A' at the spot and generated a local path to steer away from 'A', which was also a piece-wise straight-line path. The robot then followed the current path successfully to reach the point 'a'. While the robot moved from 'a' to 'b', it detected an unknown or new object 'B'. Now the EN updated the map, and again, the NEG generated a local path with the knot point 'd' (Figure 11.3a). As the robot moved from 'd' towards 'b', it became too close to the object 'B': consequently, another local path was generated as represented by the knot point 'e' (Figure 11.3b). The robot then moved from 'd' to 'e' and finally reached the subgoal 'b'. The next step was to move from 'b' towards the goal; as shown in Figure 11.3c, the path segment was too close to the lower right corner of the object 'C'. Therefore, another local path was generated as represented by the knot point 'f' and then to the 'goal'. Figure 11.3d shows the original global path and the actual path traveled. Note that the navigation process terminates when the robot arrives at the goal or a failure condition is reported, i.e., when the EN fails to find a feasible path in certain time period (i.e., within specified number of generations of the NEG).

As we already mentioned, the EN combines off-line and on-line planning with the same data structure and the same planning algorithm. That is, the only difference between FEG and NEG is in the parameters they use: population size pop_size_g, number of generations T_g, maximum length of a chromosome n_g, etc. for FEG, and pop_size_l, T_l, n_l, etc. for NEG. Note that both FEG and NEG do *global planning*; even if NEG usually generates a local path, it operates on the updated *global map*. Moreover, if no object is initially known in the environment, or no initially known object is between the start location and the goal location, then FEG will generate a straight-line path with just two nodes:

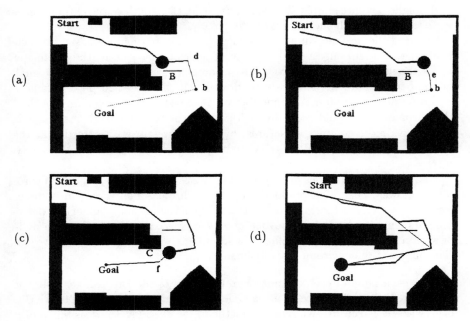

Fig. 11.3. An actual path traveled

the start and the goal locations. It will solely depend on the NEG to lead the robot towards the goal while avoiding unknown obstacles.

In the following, we discuss components of FEG and NEG in detail.

Chromosomes are ordered lists of path nodes as shown in Figure 11.4. Each of the path nodes, apart from the pointer to the next node, consists of x and y coordinates of an intermediate knot point along the path, and a Boolean variable b, which indicates whether the given node is feasible or not.

Fig. 11.4. Chromosome representing a path

The length of the chromosomes (the number of the path nodes represented in a chromosome) is variable. In off-line planning, the maximum length of a chromosome is set to be the number n_g of vertices representing known objects in the environment. It is unlikely that all feasible paths would require a large number of (e.g., n_g) intermediate nodes: even in complex environments a feasible path might be quite simple. Therefore, we make the length of the chromosomes variable to deal with such situations gracefully.

During the on-line planning, the local path for getting around an obstacle is likely to contain only a small number of nodes, consequently, the parameter n_l, as the maximum length of the chromosome in this phase, is relatively small at the beginning of the local search. However, if the evolution process fails to find a feasible path after some number of generations, the maximum length of the chromosome should grow: in such situation it is quite likely that feasible paths have more complex structures. In the EN system we assumed that the parameter n_l was a function of the current generation number t, more precisely, $n_l(t) = t$.

The initial populations of (pop_size_g for FEG, and pop_size_l for NEG) chromosomes were generated randomly. For each chromosome, a random number was generated from the range $2..max(2, n_g)$ (for the off-line planner) to determine its length. The coordinates x and y were created randomly for each node of such a chromosome (the values of coordinates were restricted to be within the confine of the environment, of course).

For each node of each chromosome, the value of the Boolean variable b is determined (feasibility check). If the node is feasible, its b value is set to TRUE, otherwise, it is set to FALSE. The methods for checking the feasibility of a node (i.e., location validity, clearance from nearby objects, and connectivity) are relatively simple and based on algorithms described by Pavlidis [310].

The fitness (the total path cost) of a chromosome $p = \langle m_1, m_2, \ldots, m_n \rangle$ is determined by two separate evaluation functions (for feasible and infeasible individuals):

- for a feasible path p:

$$\text{Path\_Cost}(p) = w_d \cdot dist(p) + w_s \cdot smooth(p) + w_c \cdot clear(p),$$

  where the weights w_d, w_s, and w_c normalize the total cost of a path, and

  - $dist(p) = \sum_{i=1}^{n-1} d(m_i, m_{i+1})$, where $d(m_i, m_{i+1})$ is the distance between knot points m_i and m_{i+1}; i.e., the function $dist(p)$ returns the total length of the path p.
  - $smooth(p) = \max_{i=2}^{n-1} s(m_i)$, where

$$s(m_i) = \frac{\theta_i}{\min\{d(m_{i-1}, m_i), d(m_i, m_{i+1})\}};$$

    i.e., the function $smooth(p)$ returns the largest curvature of p at a knot point.
  - $clear(p) = \max_{i=1}^{n-1} c_i$, where

$$c_i = \begin{cases} d_i - \tau & \text{if } d_i \geq \tau \\ a(\tau - d_i) & \text{otherwise,} \end{cases}$$

d_i is the minimum distance between the segment (m_i, m_{i+1}) of the path and all known objects, τ defines a safe distance, and a is a co-efficient; i.e., the function $clear(p)$ returns the largest number which measures clearance between all segments of p and the objects.

- for an infeasible path p:

$$\text{Path\_Cost}(p) = \alpha + \beta + \gamma,$$

where α is the number of intersections of the path p with all walls of the objects, β is the average number of intersections per infeasible segment, and γ provides the cost of the worst feasible path in the current popula-tion; because of this last variable, any feasible path in the population is better than any infeasible one (see also section 15.3, part C).

Several operators (crossover, two mutations, insertion, deletion, smooth, and swap) were included in the FEG and NEG. We discuss them in turn.

Crossover. This operator is similar to the classical one-point crossover widely used in genetic algorithms. It recombines "good" parts of the paths present in both parents to produce hopefully better path represented by the offspring. Two selected chromosomes are cut in some positions and glued together: the first part of the first chromosome with the second part of the second chromosome, and the first part of the second chromosome with the second part of the first chromosome. However, the crossing points in both chromosomes are not selected randomly: if infeasible nodes are present in the chromosome, the crossing points fall after one of them.

Mutation_1. This mutation is responsible for fine tuning values of coordinates of the nodes listed in the chromosome. If a node of a chromosome is selected for this mutation, its coordinates are modified. For example, the coordinate $x \in \langle a, b \rangle$ (as well as coordinate y) is changed in the following way:

$$x' = \begin{cases} x - \delta(t, x - a), & if \ r = 0 \\ x + \delta(t, b - x), & if \ r = 1 \end{cases}$$

where r is a random bit, and the function $\delta(t, z)$ returns a value in the range of $[0..z]$ such that the probability of $\delta(t, z)$ being close to zero increases as t increases (t is the current generation number of the evolution process). The operator is modeled on non-uniform mutation used in evolutionary systems for nonlinear optimization (Chapter 7). This mutation is responsible for "smoothing over" the shape of the path.

Mutation_2. This mutation is useful in cases when a larger change in a value is required (this situation occurs often during the on-line planning phase, when an obstacle is blocking the path). If a node of a chromosome is selected for this mutation, its coordinates are modified. For example, the coordinate $x \in \langle a, b \rangle$ (as well as coordinate y) is changed in the following way:

$$x' = \begin{cases} x - \Delta(t, x - a), & if \ \ r = 0 \\ x + \Delta(t, b - x), & if \ \ r = 1 \end{cases}$$

where r is a random bit, and the function $\Delta(t, z)$ returns a value in the range of $[0..z]$ such that the probability of $\Delta(t, z)$ being close to z increases as generation number t increases.

Insertion. This operator inserts a new node into the existing path; every place between two nodes has the same probability of such insertion.

Deletion. This operator deletes a node from the path; every node has the same probability for such deletion.

Smooth. This operator smooths a part of the path by cutting sharp turns. For selected knot point m_i (with a high curvature), the operator selects two new knot points k_1 and k_2 (from segments (m_{i-1}, m_i) and (m_i, m_{i+1}), respectively), inserts them into the path, removes m_i; so it creates a new path p':

$$p' = (m_1, \ldots, m_{i-1}, k_1, k_2, m_{i+1}, \ldots, m_n).$$

Swap. This operator splits the selected chromosome into two parts (the splitting point is determined at random) and swaps these parts.

Based on the preliminary experimental results, the EN has proved to be efficient and effective in comparison with navigators using traditional approaches (e.g., [130]). Results of the current version of the system on two different environments are presented in Figures 11.5 and 11.6.

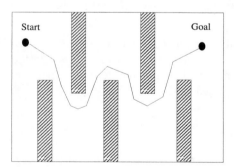 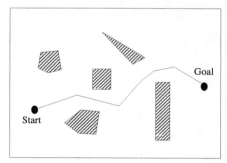

Fig. 11.5. Results of the EN on two environments

Of course, there is a need to explore the EN's potential by conducting more tests under different environments, most importantly, by implementation of the EN on a real robot. At the same time, several issues of such evolutionary navigators remain to be resolved; these include (1) design of smarter termination conditions for FEG and NEG to better realize the optimization goals (currently the algorithms terminate either when a feasible path is found or some fixed number of generations have elapsed), (2) introduction of adaptive frequencies

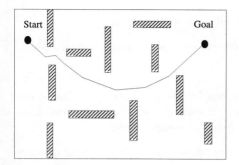 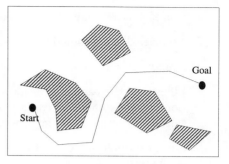

Fig. 11.6. Results of the EN on another two environments

of the genetic operators, as opposed to the constants in the current version of the system (this modification should enhance the performance of the system and is based on the simple observation that different operators may play different roles at different stages of the evolution process), (3) extension of the EN to operate in an environment with non-polyhedral objects, (4) incorporation of the knowledge of the current stage of the search into workings of operators (e.g., it might be more meaningful to cross two paths at infeasible knot points), and (5) exploration of some learning mechanism so that the EN can take advantage of past experiences.

Despite its efficiency and effectiveness in many cases, however, the EN has a major limitation: it assumes that a feasible and sufficiently good actual path can be obtained by minor perturbation from the current best path; the system is not designed to be able to replace the current global path, at some stage of the traversal, by another (possibly better) global path entirely. Thus it might be worthwhile to experiment with other solutions; for example, an adaptive navigator (AN) is currently under construction. Unlike the EN, which consists of off-line and on-line planners, an adaptive navigator would be an on-line planner completely; it would constantly adapt the path connecting the current location of the robot and the goal based on newly gathered sensing information.

11.5 Remarks

In this final section we discuss briefly a few relatively recent applications of evolutionary techniques, which, for various reasons, are interesting (from the perspective of constructing an evolution program). We discuss them in turn.

There are some applications (e.g., network design problems), in which a solution is a graph. The problem of representing graphs in genetic algorithms is quite interesting as such. Recently, Palmer and Kershenbaum [304] reported on experiments with various ways of representing trees. They identified desirable properties for a good representation; these include:

1. Completeness: ability to represent all possible trees,

2. No bias: all trees should be represented by the same number of encodings,

3. Soundness: representing only trees,

4. Efficiency: the ease of transformations between the encoded representation of the tree and the tree's representation in a more conventional form suitable for evaluating the fitness function and constraints,

5. Locality: small changes in the representation of a tree make small changes in the tree.

Of course, an ideal representation of a tree should have all of the above properties. However, this is not the case with most of the representations. In [304] the authors consider a few representations. These include

- characteristic vector representation, where a tree is represented as a binary vector of the length equal to the number of edges in the underlying graph,

- predecessors representation, where a node is designated as a root and the predecessor of each node is recorded: here, a tree is encoded as an integer vector of the length equal to the number of nodes,

- Prüfer numbers representation, where a tree is coded as a $n - 2$-"digit" number (n is the number of nodes in a tree), where each "digit" is integer determined by a special algorithm (for details, see [304].

The authors proposed also a new representation, which was based on a simple observation, that certain nodes should be interior nodes and others should be leaf nodes. In this representation the chromosome holds a bias value for each node and each possible edge (thus a tree is represented as a vector of $n + \frac{n(n-1)}{2}$ numbers); the biases modify the cost matrix C_{ij} of the graph:

$$C'_{ij} = C_{ij} + P_1(C_{max})b_{ij} + P_2(C_{max})(B_i + b_j),$$

where P_1 and P_2 are parameters of the method, and C_{max} is the maximum link cost. The tree that the chromosome represents is found by applying Prim's algorithm to find a minimum spanning tree over nodes using the biased cost matrix. This representation also can encode any tree given suitable values of the b_i. It has quite interesting properties: for a full discussion the reader is referred to [304].

Another paper by Abuali et al. [3] investigated a new encoding scheme for representing spanning trees (for the probabilistic minimum spanning tree problem). This encoding scheme is based on so-called determinant codes, which are vectors of $n-1$ integers; the i-th number k in a determinant code corresponds to an edge from vertex k to $i + 1$.[6] For experimental results and comparison of this encoding with other methods, see [3].

---

[6]The determinant encoding schemes may produce graphs which are not spanning trees; in the work reported in [3], a repair algorithm is used.

Esbensen [101] reports on a genetic algorithm for finding optimal Steiner trees. The Steiner tree problem (its decision version is NP-complete) is formulated as follows: given a graph and a specified subset of vertices, the task is to find a minimum cost subgraph spanning the specified vertices. The author used a deterministic decoder which interprets any set of selected Steiner vertices as a valid Steiner tree. Thus each chromosome is a binary string, where each bit corresponds to a vertex; such a binary string represents a subset of Steiner vertices.[7] This subset of vertices together with specified vertices constitutes the starting point for the decoder, which (1) constructs the subgraph induced by these vertices; (2) computes a minimum spanning tree for this subgraph; (3) constructs (from this minimum spanning tree) another subgraph by substituting each edge by the corresponding shortest path in the original graph; (4) computes a minimum spanning tree for the resulting subgraph; and (5) computes the Steiner tree by repeatedly deleting (from the latest minimum spanning tree) all vertices not included in the original list of vertices of degree 1. For experimental results (which include cases of graphs having up to 2500 vertices), see [101].

The set covering problem (SCP) is the problem of covering the rows of a $n \times k$ binary matrix $A = (a_{ij})$ by a subset of the columns at minimal cost; the problem has many practical applications (location of facilities, crew scheduling, etc). Each column $1 \leq j \leq k$ has associated cost c_j; thus the SCP can be expressed as

$$\text{minimize } \sum_{j=1}^{k} c_j x_j,$$

where x_j denotes a binary decision variable ($x_j = 1$ iff column j is selected in the cover), subject to

$$\sum_{j=1}^{k} a_{ij} x_j \geq 1,$$

for all $1 \leq i \leq n$.

Beasley [31] experimented with modified genetic algorithm on many SCPs, from 200 rows by 1,000 columns to 1,000 rows by 10,000 columns. The results were very good; the algorithm generated optimal solutions for smaller size problems and high-quality solutions for large size problems.

Note that the SCP has "an ideal" representation from the perspective of a GA: a binary string of x_j's ($1 \leq j \leq k$) represents a potential solution to the problem. No additional coding is necessary. The evaluation function (for feasible individuals) is also straightforward; it is just

$$f(\boldsymbol{x}) = \sum_{j=1}^{k} c_j x_j,$$

The major challenge in the SCP is the issue of feasibility; any operators applied to such binary strings can produce offspring which violate the problem

---

[7]Steiner vertices consist of additional vertices, which should be added to the specified set. Thus the length of the chromosome is determined by the difference between the total number of nodes and the number of specified nodes.

constraints (i.e., some rows may not be covered). Beasley [31] used a repair algorithm (see Chapter 15 for a general discussion on repair algorithms) to maintain feasibility of solutions. This repair algorithm was responsible for covering uncovered rows; the search for additional columns was based on the ratio between cost of a column and the number of uncovered rows which it covers. It it interesting to note that once the repair process is completed and an infeasible individual is converted into a feasible one, a local optimization step is performed, in which redundant columns are removed (i.e., columns which can be removed without violation of constraints).

Bui and Eppley [55] describe a hybrid genetic algorithm for the maximum clique problem and provide its experimental results on test cases up to 1500 vertices and over half million edges. The maximum clique problem is to find a complete subgraph (i.e., clique) of a given graph of maximum size (measured in number of vertices). The decision version of the problem is NP-complete [134]; not surprisingly, many various approximation methods have been proposed. Note that the task is to find a subset of vertices, hence it is possible to use straightforward binary representation: a binary vector

$$\langle b_1, \ldots, b_n \rangle$$

defines such a subset (for all n vertices arranged in an arbitrary order,[8] where all vertices were $b_i = 1$ implies selection of the i-th node). Instead of designing sophisticated operators which would maintain feasibility of solutions (i.e., which would guarantee that an offspring is a clique) or instead of designing a repair algorithm (which would correct an arbitrary solution into a clique), the authors constructed a clever objective function, which distinguishes between non-cliques and (almost)-cliques; this function was defined as

$$f(X) = \alpha|X| + \beta \frac{2e(X)}{|X|(|X|-1)},$$

where α and β are integers (called cardinality weight and completeness weight, respectively), and $e(X)$ returns the number of edges in the subgraph induced by X. It is interesting to note that the ratio β/α is variable: at the beginning of the algorithm it stays small (exploration phase, where cardinality of a solution is emphasized) and as the algorithm progresses, the ratio increases (completeness of a solution is emphasized). It is important to underline that the algorithm was extended by incorporation of a local optimization routine. This heuristic routine (1) determines whether removing a particular vertex (vertices are considered in a special order) would improve the fitness of the solution (if yes, the vertex is removed) and (2) tries also to increase the cardinality of the solution; it determines whether an addition of a vertex would increase the fitness value of the solution (if so, the vertex is added). For a full discussion of the details of the system and experimental results, see [55].

---

[8]In [55], however, the authors experimented with two different preprocessing steps; these steps order the vertices either in decreasing order of their degrees or take the order resulting from the depth first search.

An interesting application of an evolutionary technique for pallet loading was described by Juliff [215]. The pallet loading problem is a special scheduling problem which involves (1) packing of cartons onto pallets and (2) stacking of the pallets on a truck. The requirements are quite complex due to maintaining a balance of the load (at all times) during the delivery, and maximizing the efficiency of loading and unloading cartons. As for other scheduling problems (see section 11.1), it is possible to build a system with direct or indirect representations; in the latter one, a scheduler (or, in this case, load-builder), would complete the job. However, both these approaches have some disadvantages: highly problem-specific operators (direct representation) or limited search (indirect representation): usually one chromosome represents a single feature and other features are not explicitly represented (the knowledge of these additional features is just incorporated in the load-builder), so the genetic search is not fully guided.

Juliff developed a few systems for pallet loading [215] based on indirect representation and intelligent load-builders; one of these systems incorporates a multi-chromosome structure to handle various features of the problem (actually, three chromosomes were used); not surprisingly, a multi-chromosome system outperformed other, single-chromosome systems.

Recently, many interesting applications of evolutionary techniques to various problems in management science (including scheduling and timetable problems) have been reported. A recent survey by Nissen of such applications provides a complete reference [297].

Also, there is a growing interest in evolutionary techniques for industrial engineering; many papers address particular issues and provide descriptions of particular solutions in this area. For more information, see for example, a special issue on genetic algorithms and industrial engineering of Computers and Industrial Engineering Journal [135].

Let us conclude this chapter by a general observation that more and more various applications of evolutionary techniques to combinatorial optimization problems (including the TSP, see Chapter 10) incorporate some heuristic local search algorithms to improve their performance (for an excellent up-to-date survey of local search techniques, see [2]). Either this is done as a separate step of the algorithm, or such algorithms are adopted as clever (intelligent) mutations. The results reported by Yagiura and Ibaraki in their recent study [411] indicate that such genetic local search algorithms are powerful techniques which compare very well with other techniques (random multi-start local search or pure genetic algorithms).

These results emphasize also the importance of mutation operators, which should not be just background operators, but are essential components of any evolutionary system. This is in agreement (in some sense) with recent results of Jones [213], who experimented with macro-mutations and indicated their significance in binary codings (with respect to the crossover operator). Anyway, it seems that the higher the cardinality of the alphabet used for encoding of individuals, the greater the role of mutation operator(s). This was the case for algorithms where individuals were represented as floating-point vectors, integers, matrices, finite state machines, etc. It will not be surprising to see a new trend in the genetic programming approach (see Chapter 13), which would emphasize the role of various mutation operators; this may allow also a significant decrease in population sizes in genetic programming methods, hence further increasing their efficiency.

12. Machine Learning

> The real problem is not
> whether machines think,
> but whether men do.
>
> B.F. Skinner, *Contingencies of Reinforcement*

Machine learning is primarily devoted towards building computer programs able to construct new knowledge or to improve already possessed knowledge by using input information; much of this research employs heuristic approaches to learning rather than algorithmic ones. The most active research area in recent years [284] has continued to be symbolic empirical learning (SEL). This area is concerned with creating and/or modifying general symbolic descriptions, whose structure is unknown *a priori*. The most common topic in SEL is developing concept descriptions from concept examples [234], [284]. In particular, the problems in attribute-based spaces are of practical importance: in many such domains it is relatively easy to come up with a set of example events, on the other hand it is quite difficult to formulate hypotheses. The goal of a system implementing this kind of supervised learning is:

> *Given the initial set of example events and their membership in concepts, produce classification rules for the concepts present in the input set.*

Depending on the output language, we can divide all approaches to automatic knowledge acquisition into two categories: symbolic and non-symbolic. Non-symbolic systems do not represent knowledge explicitly. For example, in statistical models knowledge is represented as a set of examples together with some statistics on them; in a connectionist model, knowledge is distributed among network connections [335]. On the other hand, symbolic systems produce and maintain explicit knowledge in a high-level descriptive language. The best known examples of this category of system are AQ and ID families [281] and [314].

In this chapter we describe two genetic-based machine learning methodologies and discuss an evolution program (GIL, for Genetic Inductive Learning), proposed by Janikow [200]. The first two approaches fall somewhere between symbolic and non-symbolic systems: they use (to some extent) a high-level descriptive language, on the other hand, the operators used are not defined in that language and operate at the non-symbolic level. Even in some recent problem-oriented representations ([149],[363],[340], [91]) the operators still operate at the conservative traditional subsymbolic level. The third system GIL is an evolution program tailored to "learning from examples"; the system incorporates the problem-specific knowledge in data structures and operators.

Let us consider an example, which will be used throughout the chapter.

Example 12.1. This is taken from the world of Emerald's robots (see [217] and [407]). Each robot is described by the values of six attributes; the attributes with their domains are:

Attributes:	Values of Attributes:
Head_Shape	**R**ound, **S**quare, **O**ctagon
Body_Shape	**R**ound, **S**quare, **O**ctagon
Is_Smiling	**Y**es, **N**o
Holding	**S**word, **B**alloon, **F**lag
Jacket_Color	**R**ed, **Y**ellow, **G**reen, **B**lue
Has_Tie	**Y**es, **N**o

The boldface letters are used to identify attributes and their values, e.g., (**J** = **Y**) means "Jacket_Color is Yellow". The examples of concepts descriptions (where each concept C_i is described in terms of these six attributes and their values) are:

C_1 Head is round and jacket is red, or head is square and is holding a balloon

C_2 Smiling and holding balloon, or head is round

C_3 Smiling and not holding sword

C_4 Jacket is red and is wearing no tie, or head is round and is smiling

C_5 Smiling and holding balloon or sword

□

Attributes are of three types: nominal (their domains are sets of values), linear (their domains are linearly ordered), and structured (their domains are partially ordered). Events represent different decision classes: events from a particular class constitute its positive examples, all other events its negative examples. Learning examples are given in the form of events, and each event is a vector of attribute values.

The concept descriptions are represented in VL_1 (simplified version of the Variable Valued Logic System) [281] — a widely accepted language to represent input events for any program operating in an attribute–based space.

A description of a concept C is a disjunction of complexes

$$c_1 \vee \ldots \vee c_k \Rightarrow C;$$

each complex (c_i) is expressed as a conjunction of selectors, which are (attribute relation set_of_values) triplets (e.g., $\langle J = R \rangle$ for "Jacket_Color is red").

The concepts $C_1 - C_5$ can be expressed as

$$\langle S = R \rangle \wedge \langle J = R \rangle \vee \langle S = S \rangle \wedge \langle H = B \rangle \Rightarrow C_1$$
$$\langle I = Y \rangle \wedge \langle H = B \rangle \vee \langle S = R \rangle \Rightarrow C_2$$
$$\langle I = Y \rangle \wedge \langle H \neq S \rangle \Rightarrow C_3$$
$$\langle J = R \rangle \wedge \langle T \neq Y \rangle \vee \langle S = R \rangle \wedge \langle I = Y \rangle \Rightarrow C_4$$
$$\langle I = Y \rangle \wedge \langle H = \{B, S\} \rangle \Rightarrow C_5.$$

Note that the selector $\langle T \neq Y \rangle$ can be interpreted as $\langle T = N \rangle$, since the attribute T is a Boolean (nominal) attribute, and the selector $\langle H = \{B, S\} \rangle$ is interpreted as "Holding Balloon or Sword" (internal disjunction).

The problem is to construct a system to learn the concepts, i.e., to determine decision rules that account for all positive examples and no negative ones. We can evaluate and compare systems on all robot descriptions in terms of error rates and complexities of generated rules. The system should be able to predict a classification of previously unseen examples, or suggest (possibly more than one) classifications of partially specified descriptions.

During the past two decades there has been a growing interest in applying evolution programming techniques to machine learning (GBML systems, for genetics based machine learning systems) [88]. This was due to the attractive idea that chromosomes, representing knowledge, are treated as data to be manipulated by genetic operators, and, at the same time, as executable code to be used in performing some task. However, early applications, although partially successful (e.g.,[323],[340],[341]), also encountered many problems [86]. In general, in the GA community there are two competing approaches to address the problem. As stated by De Jong [86]:

> "To anyone who has read Holland [191], a natural way to proceed is to represent an entire rule set as a string (an individual), maintain a population of candidate rule sets, and use selection and genetic operators to produce new generations of rule sets. Historically, this was the approach taken by De Jong and his students while at the University of Pittsburgh (e.g., see Smith [363], [364]), which gave rise to the phrase 'the Pitt approach'.
>
> However, during the same time period, Holland developed a model of cognition (classifier systems) in which the members of the population are individual rules and a rule set is represented by the entire population (e.g., see Holland and Reitman, [189]; Booker [45]). This quickly became known as 'the Michigan approach' and initiated a friendly but provocative series of discussions concerning the strengths and weaknesses of the two approaches."

We believe that a third approach based on evolution programming techniques should be the most fruitful one. The idea of incorporating problem-specific knowledge, as usual by (1) careful design of appropriate data structures, and (2) problem-specific "genetic" operators, must pay off in system precision and performance. However, the Pitt approach is closer to our idea of evolution programming, since it maintains a population of complete solutions (set of rules) to the problem, whereas the Michigan approach (classifier systems), through bidding, the bucket brigade algorithm, and a genetic component in modifying rules, establishes a new methodology very different to evolution programming technique. On the other hand, implementations of the Pitt approach, even when they represent a chromosome in a high-level description language, do not use any learning methodology to modify their operators. This is the basic difference between the Pitt approach and the evolution program approach.

In the sequel, we discuss the basic principles behind the classifier systems (the Michigan approach, section 12.1), the Pitt approach (section 12.2), and Janikow's [200] evolution program (section 12.3) for inductive learning of decision rules in attribute-based examples.[1]

12.1 The Michigan approach

Classifier systems are a kind of rule-based system with general mechanisms for processing rules in parallel, for adaptive generation of new rules, and for testing the effectiveness of existing rules. Classifier systems provide a framework in which a population of rules encoded as bit strings evolves on the basis of intermittently given stimuli and reinforcement from its environment. The system "learns" which responses are appropriate when a stimulus is presented. The rules in a classifier system form a population of individuals evolving over time.

A classifier system (see Figure 12.1) consists of the following components:

- detector and effector,

- message system (input, output, and internal message lists),

- rule system (population of classifiers),

- apportionment of credit system (bucket brigade algorithm), and

- genetic procedure (reproduction of classifiers).

The environment sends a message (a move on a board, an example of a new event, etc.), which is accepted by the classifier system's detectors and placed on the input message list. The detectors decode the message into one or more (decoded) messages and place them on the (internal) message list. The messages activate classifiers; strong, activated classifiers place messages on the message list. These new messages may activate other classifiers or they can send some messages to the output message list. In the latter case, the classifier system's

---

[1]For more information on classifier systems, see [107]; a recent special issue of Evolutionary Computation journal.

ENVIRONMENT

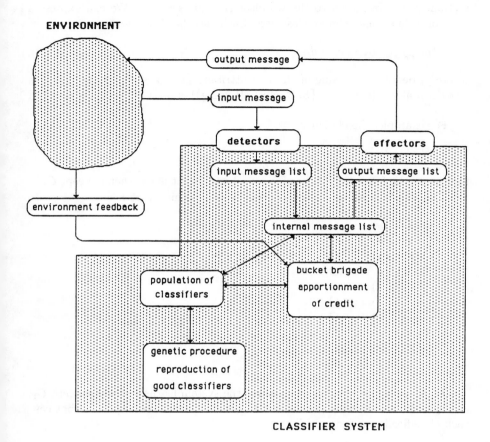

CLASSIFIER SYSTEM

Fig. 12.1. A classifier system and its environment

effectors code these messages into an output message (a move on a board, decision, etc.), which is returned to the environment. The environment evaluates the action of the system (environment feedback), and the bucket brigade algorithm updates the strengths of the classifiers.

We discuss some of these actions in more detail using the world of Emerald's robots as example. Let us provide some basic preliminaries first. A single classifier consists of two parts: (1) a condition part, and (2) a message. The condition part is a finite-length string over some alphabet; the alphabet includes the "don't care" symbol, "*". The message part is a finite length string over the same alphabet; however, it does not contain the "don't care" symbol. Each classifier has its own strength. The strength is important in the bidding process, where classifiers compete to post their messages — we discuss it later in this section. Now, we return to the world of Emerald's robots. We can express a decision rule as one or more classifiers. Each classifier has a form

$$(p_1, p_2, p_3, p_4, p_5, p_6) : d,$$

where p_i denotes the value of the i-th attribute $(1 \leq i \leq 6)$ for the domains described above (e.g., $p_4 \in \{S, B, F, *\}$ (Holding)) and $d \in \{C_1, C_2, C_3, C_4, C_5\}$.

For example, the classifier

$$(R * * * R*) : C_1$$

represents the following: "If head is round and jacket is red then concept C_1".
    The set of classifiers for five concepts given in this chapter is:

$$
\begin{array}{lcl}
(R * * * R*) & : & C_1 \\
(S * *B * *) & : & C_1 \\
(* * YB * *) & : & C_2 \\
(R * * * **) & : & C_2 \\
(* * YB * *) & : & C_3 \\
(* * YF * *) & : & C_3 \\
(* * * * RN) & : & C_4 \\
(R * Y * **) & : & C_4 \\
(* * YB * *) & : & C_5 \\
(* * YS * *) & : & C_5.
\end{array}
$$

To simplify the example, we assume the system learns a single concept C_1; any system can be easily generalized to handle multiple concepts. In that case each classifier has a form

$$(p_1, p_2, p_3, p_4, p_5, p_6) : d,$$

where $d = 1$ (membership to the concept C_1) or $d = 0$ (otherwise).
    Let us assume that at some stage of the learning process there is a small (random) population of classifiers q in the system (each classifier is given with its strength s):

$$
\begin{array}{lclcl}
q_1 = (* * *SR *) & : & 1, & s_1 = 12.3, \\
q_2 = (* * Y * *N) & : & 0, & s_2 = 10.1, \\
q_3 = (SR * * * *) & : & 1, & s_3 = 8.7, \\
q_4 = (*O * * * *) & : & 0, & s_4 = 2.3.
\end{array}
$$

Assume further that an input message arrives from the environment (a new example event):

$$(RRYSRN)$$

It is a description of a single robot with round (R) head, round (R) body, smiling (Y), holding sword (S), in red (R) jacket and without tie (N). Obviously, this robot constitutes the positive example of the concept C_1 due to its round head and red jacket.

Three classifiers are activated by this message: q_1, q_2, and q_4. These classifiers bid; each bid is proportional to the classifier's strength ($bid\_i = b \cdot s_i$). The strongest classifier q_1 wins and posts its message; since the message is decoded into the correct classification, the classifier gets a reward $r > 0$; the strength of the classifier becomes

$$s_1 := s_1 - bid_1 + r$$

(if the message was decoded into wrong answer, the "reward" r would be negative). For coefficients $b = 0.2$ and $r = 4.0$, the new strength of the classifier q_1 is $s_1 = 12.3 - 2.46 + 4.0 = 13.84$.

One of the parameters of the classifier system is the GA period, t_{ga}, which specifies the number of time steps (number of cycles just described above) between GA calls. Of course, t_{ga} can be a constant, it can be generated randomly (with the average equal to t_{ga}), or it need not be even specified and the decision of invocation of GA can be made of the basis of the performance of the system. Anyway, let us assume that the time has come to apply the genetic algorithm to the classifiers.

We consider the strengths of the classifiers as their fitness — a selection process can be performed using roulette wheel selection (Chapter 2). However, in the classifier systems we are no longer interested in the strongest (the most fit) classifier, but rather in a whole population of classifiers that perform the classification task. This implies that we should not generate the whole population and that we should be careful in selecting individuals for replacement. Usually, the *crowding factor model* (Chapter 4) is used, since it replaces similar population members.

The operators used are, again, mutation and crossover. However, some modifications are necessary. Let us consider the first attribute, Head_Shape — its domain is the set $\{R, S, O, *\}$. Thus, when mutation is called, we would change the mutated character to one of the other three characters (with equal probability):

$$R \to \{S, O, *\},$$
$$S \to \{R, O, *\},$$
$$O \to \{R, S, *\},$$
$$* \to \{R, S, O\}.$$

The strength of the offspring usually is the same as its parent.

The crossover does not require any modification. We take advantage of the fact that all classifiers are of equal length; to crossover two selected parents, say q_1 and q_2:

$$(* * *|SR*) : 1, \text{ and } (* * Y| * *N) : 0,$$

we generate a random crossover position point (say, we crossover after the third character, as marked), and the offspring are

$$(* * * * * N) \; : \; 0, \text{ and } (* * YSR*) \; : \; 1.$$

The strengths of the offspring are a (possibly weighted) average of those of the parents.

Now, the classifier system is ready to continue its learning process and start another cycle of t_{ga} steps, accepting further positive and negative events as examples, modifying the strengths of classifiers. We hope that finally the population of classifiers converges to some number of strong individuals, e.g.,

$$
\begin{aligned}
(R * * * R*) &\; : \; 1, \\
(S * *B * *) &\; : \; 1, \\
(O * * * * *) &\; : \; 0, \\
(* * *SY*) &\; : \; 0.
\end{aligned}
$$

The example discussed above was very simple: we aimed at explaining the basic components of the classifier system using a learning paradigm for the world of Emerald's robots. Note, however, that we used the simplest bidding system (e.g., we did not use *effective bid* variable $e\_bid = bid + N(0, \sigma_{bid})$, which introduces some random noise with standard deviation σ_{bid} and expected value zero), that we did not use any taxation system (usually each classifier is taxed to prevent biasing the population towards productive rules), and that we select only a single winner (the strongest classifier) at each step — in general, more then one classifier can be a winner, placing its message on the message list. Moreover, the bucket brigade algorithm was not used in the sense that the reward was available at every time step (for every provided example). There was no relationship between examples and there was no need to trace a chain of rewards to apportion credit to the classifiers whose messages activated the current (winner) classifier. For some problems, like planning problems, the length of the message part is the same as the length of the condition part. In such cases, an activated classifier (old winner) would place its message on the (internal) message list, which, in turn, may activate some other classifiers (new winners). Then the strength of the old winners is increased by a reward — a payment from the new winners.

For a full discussion on different versions of the classifier systems and the historical perspective, the reader is referred to [154].

12.2 The Pitt approach

The Michigan approach can be perceived as a computational model of cognition: the knowledge of a cognitive entity is expressed as a collection of rules which undergo modifications over a time. We can evaluate the whole cognitive unit in terms of its interaction with the environment; an evaluation of a single rule (i.e., a single individual) is meaningless.

On the other hand, the Pitt approach adopts the view that each individual in a population represents the whole set of rules (a separate cognitive entity). These individuals compete among themselves, the weak individuals die, the strong survive and reproduce: this is done by means of natural selection proportional to the their fitness, crossover, and mutation operators. In short, the Pitt approach applies the genetic algorithm to the learning problem. By doing this, Pitt approach avoids the delicate credit assignment problem, for which a heuristic method (such as the bucket brigade algorithm) should distribute (positive or negative) credit among the rules which cooperated in produced (desirable or undesirable) behavior.

However, there are some interesting issues to be addressed. The first one is representation. Should we use fixed-length binary vectors (with a fixed field format) to represent set of rules? Such a representation would be ideal for generating new rules, since we can use classical crossover and mutation for that purpose. However, it seems to be too restrictive and appropriate only for systems which work at a lower sensory level. What about genes in a chromosome representing values of attributes (i.e., number of genes equal to number of attributes)? This decision (not supported by specialized operators) does not seem to work: if the cardinalities of some domains are large, the probability that the system would converge prematurely is very high [86], even with much higher mutation rates than usual. It seems that some internal representational structure is necessary to provide punctuation marks between units in a chromosome, together with operators which are aware of chromosomal representation. Such an approach was implemented and discussed in [364] and [340].

Smith [363] went even further and experimented with variable-length individuals. He generalized many results previously valid only for GAs with fixed length strings to apply to GAs with variable length strings.

However, it seems that some other bold decisions are necessary to address the complex issues of representation and operators. Only recently [24] was the need for such decisions recognized:

"A solution to this problem is to select different genetic operators that are more appropriate to 'natural representation'. There is nothing sacred about the traditional string oriented genetic operators. The mathematical analysis of GAs shows that they work best when the internal representation encourages the emergence of useful building blocks that can be subsequently combined with other to produce improved performance. String representations are just one of many ways of achieving this."

In the next section, we present an evolution program (based on the Pitt approach), which does just this: the representation is rich and natural, with specialized, representation sensitive operators, taken directly from a learning methodology.

12.3 An evolution program: the GIL system

The implemented evolution program GIL [200] moves the genetic algorithm (the Pitt approach) closer to the symbolic level — mainly by defining specialized operators manipulating at the problem level. Again, the basic components of GIL, like any other evolution program, are data structures and genetic operators. The system was designed for learning single concepts only. However, it can be easily extended to learn in multi-concept environments by introducing multiple populations.

12.3.1 Data structures

One chromosome represents a solution to the concept description being learned; the assumption is that any event not covered by such a description belongs to the negation of the concept (does not belong to the concept). Each chromosome is a set (disjunction) of a number of complexes. The number of complexes in a chromosome can vary — the assumption that all chromosomes are of equal length (as GAs maintain populations of fixed-length strings) would be, to say the least, artificial, so it would be contrary to evolution programming technique. This decision, however, is not new in the GA community: as described in the previous section, Smith [363] has extended many formal results on genetic algorithms to variable-length strings and implemented a system that maintained a population of such (variable-length) strings.

Each complex, as defined in VL_1, is a conjunction of a number of selectors corresponding to different attributes. Each selector, in turn, is the internal disjunction from the domain of its attribute. For example, the following might be a chromosome (and therefore a description of the concept C_1):

$$(\langle S = R \rangle \wedge \langle J = R \rangle) \vee (\langle S = S \rangle \wedge \langle H = B \rangle)$$

Mostly for efficiency reasons, binary representation for selectors was used for internal representation of chromosomes. A binary 1 at position i implies the inclusion of the i-th domain value in this selector. This means that the size of the domain of an attribute is equal to the length of the binary substring corresponding to this attribute. Note that a collection of all 1s for some selector is equivalent to the *don't care* symbol for this attribute. Thus the chromosome describing concept C_1 in a notation similar to one used for classifier systems is represented as

$$\langle R * * * R * \vee S * * B * * \rangle,$$

whereas its internal representation in the GIL system is

$$\langle 100|111|11|111|1000|11 \vee 010|111|11|010|1111|11 \rangle,$$

where bars separate selectors. Note that such a representation handles internal disjunction gracefully, e.g., the concept C_5

$\langle I = Y \rangle \wedge \langle H = \{B, S\} \rangle$,

can be represented as

$\langle 111|111|10|110|1111|11 \rangle$.

12.3.2 Genetic operators

The operators of the GIL system are modeled on the methodology of inductive learning provided by Michalski [280]; the methodology describes various inductive operators that constitute the process of inductive inference. These include: *condition dropping*, e.g., dropping a selector from the concept description, *adding alternative rule* and *dropping a rule, extending reference* — extending an internal disjunction, *closing interval* — for linear domains filling up missing values between two present values, and *climbing generalization* — for structured domains climbing the generalization tree.

The GIL system defines inductive operators separately on three abstract levels: chromosome level, complex level, and selector level. We discuss some of them in turn.

Chromosome level: the operators act on the whole chromosomes:

- *RuleExchange:* the operator is similar to a crossover of the classical GA, as it exchanges selected complexes between two parent chromosomes. For example, two parents

    $\langle 100|111|11|111|1000|11 \vee 010|111|11|010|1111|11 \rangle$ and
    $\langle 111|001|01|111|1111|01 \vee 110|100|10|111|0010|01 \rangle$

    may produce the following offspring:

    $\langle 100|111|11|111|1000|11 \vee 111|001|01|111|1111|01 \rangle$ and
    $\langle 010|111|11|010|1111|11 \vee 110|100|10|111|0010|01 \rangle$.

- *RuleCopy:* the operator is similar to *RuleExchange*; however, it copies random complexes from one parent to another. For example, two parents

    $\langle 100|111|11|111|1000|11 \vee 010|111|11|010|1111|11 \rangle$ and
    $\langle 111|001|01|111|1111|01 \vee 110|100|10|111|0010|01 \rangle$

    may produce the following offspring:

    $\langle 100|111|11|111|1000|11 \vee 111|001|01|111|1111|01 \vee$
    $110|100|10|111|0010|01 \rangle$ and
    $\langle 010|111|11|010|1111|11 \rangle$.

- *NewPEvent:* this unary operator incorporates a description of a positive event into the selected chromosome. For example, for a parent

    $\langle 100|111|11|111|1000|11 \vee 010|111|11|010|1111|11 \rangle$

    and an uncovered event

⟨100|010|10|010|0010|01⟩,

the following offspring is produced:

⟨100|111|11|111|1000|11 ∨ 010|111|11|010|1111|11∨
100|010|10|010|0010|01⟩.

- *RuleGeneralization:* this unary operator generalizes a random subset of complexes. For example, for a parent

⟨100|111|11|111|1000|11 ∨ 010|111|11|010|1111|11∨
100|010|10|010|0010|01⟩

and the second and third complexes selected for generalization, the following offspring is produced:

⟨100|111|11|111|1000|11 ∨ 110|111|11|010|1111|11⟩.

- *RuleDrop:* this unary operator drops a random subset of complexes. For example, for a parent

⟨100|111|11|111|1000|11 ∨ 010|111|11|010|1111|11∨
100|010|10|010|0010|01⟩,

the following offspring might be produced:

⟨100|111|11|111|1000|11⟩.

- *RuleSpecialization:* this unary operator specializes a random subset of complexes. For example, for a parent

⟨100|111|11|111|1000|11 ∨ 010|111|11|010|1111|10∨
111|010|10|010|1111|11⟩

and the second and third complexes selected for specialization, the following offspring is produced:

⟨100|111|11|111|1000|11 ∨ 010|010|10|010|1111|10⟩.

Complex level: the operators act on complexes of the chromosomes:

- *RuleSplit:* this operator acts on a single complex splitting it into a number of complexes. For example, a parent

⟨100|111|11|111|1000|11⟩

may produce the following offspring (the operator splits the second selector):

⟨100|011|11|111|1000|11 ∨ 100|100|11|111|1000|11⟩.

- *SelectorDrop:* this operator acts on a single complex and "drops" a single selector, i.e., all values for selected selector are replaced by a string '11...1'. For example, a parent

⟨100|010|11|111|1000|11⟩

may produce the following offspring (the operator works on the fifth selector):

⟨100|010|11|111|1111|11⟩.

- *IntroSelector:* this operator acts by "adding" a complex, i.e., it eliminates a selector with a string '11...1'. For example, a parent

  ⟨100|010|11|111|1111|11⟩

  may produce the following offspring (the operator works on the fifth selector):

  ⟨100|010|11|111|0001|11⟩.

- *NewNEvent:* this unary operator incorporates a description of a negative event to the selected chromosome. For example, for a parent

  ⟨110|010|11|111|1111|11⟩

  and the covered negative event

  (100|010|10|010|0100|10),

  the following offspring is produced:

  ⟨010|010|11|111|1111|11 ∨ 110|010|01|111|1111|11∨
         110|010|11|101|1111|11∨110|010|11|111|1011|11∨
         110|010|11|111|1111|01⟩.

Selector level: the operators act on selectors:

- *ReferenceChange:* the operator adds or removes a single value (0 or 1) from the chromosome, i.e., from the domain of one of the selectors. For example, a parent

  ⟨100|010|11|111|0001|11⟩

  may produce the following offspring (note the difference in the fourth selector):

  ⟨100|010|11|110|0001|11⟩

- *ReferenceExtension:* the operator extends the domain of a selector by allowing a number of additional values. For different types of attributes (nominal, linear, structured) it uses different probabilities of selecting values. For example, a parent

  ⟨100|010|11|111|1010|11⟩

  may produce the following offspring (the operator "closes" the domain of the fifth selector):

  ⟨100|010|11|111|1110|11⟩.

- *ReferenceRestriction:* this operator removes some domain values from a selector. For example, a parent

  ⟨100|010|11|111|1011|11⟩

  may produce the following offspring:

  ⟨100|010|11|111|1000|11⟩.

The GIL system is quite complex and requires a number of parameters (e.g., the probabilities of applying these operators). For a discussion of these and other implementational issues, the reader is referred to [200]. It is interesting to note, however, that the operators are given a priori probabilities, but the actual probabilities are computed as a function of these probabilities and two other parameters: an (a priori) desired balance between specialization and generalization, and a (dynamic) measure of the current coverage. This idea is similar to one discussed earlier (last section of Chapter 8).

At each iteration all chromosomes are evaluated with respect to their completeness and consistency (and possibly cost if desired), and a new population is formed with those better ones more likely to appear. Then, the operators are applied to the new population, and the cycle repeats.

12.4 Comparison

A recent publication [407] provides an evaluation of a number of learning strategies, including a classifier system (CFS), a neural net (BpNet), a decision tree learning program (C4.5), and a rule learning program (AQ15). The systems used examples from the world of Emerald robots; the systems were supposed to learn five concepts $(C_1 - C_5)$ given at the beginning of this chapter, while seeing only a varying percentage of the positive and negative examples (there were a total of 432 different robots present, i.e., all possible combinations of the attribute values were there). The systems were compared by providing an average error in recognizing all of the 432 (seen and unseen) robots; the results (from [407]) are given in Table 12.1.

System	Learning Scenario (Positive % / Negative %)				
	6% / 3%	10% / 10%	15% / 10%	25% / 10%	100% / 10%
AQ15	22.8%	5.0%	4.8%	1.2%	0.0%
BpNet	9.7%	6.3%	4.7%	7.8%	4.8%
C4.5	9.7%	8.3%	11.3%	2.5%	1.6%
CFS	21.3%	20.3%	21.5%	19.7%	23.0%

Table 12.1. Summary of the error rates for different systems

Table 12.2 provides a recognition rate for the GIL system on the individual concepts basis (from [200]). As expected, the evolution program GIL performs much better than system CFS based on the classifier system approach; surprisingly, GIL outperformed other learning systems as well. Its superiority is most visible in the cases with a small percentage of seen and unseen examples.

For a further discussion on comparison of these systems (e.g., complexity of the generated rules), implementational issues of the GIL system, and results of

Concept	Learning Scenario (Positive % / Negative %)				
	6% / 3%	10% / 10%	15% / 10%	25% / 10%	100% / 10%
C_1	11.1%	5.3%	0.0%	0.0%	0.0%
C_2	0.0%	0.0%	0.0%	0.0%	0.0%
C_3	0.0%	0.0%	0.0%	0.0%	0.0%
C_4	10.4%	0.0%	0.0%	0.0%	0.0%
C_5	0.0%	0.0%	0.0%	0.0%	0.0%

Table 12.2. Summary of the error rates for the evolution program GIL

the other experiments (multiplexers, breast cancer, etc.), the reader is referred to [200].

12.5 REGAL

An interesting approach for inducing concept descriptions from examples was reported recently by Giordana and Saitta [139]. The developed system REGAL learns concept descriptions in disjunctive normal form:

$$c_1 \vee \ldots \vee c_k \Rightarrow C;$$

each complex (c_i) is expressed as a conjunction of selectors, which may contain internal disjunctions, e.g.,

$$\langle S = R \vee O \rangle \wedge \langle J = R \vee Y \rangle \vee \langle S = S \rangle \wedge \langle H = B \vee F \rangle \Rightarrow C.$$

The key issue is, again, representation. The REGAL system works on fixed length binary strings, hence the necessity of mapping disjunctive normal form expressions into such strings. This has been achieved by imposing limits on the formula complexity, which is defined through the language template Λ representing the maximally complex formula. Then, any other well formed formula is obtained by deleting some literal from Λ; in this way the literals in Λ can be set in correspondence with the bits of a string. The REGAL system uses 2-point and uniform crossovers as well as generalizing and specializing crossovers specifically designed for the task at hand.

The system was tested in "learning one disjunct at a time" mode and "learning many disjuncts at one time" mode, where sharing functions were used. For more detailed description of the system and the results of experiments the reader is referred to [139]; further experiments are described in [140].

13. Evolutionary Programming and Genetic Programming

The past is present, isn't it?
It's the future too.

Eugene O'Neill, *Long Day's Journey Into Night*

In this chapter we review briefly two powerful evolutionary techniques; these are evolutionary programming (section 13.1) and genetic programming (section 13.2). These two techniques were developed a quarter of a century apart from each other; they aimed at different problems; they use different chromosomal representations for individuals in the population, and they put emphasis on different operators. Yet, they are very similar from our perspective of "evolution programs": for particular tasks they aim at, they use specialized data structures (finite state machines and tree-structured computer programs) and specialized "genetic" operators. Also, both methods must control the complexity of the structure (some measure of the complexity of a finite state machine or a tree might be incorporated in the evaluation function). We discuss them in turn.

13.1 Evolutionary programming

The original evolutionary programming (EP) techniques were developed by Lawrence Fogel [126]. They aimed at evolution of artificial intelligence in the sense of developing ability to predict changes in an environment. The environment was described as a sequence of symbols (from a finite alphabet) and the evolving algorithm supposed to produce, as an output, a new symbol. The output symbol should maximize the payoff function, which measures the accuracy of the prediction.

For example, we may consider a series of events, marked by symbols $a_1, a_2, \ldots$; an algorithm should predict the next (unknown) symbol, say a_{n+1} on the basis of the previous (known) symbols, $a_1, a_2, \ldots, a_n$. The idea of evolutionary programming was to evolve such an algorithm.

Finite state machines (FSM) were selected as a chromosomal representation of individuals; after all, finite state machines provide a meaningful representation of behavior based on interpretation of symbols. Figure 13.1 provides an example of a transition diagram of a simple finite state machine for a parity check. Such transition diagrams are directed graphs that contain a node for each state and edges that indicate the transition from one state to another, input and output values (notation a/b next to an edge leading from state S_1 to the state S_2 indicates that the input value of a, while the machine is in state S_1, results in output b and the next state S_2.

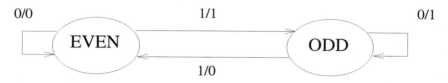

0/0   1/1   0/1

1/0

Fig. 13.1. A FSM for a parity check

There are two states EVEN and ODD (machine starts in state EVEN); the machine recognizes a parity of a binary string.

The technique of evolutionary programming is to maintain a population of finite state machines; each such individual represents a potential solution to the problem (i.e., represents a particular behavior). As already mentioned, each FSM is evaluated to give some measure of its "fitness". This is done in the following way: each FSM is exposed to the environment in the sense that it examines all previously seen symbols. For each subsequence, say, $a_1, a_2, \ldots, a_i$ it produces an output a'_{i+1}, which is compared with the next observed symbol, a_{i+1}. For example, if n symbols were seen so far, a FSM makes n predictions (one for each of the substrings a_1, a_1, a_2, and so on, until $a_1, a_2, \ldots, a_n$); the fitness function takes into account the overall performance (e.g., some weighted average of accuracy of all n predictions).

As in evolution strategies (section 8.1), the evolutionary programming technique first creates offspring and later selects individuals for the next generation. Each parent produces a single offspring; hence the size of the intermediate population doubles (as in (pop_size, pop_size)-ES). Offspring (new FSMs) are created by random mutations of the parent population (see Figure 13.2). There are five possible mutation operators: change of an output symbol, change of a state transition, addition of a state, deletion of a state, and change of the initial state (there are some additional constraints on the minimum and maximum number of states). These mutations are chosen with respect to some probability distribution (which can change during the evolutionary process); also it is possible to apply more than one mutation to a single parent (a decision on the number of mutations for a particular individual is made with respect to some other probability distribution).

The best pop_size individuals are retained for the next generation; i.e., to qualify for the next generation an individual should rank in the top 50% of the intermediate population. In the original version [126] this process was iterated several times before the next output symbol was made available. Once a new

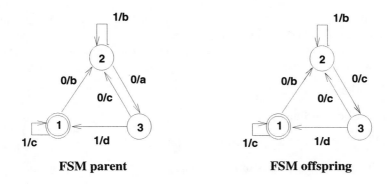

Fig. 13.2. A FSM and its offspring. Machines start in state 1

symbol is available, it is added to the list of known symbols, and the whole process is repeated.

Of course, the above procedure can be extended in many way; as stated in [121]:

> "The payoff function can be arbitrarily complex and can posses temporal components; there is no requirement for the classical squared error criterion or any other smooth function. Further, it is not required that the predictions be made with a one-step look ahead. Forecasting can be accomplished at an arbitrary length of time into the future. Multivariate environments can be handled, and the environmental process need not be stationary because the simulated evolution will adapt to changes in the transition statistics."

As mentioned in section 8.2, evolutionary programming techniques were generalized to handle numerical optimization problems; for details see [117] or [121]. For other examples of evolutionary programming techniques, see also [126] (classification of a sequence of integers into primes and nonprimes), [120] (for application of EP technique to the interated prisoner's dilemma), as well as [123, 124, 378, 254] for many other applications.

13.2 Genetic programming

Another interesting approach was developed relatively recently by Koza [228], [231]. Koza suggests that the desired program should evolve itself during the evolution process. In other words, instead of solving a problem, and instead of building an evolution program to solve the problem, we should rather search the space of possible computer programs for the best one (the most fit). Koza developed a new methodology, named Genetic Programming (GP), which provides a way to run such a search. A population of executable computer programs is created, individual programs compete against each other, weak programs die, and strong ones reproduce (crossover, mutation)...

There are five major steps in using genetic programming for a particular problem. These are

- selection of terminals,

- selection of a function,

- identification of the evaluation function,

- selection of parameters of the system,

- selection of the termination condition.

It is important to note that the structure which undergoes evolution is a hierarchically structured computer program.[1] The search space is a hyperspace of valid programs, which can be viewed as a space of rooted trees. Each tree is composed of functions and terminals appropriate to the particular problem domain; the set of all functions and terminals is selected *a priori* in such a way that some of the composed trees yield a solution.

For example, two structures e_1 and e_2 (Figure 13.3) represent expressions $2x + 2.11$ and $x \cdot \sin(3.28)$, respectively. A possible offspring e_3 (after crossover of e_1 and e_2) represents $x \cdot \sin(2x)$.

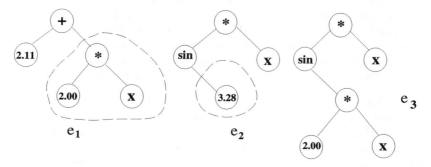

Fig. 13.3. Expression e_3: an offspring of e_1 and e_2. Broken line includes areas being exchanged during the crossover operation

The initial population is composed of such trees; construction of a (random) tree is straightforward. The evaluation function assigns a fitness value which evaluates the performance of a tree (program). The evaluation is based on a preselected set of test cases; in general, the evaluation function returns the sum of distances between the correct and obtained results on all test cases. The selection is proportional; each tree has a probability of being selected to the

---

[1]Actually, Koza has chosen LISP's S-expressions for all his experiments. Currently, however, there are implementations of GP in C and other programming languages.

next generation proportional to its fitness. The primary operator is a crossover that produces two offspring from two selected parents. The crossover creates offspring by exchanging subtrees between two parents. There are other operators as well: mutation, permutation, editing, and a define-building-block operation [228]. For example, a typical mutation selects a node in a tree and generates a new (random) subtree which originates in the selected node.

In addition to five major steps for building a genetic program for a particular problem, Koza [232] recently considered the advantages of adding a further feature: a set of procedures. These procedures are called Automatically Defined Functions (ADF). It seems that this is an extremely useful concept for genetic programming techniques, which makes its major contribution in the area of code reusability. ADFs discover and exploit the regularities, symmetries, similarities, patterns, and modularities of the problem at hand, and the final genetic program may call these procedures at different stages of its execution.

Probably it would be a mistake to classify genetic programming as another version of evolution programming which special chromosomal representation of individuals. The fact that genetic programming operates on computer programs has a few interesting aspects. For example, the operators can be viewed also as programs, which can undergo a separate evolution during the run of the system. Additionally, a set of functions can consist of several programs which perform complex tasks; such functions can evolve further during the evolutionary run (e.g., ADF). Clearly, it is one of the most exciting areas of the current development in the evolutionary computation field, and has already accumulated a significant amount of experimental data (see, for example, apart from [231] and [232], also [225] and [8]).

14. A Hierarchy of Evolution Programs

In this book we discussed different strategies, called Evolution Programs, which might be applied to hard optimization problems and which were based on the principle of evolution. Evolution programs borrow heavily from genetic algorithms. However, they incorporate problem-specific knowledge by using "natural" data structures and problem-sensitive "genetic" operators. The basic difference between GAs and EPs is that the former are classified as weak, problem-independent methods, which is not the case for the latter.

The boundary between weak and strong methods is not well defined. Different evolution programs can be built which display a varying degree of problem dependence. For a particular problem P, in general, it is possible to construct a family of evolution programs EP_i, each of which would 'solve' the problem (Figure 14.1). The term 'solve' means 'provide a reasonable solution', i.e., a solution which need not, of course, be optimal, but is feasible (it satisfies problem constraints).

The evolution program EP_5 (Figure 14.1) is the most problem specific and it addresses the problem P only. The system EP_5 will not work for any modified version of the problem (e.g., after adding a new constraint or after changing the size of the problem). The next evolution program, EP_4, can be applied to some (relatively small) class of problems, which includes the problem P; other evolution programs EP_3 and EP_2 work on larger domains, whereas EP_1 is domain independent and can be applied to any optimization problem.

We have already seen a part of such a hierarchy in various places of the book. Let us consider a particular 20×20 nonlinear transportation problem, P. There are 400 variables with $20 + 20 = 40$ equations (of which 39 are independent). Additional constraints require that the variables must take nonnegative values. In principle, it is possible to construct an evolution program, say EP_5, which would solve this particular problem. It might be a genetic algorithm with 39 penalty functions tuned very carefully for these constraints or with decoders and/or repair algorithms. Any change in the size of the problem (moving from

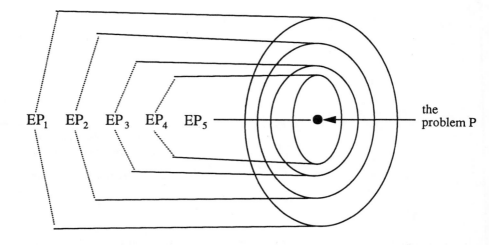

Fig. 14.1. A hierarchy of evolution programs

20×20 to 20×21) or any change in a transportation cost from one source to one destination would result in a failure of the EP_5 system.

There is also an evolution program, GENETIC-2 (Chapter 9), which can be applied to any transportation problem. Let us call this system EP_4. This system still belongs to the class of strong methods, since it can be applied only for nonlinear transportation problems. However, it is much weaker than EP_5, since it can handle any transportation problem.

Another evolution program, say EP_3, applicable to the problem P, is GENOCOP (Chapter 7). It optimizes any function with the presence of any set of linear constraints, which is the case for the transportation problem, P. Obviously, EP_3 is a weaker method than EP_4. However, it can still be considered as a relatively strong method, since it can be applied only to numerical optimization problems with linear constraints.

Yet another evolution program (let us call it EP_2) can be applied to our 20×20 transportation problem, P: an evolution strategy (Chapter 8). Evolution strategies can be applied to any numerical optimization problem with (not necessarily linear) inequality constraints. Clearly, the problem P belongs to the domain of EP_2; also, EP_2 is a weaker method than EP_3, since it handles any type of inequalities (for the problem P, equalities can be easily replaced by inequalities using the method discussed in Chapter 7).

We can also construct a general purpose evolution program, EP_1, which might be just a classical genetic algorithm with a standard set of penalty functions; each penalty function would correspond to one of the problem's constraints. The system EP_1 is domain independent — it can approach any optimization problem with any set of constraints. For numerical optimization problems, the constraints may include nonlinear equalities, which makes this system a weaker method than EP_2, which is restricted to inequalities only. Moreover, EP_1 can be applied also to other (non-numerical) problems as well. Here we assume that the program EP_1 always return a feasible solution. We can enforce it easily, if the initial population consists of feasible solutions and penalty functions, decoders, or if repair algorithms keep individuals within the search space.

Let us denote by $dom(EP_i)$ a set of all problems to which the evolution program EP_i can by applied, i.e., the program returns a feasible solution. Clearly,

$$dom(EP_5) \subseteq dom(EP_4) \subseteq dom(EP_3) \subseteq dom(EP_2) \subseteq dom(EP_1).$$

Obviously, the above example is by no means complete: it is possible to create other evolution programs which would fit between EP_i and EP_{i+1} for some $1 \le i \le 4$. Of course, there might be also other evolution programs which overlap with others in the above hierarchy. For example, we can build systems to optimize transportation problems with the cost functions restricted to polynomials, or to optimize problems restricted to a convex search space, or problems with constraints in the form of nonlinear equations. In other words, the set of evolution programs is partially ordered; we denote the ordering relation by $\prec$ with the following meaning: if $EP_p \prec EP_q$ then the evolution program EP_p is a weaker method than EP_q, i.e., $dom(EP_q) \subseteq dom(EP_p)$. Referring to our example of a transportation problem, P, and a hierarchy of evolution programs, EP_i,:

$$EP_1 \prec EP_2 \prec EP_3 \prec EP_4 \prec EP_5.$$

The hypothesis is that if $EP_p \prec EP_q$, then the stronger method, EP_q, should in general perform better than a weaker system, EP_p. We do not have any proof of this hypothesis, of course, since it is based solely on a number of experiments and the simple intuition that problem-specific knowledge enhances an algorithm in terms of performance (time and precision) and at the same time narrows its applicability. We have already seen the better performance of GENETIC-2 against GENOCOP, and we discussed how GENOCOP outperforms classical GA on a particular class of problems. If this hypothesis is true, GENOCOP should give better results than an evolution program based on evolution strategy for problems with linear constraints, since ES is a weaker method than GENOCOP. Some other researchers support the above hypothesis; let us cite from Davis [77]:

> "It is a truism in the expert system field that domain knowledge leads to increased performance in optimization, and this truism has certainly been borne out of my experience applying genetic algorithms to industrial problems. Binary crossover and binary mutation are knowledge-blind operators. Hence, if we resist adding knowledge to our genetic algorithms, they are likely to underperform nearly

any reasonable optimization algorithm that does take into account such domain knowledge."

Goldberg [156], [154], provides an additional perspective. Let us quote from [156]:

"Certainly humans have developed very efficient search procedures for narrow classes of problems — genetic algorithms are unlikely to beat conjugate direction or gradient methods on continuous, quadratic optimization problems — but this misses the point. [...] The breadth combined with relative — if not peak — efficiency defines the primary theme of genetic search: robustness."

We visualize this observation in Figure 14.2, where a (classical) method, Q, works well for a problem, P, and nowhere else, whereas GAs perform reasonably across the spectrum. (Figure 14.2 is a simplification of similar figures given in [156] and [154].)

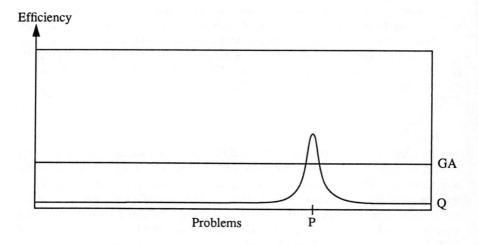

Fig. 14.2. Efficiency/problem spectrum and GAs

However, in the presence of nontrivial, hard constraints, the performance of GAs deteriorates quite often. On the other hand, evolution programs, by incorporating some problem-specific knowledge, may outperform even classical methods (Figure 14.3).

We should emphasize, again, that most evolution programs presented in the book do not have much theoretical support. There is neither a Schema Theorem (as for classical genetic algorithms) nor are there convergence theorems (as for evolution strategies). It is also important to underline that evolution programs are generally much slower than other optimization techniques. On

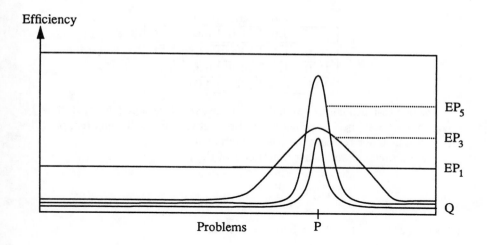

Fig. 14.3. Efficiency/problem spectrum and EPs

the other hand, their time complexity quite often grows in a linear (or $n \log n$) manner together with the problem size — which is not the case for most of the other techniques. Also, recent work by Nick Radcliff [315, 316, 317] on formal analysis and properties of genetic operators (e.g., respectful recombinations) applied to arbitrary data structures, is an important step towards providing some theoretical justification for evolution programs.

The above discussion was supported recently [261] by a series of experiments. The idea of stronger and weaker evolution programs was tested on one particular problem P (nonlinear transportation problem) and five evolution programs EP_i $(i = 1, \ldots, 5)$. We discuss these experiments in the following paragraphs.

Let us define a particular nonlinear balanced transportation problem P. Assume 3 sources and 4 destinations. The supply is:

$$source(1) = 10, \; source(2) = 15, \text{ and } source(3) = 20.$$

The demand is:

$$dest(1) = 3, \; dest(2) = 20, \; dest(3) = 5, \text{ and } dest(4) = 17.$$

The total flow in the problem P is 45. As discussed in Chapter 9, the optimum solution for the nonlinear transportation problem may contain neither zeros nor integer values (as it is the case in the linear transportation problem). For example, for some transportation cost functions f_{ij} the following solution might be optimal:

Amount transported

	3.0	20.0	5.0	17.0
10.0	1.34	1.52	0.01	7.13
15.0	1.15	10.39	0.39	3.07
20.0	0.51	8.09	4.60	6.80

For our test problem P we have used the same function f for each flow f_{ij}; a cost-matrix was used to provide variation between flows. The matrix provides the c_{ij}'s which act to scale the basic function shape.

We adopted the following function f of the flows x_{ij}:

$$f(x_{ij}) = \begin{cases} 0 & \text{if } x_{ij} = 0, \\ d + c_{ij} \cdot \sqrt{x_{ij}} & \text{otherwise,} \end{cases}$$

for $i = 1, 2, 3$, $j = 1, 2, 3, 4$, where $d = 5.0$, and

$$\begin{array}{llll} c_{11} = 0.0 & c_{12} = 21.0 & c_{13} = 50.0 & c_{14} = 62.0 \\ c_{21} = 21.0 & c_{22} = 0.0 & c_{23} = 17.0 & c_{24} = 54.0 \\ c_{31} = 50.0 & c_{32} = 17.0 & c_{33} = 0.0 & c_{34} = 60.0. \end{array}$$

So the problem P is to minimize

$$\sum_{i=1}^{3} \sum_{j=1}^{4} f(x_{ij}),$$

subject to the following constraints:

$$\begin{aligned} x_{11} + x_{12} + x_{13} + x_{14} &= 10 \\ x_{21} + x_{22} + x_{23} + x_{24} &= 15 \\ x_{31} + x_{32} + x_{33} + x_{34} &= 20 \\ x_{11} + x_{21} + x_{31} &= 3 \\ x_{12} + x_{22} + x_{32} &= 20 \\ x_{13} + x_{23} + x_{33} &= 5 \\ x_{14} + x_{24} + x_{34} &= 17. \end{aligned}$$

We solved the above problem P using GAMS (see Chapter 6). The GAMS' best solution was:

$$\sum_{i=1}^{3} \sum_{j=1}^{4} f(x_{ij}) = 430.64,$$

which was achieved for

$$\begin{array}{llll} x_{11} = 3.0 & x_{12} = 0.0 & x_{13} = 0.0 & x_{14} = 7.0 \\ x_{21} = 0.0 & x_{22} = 5.0 & x_{23} = 0.0 & x_{24} = 10.0 \\ x_{31} = 0.0 & x_{32} = 15.0 & x_{33} = 5.0 & x_{34} = 0.0. \end{array}$$

This result would serve us as a convenient reference point in evaluation of evolution programs presented here. We refer to GAMS as a classical (gradient–based) method Q for a problem P (see Figure 14.3).

For a fair comparison of the evolution programs EP_i ($i = 1, \ldots, 5$), we set population size to 70 and the number of generations to 5,000 for all our experiments. Each experiment was repeated 20 times; all averages for a particular experiment reported in the following subsections refer to averages obtained from these 20 runs. It is also important to point out that the presented evolution programs use different initialization techniques, however, we discuss them little bit later.

Evolution Program EP_1

The weakest evolution program EP_1 used in the experiments was the GENESIS 1.2ucsd system[1] developed by Nicol Schraudolph at the University of California, San Diego (the system is based on GENESIS 4.5, a genetic algorithm package written by John Grefenstette). In principle, one can use such generic tool to optimize a variety of problems and the $dom(EP_1)$ is virtually unlimited.

Let us exercise the usefulness of this evolution program on our test case, problem P. It is clear that the system will not provide any useful solutions if constraints are not incorporated by means of penalty functions. For example, we performed several runs of EP_1, defining only a domain for each of the twelve variables. Here we did not have much choice—the domain for each variable was selected as a range from zero to the smaller marginal sum for a given row and column:

$$0.0 \le x_{11} \le 3.0 \quad 0.0 \le x_{12} \le 10.0 \quad 0.0 \le x_{13} \le 5.0 \quad 0.0 \le x_{14} \le 10.0$$
$$0.0 \le x_{21} \le 3.0 \quad 0.0 \le x_{22} \le 15.0 \quad 0.0 \le x_{23} \le 5.0 \quad 0.0 \le x_{24} \le 15.0$$
$$0.0 \le x_{31} \le 3.0 \quad 0.0 \le x_{32} \le 20.0 \quad 0.0 \le x_{33} \le 5.0 \quad 0.0 \le x_{34} \le 17.0.$$

Obviously, none of the solutions found by the program satisfied constraints of the problem; a typical output is given below:

$$x_{11} = 2.05 \quad x_{12} = 0.00 \quad x_{13} = 0.00 \quad x_{14} = 0.00$$
$$x_{21} = 0.00 \quad x_{22} = 10.65 \quad x_{23} = 0.00 \quad x_{24} = 0.00$$
$$x_{31} = 0.00 \quad x_{32} = 0.00 \quad x_{33} = 0.00 \quad x_{34} = 0.00.$$

As expected, the above nonfeasible solution is without any value for the user. It can be "improved" even further: a solution $x_{ij} = 0.0$ for all $1 \le i \le 3, 1 \le j \le 4$ yields the optimum transportation cost (zero)!

Clearly, it is necessary to incorporate some penalties on constraints. Since the evolution program EP_1 should not depend on the problem to be solved, we experimented only with some standard penalty functions. We have considered two sets of such penalty functions. The first one (p_i's, moderate penalties) measures each penalty as a linear function of the violation of the constraint, the other set (q_i's, high penalties) squares the violation of the constraint. For our problem P with seven linear equalities, these functions are given below:

---

[1]The system was run with the dynamic parameter encoding option (Schraudolph, & Belew, 1992); however, this option did not improve the performance of the system because the precision was not the issue here. The same comment applies to the evolution program EP_5 discussed later.

$$p_1 = c \cdot |x_{11} + x_{12} + x_{13} + x_{14} - 10|,$$
$$p_2 = c \cdot |x_{21} + x_{22} + x_{23} + x_{24} - 15|,$$
$$p_3 = c \cdot |x_{31} + x_{32} + x_{33} + x_{34} - 20|,$$
$$p_4 = c \cdot |x_{11} + x_{21} + x_{31} - 3|,$$
$$p_5 = c \cdot |x_{12} + x_{22} + x_{32} - 20|,$$
$$p_6 = c \cdot |x_{13} + x_{23} + x_{33} - 5|,$$
$$p_7 = c \cdot |x_{14} + x_{24} + x_{34} - 17|,$$

and $q_i = p_i^2/c$ $(i = 1, \ldots, 7)$. In all experiments we used $c = 10.0$; for this number the penalties constitute a significant percentage of the total cost (which, as indicated by the results of the GAMS system, is around 400). The results of the experiments were quite interesting.

The following point represents a *typical* output for experiments with penalties p_i:

$$
\begin{array}{llll}
x_{11} = 3.00 & x_{12} = 3.77 & x_{13} = 0.00 & x_{14} = 0.00 \\
x_{21} = 0.00 & x_{22} = 1.23 & x_{23} = 0.00 & x_{24} = 13.77 \\
x_{31} = 0.00 & x_{32} = 15.00 & x_{33} = 5.00 & x_{34} = 0.00.
\end{array}
$$

The above solution is just 'typical': we are unable to provide the *best* output due to the fact that it is relatively hard to evaluate the goodness of nonfeasible solutions. To get a feasible solution from a nonfeasible one, we have to make a few adjustments and the final transportation cost depends on these. For example, the above solution may be corrected into the following feasible solution:

$$
\begin{array}{llll}
x_{11} = 3.00 & x_{12} = 3.77 & x_{13} = 0.00 & x_{14} = 3.23 \\
x_{21} = 0.00 & x_{22} = 1.23 & x_{23} = 0.00 & x_{24} = 13.77 \\
x_{31} = 0.00 & x_{32} = 15.00 & x_{33} = 5.00 & x_{34} = 0.00,
\end{array}
$$

which yields the total transportation cost of 453.43. Of course, some other corrections yield better or worse transportations costs. (The above correction was done manually. It was based on a simple observation that the totals of the first row and the fourth column are smaller than the corresponding marginal sums by 3.23; hence we added 3.23 to x_{14}).

In the above example, a manual correction of the nonfeasible solution resulted in a respectable value 453.43. However, it is important to stress that it was possible only because of low dimensions of the problem. The process of finding a 'good' correction of a nonfeasible solution for a 20×20 transportation problem might be as difficult as solving the original problem. It seems that stronger penalties should be used to force the solution into a feasible region.

Indeed, the approach of stronger penalties provided solutions which were "almost" feasible. The following point represents the best output for experiments with penalties q_i:

$$
\begin{array}{llll}
x_{11} = 3.00 & x_{12} = 6.98 & x_{13} = 0.00 & x_{14} = 0.00 \\
x_{21} = 0.00 & x_{22} = 0.00 & x_{23} = 3.06 & x_{24} = 11.93 \\
x_{31} = 0.00 & x_{32} = 13.02 & x_{33} = 1.93 & x_{34} = 5.03.
\end{array}
$$

The above solution can be transformed easily (manual rounding) into a feasible solution:

$$
\begin{array}{llll}
x_{11} = 3.00 & x_{12} = 7.00 & x_{13} = 0.00 & x_{14} = 0.00 \\
x_{21} = 0.00 & x_{22} = 0.00 & x_{23} = 3.00 & x_{24} = 12.00 \\
x_{31} = 0.00 & x_{32} = 13.00 & x_{33} = 2.00 & x_{34} = 5.00,
\end{array}
$$

which yields 502.53 as the total transportation cost. This cost is worse than the cost of 453.43 we obtained from the moderate penalties approach, however, it should be stressed again that the process of finding a 'good' correction in the moderate penalties approach can be quite complex for high dimensional problems. We can think about this step as a process of solving a new transportation problem with modified marginal sums (which represent differences between actual and required totals), where variables, say, δ_{ij}, represent respective corrections to original variables x_{ij}. Thus, in general, stronger penalties provide better results. At the same time these results are still worse than the results obtained from the commercial software GAMS (system Q in Figure 14.3). Also, it should be pointed out that 'very strong' penalties do not improve the performance of the program. In extreme, if we assign zero fitness to individuals which violate a constraint, very often the system would settle for the first feasible solution found.

The final (and predictable) conclusion from experiments with EP_1 is that the use of penalty functions does not guarantee feasible solutions and that a 'good' repair may be expensive.

Evolution Program EP_2

As discussed in Chapter 8, evolution strategies assume a set of $q \geq 0$ inequalities,

$$g_1(x) \geq 0, \ldots, g_q(x) \geq 0,$$

as part of the optimization problem. If during some iteration an offspring does not satisfy all of these constraints, then the offspring is disqualified, i.e., it is not placed in a new population. If the rate of occurrence of such illegal offspring is high, the ESs adjust their control parameters, e.g., by decreasing the components of the vector σ.

We have used KORR 2.1, Hans-Paul Schwefel and Frank Hoffmeister's implementation of a $(\mu+\lambda)$–ES and (μ, λ)–ES, as our next evolution program, EP_2. Clearly, evolution strategies are applicable to parameter optimization problems, hence $dom(EP_2) \subseteq dom(EP_1)$ and consequently, $EP_1 \prec EP_2$.

As stated earlier, EP_2 handles only inequality constraints. Because of that the problem P was rewritten to eliminate the equalities. As a result the objective function has only six variables: y_1, y_2, y_3, y_4, y_5, and y_6, and the transportation problem P is given as:

$$
\begin{aligned}
\min\, & f(y_1)+f(y_2)+f(y_3)+f(10.0-y_1-y_2-y_3)+f(y_4)+f(y_5)+f(y_6)+ \\
& f(15.0 - y_4 - y_5 - y_6) + f(3.0 - y_1 - y_4) + f(20.0 - y_2 - y_5)+ \\
& f(5.0 - y_3 - y_6) + f(y_1 + y_2 + y_3 + y_4 + y_5 + y_6 - 8.0),
\end{aligned}
$$

where

$$y_1 = x_{11},\ y_2 = x_{12},\ y_3 = x_{13},\ y_4 = x_{21},\ y_5 = x_{22},\ y_6 = x_{23},$$

and the following eighteen constraints hold:

$g_1:$ $y_1 \geq 0$ (i.e., $x_{11} \geq 0$),
$g_2:$ $y_2 \geq 0$ (i.e., $x_{12} \geq 0$),
$g_3:$ $y_3 \geq 0$ (i.e., $x_{13} \geq 0$),
$g_4:$ $y_4 \geq 0$ (i.e., $x_{21} \geq 0$),
$g_5:$ $y_5 \geq 0$ (i.e., $x_{22} \geq 0$),
$g_6:$ $y_6 \geq 0$ (i.e., $x_{23} \geq 0$),
$g_7:$ $10.0 - y_1 - y_2 - y_3 \geq 0$ (i.e., $x_{14} \geq 0$),
$g_8:$ $15.0 - y_4 - y_5 - y_6 \geq 0$ (i.e., $x_{24} \geq 0$),
$g_9:$ $3.0 - y_1 - y_4 \geq 0$ (i.e., $x_{31} \geq 0$),
$g_{10}:$ $20.0 - y_2 - y_5 \geq 0$ (i.e., $x_{32} \geq 0$),
$g_{11}:$ $5.0 - y_3 - y_6 \geq 0$ (i.e., $x_{33} \geq 0$),
$g_{12}:$ $y_1 + y_2 + y_3 + y_4 + y_5 + y_6 - 8.0 \geq 0$ (i.e., $x_{34} \geq 0$),
$g_{13}:$ $3.0 - y_1 \geq 0$ (i.e., $x_{11} \leq 3$),
$g_{14}:$ $10.0 - y_2 \geq 0$ (i.e., $x_{12} \leq 10$),
$g_{15}:$ $5.0 - y_3 \geq 0$ (i.e., $x_{13} \leq 5$),
$g_{16}:$ $3.0 - y_4 \geq 0$ (i.e., $x_{21} \leq 3$),
$g_{17}:$ $15.0 - y_5 \geq 0$ (i.e., $x_{22} \leq 15$),
$g_{18}:$ $5.0 - y_6 \geq 0$ (i.e., $x_{23} \leq 5$).

The average value of the best transportation cost found (out of 20 independent runs) by EP_2 was 460.75, whereas the best solution found (which yields the total value of 420.74) was

$x_{11} = 3.00$	$x_{12} = 2.00$	$x_{13} = 5.00$	$x_{14} = 0.00$
$x_{21} = 0.00$	$x_{22} = 0.00$	$x_{23} = 0.00$	$x_{24} = 15.00$
$x_{31} = 0.00$	$x_{32} = 18.00$	$x_{33} = 0.00$	$x_{34} = 2.00.$

As expected, the results of EP_2 are better than results from the previous evolution program EP_1. An additional point for EP_2 is that there is no need for correcting the results to move them into the feasible region. On the other hand it seems that the performance of EP_2 depends on a starting point in the search space (which is given by the user). For that reason, it is quite hard to provide a complete analysis of the system.

Evolution Program EP_3

The third evolution program EP_3 described here is GENOCOP (Chapter 7). Since the GENOCOP (as our evolution program EP_3) can handle only linear constraints, it is clear that $dom(EP_3) \subseteq dom(EP_2)$ and consequently, $EP_2 \prec EP_3$.

The transportation problem P is a problem with $m = 12$ variables; each chromosome is coded as a vector of twelve floating point numbers $\langle y_1, \ldots, y_{12} \rangle$. Then, the problem P is

$$\min \sum_{i=1}^{12} f(y_i),$$

where

$$y_1 = x_{11}, \quad y_2 = x_{12}, \quad y_3 = x_{13}, \quad y_4 = x_{14},$$
$$y_5 = x_{21}, \quad y_6 = x_{22}, \quad y_7 = x_{23}, \quad y_8 = x_{24},$$
$$y_9 = x_{31}, \quad y_{10} = x_{32}, \quad y_{11} = x_{33}, \quad y_{12} = x_{34},$$

with six independent linear constraints:

$$y_1 + y_2 + y_3 + y_4 = 10$$
$$y_5 + y_6 + y_7 + y_8 = 15$$
$$y_9 + y_{10} + y_{11} + y_{12} = 20$$
$$y_1 + y_5 + y_9 = 3$$
$$y_2 + y_6 + y_{10} = 20$$
$$y_3 + y_7 + y_{11} = 5$$

(the seventh equation, $y_4 + y_8 + y_{12} = 10$, is unnecessary, as is linearly dependent on the given six equations); additional linear inequalities are

$$y_i \geq 0, \text{ for } i = 1, \dots, 12.$$

We performed 20 runs of GENOCOP. The values of the total transportation cost varied from 420.74 (the worst case) for the following solution (rounded to the second digit after the decimal point):

$$x_{11} = 3.00 \quad x_{12} = 4.38 \quad x_{13} = 2.62 \quad x_{14} = 0.00$$
$$x_{21} = 0.00 \quad x_{22} = 15.00 \quad x_{23} = 0.00 \quad x_{24} = 0.00$$
$$x_{31} = 0.00 \quad x_{32} = 0.62 \quad x_{33} = 2.38 \quad x_{34} = 17.00,$$

to the value of 356.98 (the best case) for a solution:

$$x_{11} = 3.00 \quad x_{12} = 7.00 \quad x_{13} = 0.00 \quad x_{14} = 0.00$$
$$x_{21} = 0.00 \quad x_{22} = 13.00 \quad x_{23} = 2.00 \quad x_{24} = 0.00$$
$$x_{31} = 0.00 \quad x_{32} = 0.00 \quad x_{33} = 3.00 \quad x_{34} = 17.00.$$

The average (out of 20 runs) transportation cost returned by the GENOCOP system was 405.45. Of course, all obtained solutions were feasible. Clearly, GENOCOP as a more problem-specific system performed much better than evolution strategies EP_2.

Evolution Program EP_4

The next evolution program EP_4 described here is GENETIC-2. As discussed in Chapter 9, the system was built to optimize any nonlinear transportation problem, so clearly $dom(EP_4) \subseteq dom(EP_3)$ and consequently, $EP_3 \prec EP_4$. In GENETIC-2 a matrix represents a potential solution; appropriate operators were defined for this representation.

We performed 20 runs of GENETIC-2. The values of the total transportation cost varied from 397.02 (the worst case) for a solution (rounded to the second digit after the decimal point):

$$\begin{aligned}
x_{11} = 3.00 \quad & x_{12} = 5.00 \quad & x_{13} = 0.00 \quad & x_{14} = 2.00 \\
x_{21} = 0.00 \quad & x_{22} = 15.00 \quad & x_{23} = 0.00 \quad & x_{24} = 0.00 \\
x_{31} = 0.00 \quad & x_{32} = 0.00 \quad & x_{33} = 5.00 \quad & x_{34} = 15.00,
\end{aligned}$$

to the value of 356.98 (the best case) for a solution:

$$\begin{aligned}
x_{11} = 3.00 \quad & x_{12} = 7.00 \quad & x_{13} = 0.00 \quad & x_{14} = 0.00 \\
x_{21} = 0.00 \quad & x_{22} = 13.00 \quad & x_{23} = 2.00 \quad & x_{24} = 0.00 \\
x_{31} = 0.00 \quad & x_{32} = 0.00 \quad & x_{33} = 3.00 \quad & x_{34} = 17.00
\end{aligned}$$

(the same solution found by GENOCOP). However, the average (again, out of 20 runs) transportation cost returned from the GENETIC-2 system was 391.65, much better than 405.45 of GENOCOP. Again, all solutions were feasible. Clearly, GENETIC-2 (EP_4) as a more problem-specific system performed better than GENOCOP (EP_3).

Evolution Program EP_5

The final evolution program EP_5 described here is based again on GENESIS 1.2ucsd system, the very same system we used for experiments described earlier. This time, however, we tried to "tune up" the set of penalty functions to focus the system just on problem P. Additionally, we eliminated all equations: the intuition being that it should be easier to maintain inequality than equality constraints.

So again, the problem P was rewritten as:

$$\begin{aligned}
\min \; f(y_1)+f(y_2)+f(y_3)+f(10.0-y_1-y_2-y_3)+f(y_4)+f(y_5)+f(y_6)+ \\
f(15.0 - y_4 - y_5 - y_6) + f(3.0 - y_1 - y_4) + f(20.0 - y_2 - y_5)+ \\
f(5.0 - y_3 - y_6) + f(y_1 + y_2 + y_3 + y_4 + y_5 + y_6 - 8.0),
\end{aligned}$$

where $y_1 = x_{11}, y_2 = x_{12}, y_3 = x_{13}, y_4 = x_{21}, y_5 = x_{22}, y_6 = x_{23}$, and

$$\begin{aligned}
0.0 \le y_1 \le 3.0, \\
0.0 \le y_2 \le 10.0, \\
0.0 \le y_3 \le 5.0, \\
0.0 \le y_4 \le 3.0, \\
0.0 \le y_5 \le 15.0, \\
0.0 \le y_6 \le 5.0.
\end{aligned}$$

The six penalty functions we tried to tune were:

$$p_1 = \begin{cases} w_1 \cdot (c_1 + y_1 + y_2 + y_3 - 10.0)^2 & \text{if } 10.0 - y_1 - y_2 - y_3 < 0.0 \\ 0.0 & \text{otherwise} \end{cases}$$

$$p_2 = \begin{cases} w_2 \cdot (c_2 + y_4 + y_5 + y_6 - 15.0)^2 & \text{if } 15.0 - y_4 - y_5 - y_6 < 0.0 \\ 0.0 & \text{otherwise} \end{cases}$$

$$p_3 = \begin{cases} w_3 \cdot (c_3 + y_1 + y_4 - 3.0)^2 & \text{if } 3.0 - y_1 - y_4 < 0.0 \\ 0.0 & \text{otherwise} \end{cases}$$

$$p_4 = \begin{cases} w_4 \cdot (c_4 + y_2 + y_5 - 20.0)^2 & \text{if } 20.0 - y_2 - y_5 < 0.0 \\ 0.0 & \text{otherwise} \end{cases}$$

$$p_5 = \begin{cases} w_5 \cdot (c_5 + y_3 + y_6 - 5.0)^2 & \text{if } 5.0 - y_3 - y_6 < 0.0 \\ 0.0 & \text{otherwise} \end{cases}$$

$$p_6 = \begin{cases} w_6 \cdot (c_6 + 8.0 - y_1 - y_2 - y_3 - y_4 - y_5 - y_6)^2 \\ \quad \text{if } y_1 + y_2 + y_3 + y_4 + y_5 + y_6 - 8.0 < 0.0 \\ 0.0 \quad \text{otherwise,} \end{cases}$$

where w_i's and c_i's are additional weights.

As usual, all penalties are added to the objective function. After many experiments (during which we increased and decreased the corresponding weights for constraints which were violated or satisfied, respectively), we arrived at the following set:

$$\begin{array}{ll} c_1 = 2.5, & w_1 = 2.0, \\ c_2 = 0.3, & w_2 = 1.3, \\ c_3 = 5.0, & w_3 = 2.5, \\ c_4 = 5.0, & w_4 = 2.0, \\ c_5 = 0.2, & w_5 = 1.3, \\ c_6 = 0.1, & w_6 = 2.0. \end{array}$$

We do not claim, of course, the the above set of weights represents the optimal configuration: the tuning was done just "by hand"; if some constraint was not satisfied, we gradually increased the corresponding weights. However, we can make the following two observations:

- the system EP_5 with the above weights performs quite well on the problem P, and

- if we change the problem P by adding another source or destination, or just by changing the problem specific weights c_{ij}, the evolution program EP_5 would not produce meaningful results.

It is clear then that $dom(EP_5) \subseteq dom(EP_4)$ and consequently, $EP_4 \prec EP_5$.

One of the runs of the EP_5 system gave the following solution:

$$\begin{array}{llll} x_{11} = 2.93 & x_{12} = 6.91 & x_{13} = 0.16 & x_{14} = 0.00 \\ x_{21} = 0.07 & x_{22} = 13.09 & x_{23} = 1.84 & x_{24} = 0.00 \\ x_{31} = 0.00 & x_{32} = 0.00 & x_{33} = 3.00 & x_{34} = 17.00. \end{array}$$

Note that all constraints are satisfied, and the value of the objective function is 391.2. The above solution can be manually corrected into the best solution found by GENOCOP and GENETIC-2:

$$\begin{array}{llll} x_{11} = 3.00 & x_{12} = 7.00 & x_{13} = 0.00 & x_{14} = 0.00 \\ x_{21} = 0.00 & x_{22} = 13.00 & x_{23} = 2.00 & x_{24} = 0.00 \\ x_{31} = 0.00 & x_{32} = 0.00 & x_{33} = 3.00 & x_{34} = 17.00. \end{array}$$

However, the system EP_5 found also a solution with a better value than 391.2, namely 378.25, for the following transportation plan:

$$x_{11} = 2.53 \quad x_{12} = 7.47 \quad x_{13} = 0.00 \quad x_{14} = 0.00$$
$$x_{21} = 0.47 \quad x_{22} = 12.53 \quad x_{23} = 2.00 \quad x_{24} = 0.00$$
$$x_{31} = 0.00 \quad x_{32} = 0.00 \quad x_{33} = 3.00 \quad x_{34} = 17.00,$$

which is much harder to correct (remember that the optimal solution need not consist of integers; for example, one of the solutions we obtained from GENETIC-2 was

$$x_{11} = 3.00 \quad x_{12} = 7.00 \quad x_{13} = 0.00 \quad x_{14} = 0.00$$
$$x_{21} = 0.00 \quad x_{22} = 12.25 \quad x_{23} = 2.75 \quad x_{24} = 0.00$$
$$x_{31} = 0.00 \quad x_{32} = 0.75 \quad x_{33} = 2.25 \quad x_{34} = 17.00,$$

with the total transportation cost equal to 380.86).

In general, it should be possible to construct a "perfect" evolution program which is tailored to the problem P. We can add further knowledge to such a system by incorporating the transportation costs c_{ij}, characteristics of six independent constraints, possibly with some additional heuristic to modify a feasible solution. Additional constraints can be added "to guide" the system in a desirable direction. However, it should be noted that the difficulty in constructing such a system grows with the dimensions of the problem, and its usefulness would be quite limited (to the problem P only).

The experimental results presented earlier confirmed the intuitive hypothesis that the problem-specific knowledge enhances the performance of the algorithm, narrowing its applicability.

As mentioned earlier, for a fair comparison of the evolution programs we set population size to 70 and the number of generations to 5,000 for all our experiments, and all runs were repeated 20 times. However, these evolution programs used different techniques in their initialization steps. The first evolution program EP_1 generates its population in a way that the individuals need not be feasible (since the constraints include equations, it would be very surprising if even one generated individual was feasible). The second evolution program EP_2 uses a single (feasible) individual as its starting point; twenty different initial feasible points were generated for these tests. The third program EP_3 makes some number (which is a parameter of the system) of attempts to find an initial feasible individual in the search space. If successful, the initial population would consist of *population_size* identical copies of the found individual. If unsuccessful, the system would prompt the user for a feasible initial point; the set of initial feasible points for these runs was the same one as used for EP_2. The fourth program EP_4 generates and maintains a population of feasible individuals, whereas EP_5 (like EP_1) generates an initial population of (possibly) nonfeasible individuals.

In comparing our evolution programs it is important to know about these differences in initialization techniques; however, the results of our experiments indicated that the influence of a particular initialization technique on the system performance was negligible. This is not surprising: for a highly constrained problem in general (and for the transportation problem in particular), a 'feasible' point in the search space does not mean a 'good' point. A heuristic initialization works only in cases where a user has a *good* heuristic to incorporate in the system (and even then it must be done carefully to avoid premature convergence!). There was no improvement in EP_1 or EP_5 when they were initialized by feasible individuals unless one feasible individual was really good. The 'clever' initialization presented in connection with the evolution program EP_4 generated a set of feasible points with average evaluation of 456. This initialization did not enhance the algorithm: it was simply necessary to start with a feasible population, since the operators of EP_4 just maintain the feasibility. Other programs (EP_2 and EP_3) used a collection of relatively poor feasible points with fitness values from the range $\langle 493, 610 \rangle$ (with the average of 562).

We conclude that the initialization process has not influenced the presented results.

After these introductory remarks, we are ready to address two practical issues connected with evolution programming: For a given problem, P,

(1) how weak (or strong) should an evolution program be?

(2) how should we proceed to construct an evolution program?

There are no easy answers for these questions. In the sequel, we provide some general comments and intuition developed on the basis of various experiments, mixed with a dose of wishful thinking.

The first question is on optimizing the selection of an evolution program to be constructed. For a given problem, P, how weak (or strong) should an evolution program be? In other words, for a given problem, P, should we construct EP_2, or rather EP_4? Our hypothesis suggests that incorporation of problem-specific knowledge gives better results in terms of precision. However, as indicated in the Introduction, the development of a stronger, high-performance system may take a long time if it involves extensive problem analysis to design specialized representation, operators, and performance enhancements. On the other hand, we may have already some standard packages, like Grefenstette's GENESIS, Whitley's GENITOR, Davis's OOGA, Schraudolph's GENESIS 1.2ucsd, or one of Schwefel's evolution strategy systems. And if we try to find an effective binary representation for a given problem, this may result in little or no software adaptation!

Well, sometimes yes, sometimes no. If one is solving a transportation problem with hard constraints (i.e., constraints which must be satisfied), there is very little chance that some standard package would produce any feasible solution, or, if we start with a population of feasible solutions and force the system to maintain them, we may get no progress whatsoever — in such cases the system does not perform better than a random search routine. On the other hand, for some other problems such standard packages may produce quite satisfactory results. In short, the responsibility for making a decision on question (1) lies with the user; the decision is a function of many factors, which include the demands of the precision of the required solution, time complexity of the algorithm, cost of developing a new system, feasibility of the solution found (i.e., importance of the problem constraints), frequency of using the developed system, and others.

Assume, then, that (for some reason) we have to (or wish to) build a new system to solve a nontrivial optimization problem. This might be the case when standard GA packages do not provide acceptable feasible solutions and there are no computational packages available appropriate for the problem. Then we have to make a choice: either we can try to construct an evolution program or we can approach the problem using some traditional (heuristic) methods. It is interesting to note that in a traditional approach it usually takes three steps to solve an optimization problem:

1. understand the problem,

2. solve the problem,

3. implement the algorithm found in the previous step.

In the traditional approach, a programmer should **solve** the problem — only then may a correct program be produced. However, very often an algorithmic solution of a problem is not possible, or at least is very hard. On the top of that, for some applications, it is not important to find the optimal solution — any solution with a reasonable margin of error (relative distance from the optimum value) will do. For example, in some transportation problem one may look only for a *good* transportation plan — finding the optimal value is not required. In our experiments some evolution systems which were developed proposed (relatively quickly) a solution with a value, say, 1109, where the optimum value was 1102. In such a case (error less than 1%), the approximate solution might be more desirable.

An evolution programming approach usually eliminates the second, most difficult step. Just after we understand the problem, we can move to the implementational issues. The major task of a programmer in constructing an evolution program is a selection of appropriate data structures as well as "genetic" operators to operate on them (the rest is left for the evolution process). This task need not be trivial, since apart from the variety of data structures which can be used for chromosomal representation, each data structure may provide a wide selection of genetic operators. This would involve an understanding of the nature of the problem by a programmer; however, there is no need to solve the problem first. To construct an evolution program, a programmer would follow five basic steps:

1. First, (s)he selects a genetic representation of solutions to the problem. As we have already observed, this requires some understanding of the nature of the problem. However, there is no need to solve the problem.

Selected representation of solutions to the problem should be "natural" and this decision is left to the programmer (note that in the current programming environments, programmers select the appropriate data structures on their own). It seems that this is *the most* important step, which influences the remaining components of an evolution program. The representation should be able to carry all important information about the solution; unfortunately, there are no ready guidelines for such selection. Note that the basic differences between many paradigms of evolutionary computation concentrate on this issue (binary strings, vectors of floating point numbers, finite state machines, computer programs, etc.).

2. The second task of a programmer is the creation of an initial population of solutions. This can be done in many ways (random, output from some heuristic algorithm, etc.). Some care should be taken in the cases when a set of problem-specific constraints must be satisfied; often a population of *feasible* solutions is accepted as a good starting point of an evolution program (see next chapter). However, in many cases we do not know any feasible solution to the problem; sometimes, to repair an infeasible solution is as hard as to find a feasible one (e.g., heavily constrained timetable problems).

3. Selection of an evaluation function (which rates solutions in terms of their fitness) should not pose a serious difficulty for many optimization problems. However, sometimes this task is far from trivial (see next chapter).

4. The 'genetic' operators should be designed carefully — the design should be based on the problem itself and its constraints. Here, it is essential to investigate the *meaning* of the information transmitted by the operators. If we search only a feasible part of the search space, the operators should transform a feasible solution(s) into another feasible solution. It is also possible to use repair algorithms, penalty functions, or other methods (see next chapter) to handle problem-specific constraints. For many real-world problems, it is more than helpful to incorporate local search heuristics into some of the operators.

5. Values for various parameters that the program uses may be provided by a programmer. However, in the more advanced versions of evolution programming environments, these may be controlled by a supervisor meta-process (like one discussed in [169]), whose only task is to tune all parameters. More and more research is directed into self-adaptation of parameters of evolution programs.

As seen in the above recipe for constructing an evolution program, the general idea behind it lies within "natural" data structures and problem-sensitive "genetic operators". However, it still might be difficult to build such a system (for a particular problem) from scratch. An experienced programmer would manage this task; however, the resulting program might be quite inefficient. To assist the user in this task, it might be worthwhile to create a new programming methodology supported by software (special programming languages) and hardware (parallel computers) — this is where our wishful thinking starts.

Currently, there are a number of different programming methodologies in computer science: structured programming, logic programming, object-oriented programming, functional programming. None of these fully support the construction of an evolution program. The goal of the new methodology would be the creation of appropriate tools for learning (here optimization is understood as a learning process) using a parallel processor architecture.

We hope to develop this idea to design a programming language PROBIOL (for PROgramming in BIOLogy) to support an EVA programming environment (EVA for EVolution progrAmming). An important issue in this methodology is an implementation of programs to control the evolution of the evolution process occurring in the "evolution engine" — the evolution engine is represented by a "society of microprocessors" [296]; some of these issues were briefly discussed in [204]. The key motivation of the new programming environment is to provide programming tools based on a parallel architecture. This is significant; as stated in [6]:

> "Parallelism is sure to change the way we think about and use computers. It promises to put within our reach solutions to problems and frontiers of knowledge never dreamed of before. The rich variety of architectures will lead to the discovery of novel and more efficient solutions to both old and new problems."

15. Evolution Programs and Heuristics

'The time has come,' the Walrus said,
'To talk of many things:
Of shoes—and ships—and sealing wax—
Of cabbages—and kings—
And why the sea is boiling hot—
And whether pigs have wings.'

Lewis Carroll, *Through the Looking-Glass*

As we already discussed in the previous chapters, the best known evolution programs include genetic algorithms, evolutionary programming, evolution strategies, and genetic programming. There are also many hybrid systems which incorporate various features of the above paradigms, and consequently are hard to classify; anyway, we refer to them just as evolution programs (or evolutionary algorithms, or evolutionary computation techniques).

As we mentioned a few times in this text, it is generally accepted that any evolutionary algorithm to solve a problem must have five basic components:

- a genetic representation of solutions to the problem,

- a way to create an initial population of solutions,

- an evaluation function (i.e., the environment), rating solutions in terms of their 'fitness',

- 'genetic' operators that alter the genetic composition of children during reproduction, and

- values for the parameters (population size, probabilities of applying genetic operators, etc.).

It is common knowledge that for a successful implementation of an evolutionary technique for a particular real-world problem, the basic components listed above require some additional heuristics. These heuristic rules apply to genetic representation of solutions, to 'genetic' operators that alter their composition, to values of various parameters, to methods for creating an initial population. It seems that one item only from the above list of five basic components of the evolutionary algorithm—the evaluation function—usually is taken "for granted" and does not require any heuristic modifications. Indeed, in many cases the process of selection of an evaluation function is straightforward (e.g., classical numerical and combinatorial optimization problems). Consequently, during the last two decades, many difficult functions have been examined; often they served as test-beds for different selection methods, various operators, different representations, and so forth. However, the process of selection of an evaluation function might be quite complex by itself, especially, when we deal with feasible and infeasible solutions to the problem; several heuristics usually are incorporated in this process. In this section we examine some of these heuristics and discuss their merits and drawbacks.

As indicated in the Introduction, all evolution programs have the same structure (Figure 0.1, Introduction) but there are also many differences between them (often hidden on a lower level of abstraction). They use different data structures for their chromosomal representations, consequently, the 'genetic' operators are different as well. They may or may not incorporate some other information (to control the search process) in their genes. There are also other differences; for example, the two lines of Figure 0.1:

> select $P(t)$ from $P(t-1)$
> alter $P(t)$

can appear in the reverse order: in evolution strategies first the population is altered and later a new population is formed by a selection process. Moreover, even within a particular technique, say, within genetic algorithms, there are many flavors and twists. For example, there are many methods for selecting individuals for survival and reproduction. As discussed in Chapter 4, these methods include (1) proportional selection, where the probability of selection is proportional to the individual's fitness, (2) ranking methods, where all individuals in a population are sorted from the best to the worst and probabilities of their selection are fixed for the whole evolution process,[1] and (3) tournament selection, where some number of individuals (usually two) compete for selection to the next generation: this competition (tournament) step is repeated population-size number of times. Within each of these categories there are further important details. Proportional selection may require the use of scaling windows or truncation methods, there are different ways for allocating probabilities in ranking methods (linear, nonlinear distributions), the size of a tournament plays a significant role in tournament selection methods. It is also important to decide on a generational policy. For example, it is possible to replace the whole population by a population of offspring, or it is possible to select the best individuals from two populations (population of parents and population of offspring)—this selection can be done in a deterministic or nondeterministic way. It is also possible to produce few (in particular, a single) offspring, which replace some

---

[1]For example, the probability of selection of the best individual is always 0.15 regardless its precise evaluation; the probability of selection of the second best individual is always 0.14, etc. The only requirements are that better individuals have larger probabilities and the total of these probabilities equals to one.

(the worst?) individuals (systems based on such a generational policy are called 'steady state'). Also, one can use an 'elitist' model which keeps the best individual from one generation to the next[2]; such a model is very helpful for solving many kinds of optimization problems. Recently, Ronald [333] experimented with an extension of a selection process by allowing fit individuals to choose their mates (so called selection-seduction approach). In this approach, individuals are still selected on the basis of their fitness, however, at the time of breeding an individual is allowed to select its mate on the basis of its own preferences (these preferences are formulated in terms of phenotypic characteristics and can constitute a part of the phenotype).

For a particular chromosomal representation there is a variety of different genetic operators. In this text we considered various types of mutation; in some of these types the probability of mutation depends on generation number and/or location of a bit. As indicated in Chapters 10 and 11, many 'mutation' operators incorporate some heuristic local-search algorithm to enhance the performance of the evolutionary algorithm. Also, apart from 1-point crossover, we have 2-point, 3-point, etc. crossovers, which exchange an appropriate number of segments between parent chromosomes, as well as 'uniform crossover', which exchanges single genes from both parents. When a chromosome is a permutation of integer numbers $1, \ldots, n$, there are also many ways to mutate such chromosome and crossover two chromosomes (e.g., PMX, OX, CX, ER, EER crossovers).[3] Recently, Bui and Moon [56] formally generalized linear-string crossovers to n-dimensional binary encodings.

The variety of structures, operators, selection methods, etc. indicate clearly that some versions of evolutionary algorithms perform better than other versions on particular problems; many comparisons of different sort have been reported in the literature (e.g., evolutionary strategies versus genetic algorithms, 1-point crossover versus 2-point crossover versus uniform crossover, etc.) As a result, in building a successful evolutionary algorithm for a particular problem (or class of problems) the user uses 'common knowledge': a set of heuristic rules which emerged during the last two decades as a summary of countless experiments with various systems and various problems. The next section describes briefly some heuristics for selecting appropriate components of an evolutionary algorithm, whereas section 15.3 provides a detailed discussion of heuristics used for evaluating an individual in a population.

15.1 Techniques and heuristics: a summary

The data structure used for a particular problem and a set of 'genetic' operators constitute the most essential components of any evolutionary algorithm. For

---

[2]This means that if the best individual from a current generation is lost due to selection or genetic operators, the system forces it into the next generation anyway.

[3]In most cases, crossover involves just two parents; however, this need not be so; see Chapter 4.

example, the original genetic algorithms devised to model *adaptation processes* mainly operated on binary strings and used a recombination operator with mutation as a background operator. Mutation flips a bit in a chromosome and crossover exchanges genetic material between two parents: if the parents are represented by five-bits strings, say $(0, 0, 0, 0, 0)$ and $(1, 1, 1, 1, 1)$, crossing the vectors after the second component would produce the offspring $(0, 0, 1, 1, 1)$ and $(1, 1, 0, 0, 0)$.

Evolution strategies were developed as a method to solve parameter optimization problems; consequently, a chromosome represents an individual as a pair of float-valued vectors, i.e., $v = (x, \sigma)$. Here, the first vector x represents a point in the search space; the second vector σ is a vector of standard deviations: mutations are realized by replacing v by (x', σ'), where

$$\sigma' = \sigma \cdot e^{N(0, \Delta\sigma)} \text{ and }$$
$$x' = x + N(0, \sigma'),$$

where $N(0, \sigma)$ is a vector of independent random Gaussian numbers with a mean of zero and standard deviations σ and $\Delta\sigma$ is a parameter of the method.

The original evolutionary programming techniques aimed at evolution of artificial intelligence and finite state machines were selected as a chromosomal representation of individuals. Offspring (new FSMs) are created by random mutations of parent population. There are five possible mutation operators: change of an output symbol, change of a state transition, addition of a state, deletion of a state, and change of the initial state (with some additional constraints on the minimum and maximum number of states).

Genetic programming techniques provide a way to run a search of the space of possible computer programs for the best one (the most fit).

Many researchers further modified evolutionary algorithms by 'adding' problem-specific knowledge to the algorithm. Several papers have discussed initialization techniques, different representations, decoding techniques (mapping from genetic representations to 'phenotypic' representations), and the use of heuristics for genetic operators. Such hybrid/nonstandard systems enjoy a significant popularity in the evolutionary computation community. Very often these systems, extended by the problem-specific knowledge, outperform other classical evolutionary methods as well as other standard techniques (as discussed in the previous chapter).

There are few heuristics to guide a user in selection of appropriate data structures and operators for a particular problem. It is common knowledge that for numerical optimization problem one should use an evolutionary strategy[4] or genetic algorithm with floating point representation, whereas some versions of genetic algorithms would be best to handle combinatorial optimization problems. Genetic programs are great in discovery of rules given as a computer program, and evolutionary programming techniques can be used successfully to model the behavior of a system (e.g., the prisoner's dilemma problem, see

---

[4]Evolutionary programming techniques have been generalized also to handle numerical optimization problems, see [117].

[120]). An additional popular heuristic in applying evolutionary algorithms to real-world problems is based on modifying the algorithm by the problem-specific knowledge; this problem-specific knowledge is incorporated in chromosomal data structures and specialized genetic operators For example, a system GENETIC-2 (Chapter 9) constructed for the nonlinear transportation problem used a matrix representation for its chromosomes, a problem-specific mutation (main operator, used with probability 0.4) and arithmetical crossover (background operator, used with probability 0.05). It is hard to classify this system: it is not really a genetic algorithm, since it can run with mutation operator only without any significant decrease of quality of results. Moreover, all matrix entries are floating point numbers. It is not an evolution strategy, since it did not encode any control parameters in its chromosomal structures. Clearly, it has nothing to do with genetic programming or evolutionary programming approaches. It is just an evolutionary technique aimed at a particular problem.

Another possibility is based on hybridization; this technique [78] incorporates existing algorithms to enhance the results of the evolutionary system. This can be done by using the output of other algorithms to seed the initial population of evolutionary system, by incorporating some local search operators into 'genetic' operators, or by 'borrowing' some encoding strategy.

Some of these ideas were embodied earlier in the evolutionary procedure called *scatter search* (Chapter 8). The process generates initial populations by screening good solutions produced by heuristics. The points used as parents are then joined by linear combinations with context-dependent weights, where such combinations may apply to multiple parents simultaneously. The linear combination operators are further modified by adaptive rounding processes to handle components required to take discrete values. (The vectors operated on may contain both real and integer components, as opposed to strictly binary components.) Finally, preferred outcomes are selected and again subjected to heuristics, whereupon the process repeats. The approach has been found useful for mixed integer and combinatorial optimization.

There are a few heuristics available for creating an initial population: one can start from a randomly created population, or use an output from some deterministic algorithm to initialize it (with many other possibilities in between these extremes). There are also some general heuristic rules for determining values for the various parameters; for many genetic algorithms applications, population size stays between 50 and 100, probability of crossover—between 0.65 and 1.00, and probability of mutation—between 0.001 and 0.01. Additional heuristic rules are often used to vary the population size or probabilities of operators during the evolution process.

It seems that neither of the evolutionary techniques is perfect (or even robust) across the problem spectrum; only the whole family of algorithms based on evolutionary computation concepts (i.e., evolutionary algorithms) have this property of robustness. But the main key to successful applications is in heuristic methods which are mixed skillfully with evolutionary techniques.

15.2 Feasible and infeasible solutions

In evolutionary computation methods the evaluation function serves as the only link between the problem and the algorithm. The evaluation function rates individuals in the population: better individuals have better chances for survival and reproduction. Hence it is essential to define an evaluation function which characterize the problem in a 'perfect way'. In particular, the issue of handling feasible and infeasible individuals should be addressed very carefully: very often a population contains infeasible individuals but we search for a *feasible* optimal. Finding a proper evaluation measure for feasible and infeasible individuals is of great importance; it directly influences the outcome (success or failure) of the algorithm.

The issue of processing infeasible individuals is very important for solving constrained optimization problems using evolutionary techniques. For example, in continuous domains, the general nonlinear programming problem[5] is to find x so as to

$$\text{optimize } f(x), \ x = (x_1, \ldots, x_n) \in R^n,$$

where $x \in \mathcal{F} \subseteq \mathcal{S}$. The set $\mathcal{S} \subseteq R^n$ defines the search space and the set $\mathcal{F} \subseteq \mathcal{S}$ defines a *feasible* search space. Usually, the search space $\mathcal{S}$ is defined as a n-dimensional rectangle in R^n (domains of variables defined by their lower and upper bounds):

$$l(i) \leq x_i \leq u(i), \quad 1 \leq i \leq n,$$

whereas the feasible set $\mathcal{F}$ is defined by an intersection of $\mathcal{S}$ and a set of additional $m \geq 0$ constraints:

$$g_j(x) \leq 0, \text{ for } j = 1, \ldots, q, \text{ and } h_j(x) = 0, \text{ for } j = q+1, \ldots, m.$$

Most research on applications of evolutionary computation techniques to nonlinear programming problems was concerned with complex objective functions with $\mathcal{F} = \mathcal{S}$. Several test functions used by various researchers during the last 20 years consider only domains of n variables; this was the case with five test functions F1–F5 proposed by De Jong [82], as well as with many other test cases proposed since then.

In discrete domains the problem of constraints was acknowledged much earlier. The knapsack problem, set covering problem, and all types of scheduling and timetabling problems are constrained. Several heuristic methods emerged to handle constraints; however, these methods have not been studied in a systematic way.

In general, a search space $\mathcal{S}$ consists of two disjoint subsets of feasible and infeasible subspaces, $\mathcal{F}$ and $\mathcal{U}$, respectively (see Figure 15.1). We do not make any assumptions about these subspaces; in particular, they need not be convex and they need not be connected (e.g., as is the case in the example in Figure 15.1

---

[5]We consider here only continuous variables.

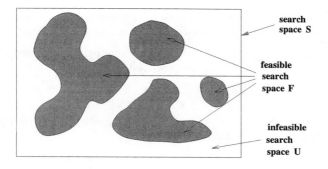

Fig. 15.1. A search space and its feasible and infeasible parts

where the feasible part $\mathcal{F}$ of the search space consists of four disjoined subsets). In solving optimization problems we search for a *feasible* optimum. During the search process we have to deal with various feasible and infeasible individuals; for example (see Figure 15.2), at some stage of the evolution process, a population may contain some feasible (b, c, d, e, i, j, k, p) and infeasible individuals (a, f, g, h, l, m, n, o), while the (global) optimum solution is marked by 'X'.

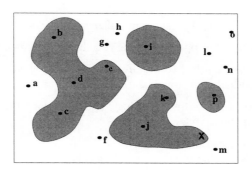

Fig. 15.2. A population of 16 individuals, a – o

The presence of feasible and infeasible individuals in the population influences other parts of the evolutionary algorithm; for example, should the elitist selection method consider a possibility of preserving the best *feasible* individual, or just the best individual overall? Further, some operators might be applicable to feasible individuals only. However, the major aspect of such a scenario is the need for evaluation of feasible and infeasible individuals. The problem of how to evaluate individuals in the population is far from trivial. In general, we have to design two evaluation functions, $eval_f$ and $eval_u$, for feasible and infeasible domains, respectively. There are many important questions to be addressed (we discuss them in detail in the next section):

A. How should two feasible individuals be compared, e.g., 'c' and 'j' from Figure 15.2? In other words, how should the function $eval_f$ be designed?

B. How should two infeasible individuals be compared, e.g., 'a' and 'n'? In other words, how should the function $eval_u$ be designed?

C. How are the functions $eval_f$ and $eval_u$ related to each other? Should we assume, for example, that $eval_f(s) \succ eval_u(r)$ for any $s \in \mathcal{F}$ and any $r \in \mathcal{U}$ (the symbol $\succ$ is interpreted as 'is better than', i.e., 'greater than' for maximization and 'smaller than' for minimization problems)?

D. Should we consider infeasible individuals harmful and eliminate them from the population?

E. Should we 'repair' infeasible solutions by moving them into the closest point of the feasible space (e.g., the repaired version of 'm' might be the optimum 'X', Figure 15.2)?

F. If we repair infeasible individuals, should we replace an infeasible individual by its repaired version in the population or rather should we use a repair procedure for evaluation purpose only?

G. Since our aim is to find a feasible optimum solution, should we choose to penalize infeasible individuals?

H. Should we start with initial population of feasible individuals and maintain the feasibility of offspring by using specialized operators?

I. Should we change the topology of the search space by using decoders?

J. Should we extract a set of constraints which define the feasible search space and process individuals and constraints separately?

K. Should we concentrate on searching for a boundary between feasible and infeasible parts of the search space?

L. How do we find a feasible solution?

Several trends for handling infeasible solutions have emerged in the area of evolutionary computation. We discussed some of them in Chapter 7 in the context of numerical optimization; here we discuss them using examples from discrete and continuous domains.

15.3 Heuristics for evaluating individuals

In this section we discuss several methods for handling feasible and infeasible solutions in a population; most of these methods emerged quite recently. Only a few years ago Richardson et al. [332] claimed: "Attempts to apply GA's with constrained optimization problems follow two different paradigms (1) modification of the genetic operators; and (2) penalizing strings which fail to satisfy all the constraints." This is no longer the case as a variety of heuristics have been proposed. Even the category of penalty functions consists of several methods which differ in many important details on how the penalty function is designed and applied to infeasible solutions. Other methods maintain the feasibility of the individuals in the population by means of specialized operators or decoders, impose a restriction that any feasible solution is 'better' than any infeasible solution, consider constraints one at a time in a particular linear order, repair infeasible solutions, use multiobjective optimization techniques, are based on cultural algorithms, or rate solutions using a particular co-evolutionary model. We discuss these techniques in turn by addressing questions A – L from the previous section.

A. Design of $eval_f$

This is usually the easiest issue: for most optimization problems, the evaluation function f for feasible solutions is given. This is the case for numerical optimization problems and for most operations research problems (knapsack problems, traveling salesman problems, set covering problems, etc.) However, for some problems the selection of an evaluation function might be far from trivial. For example, in building an evolutionary system to control a mobile robot (Chapter 11) there is a need to evaluate a robot's paths. It is unclear whether path #1 or path #2 (Figure 15.3) should have better evaluation (taking into account their total distance, clearance from obstacles, and smoothness): path #1 is shorter, but path #2 is smoother. For such problems there is a need for some heuristic measures to be incorporated into the evaluation function. Note that even the subtask of measuring the smoothness or clearance of a path is not simple.

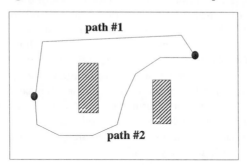

Fig. 15.3. Paths in an environment

This is also the case in many design problems, where there are no clear formulae for comparing two feasible designs. Clearly, some problem-dependent heuristics are necessary in such cases, which should provide a numerical measure $eval_f(x)$ of a feasible individual x.

One of the best examples to illustrate the problem of the need to evaluate feasible individuals is the satisfiability (SAT) problem. For a given conjunctive normal form formula, say

$$F(x) = (x_1 \vee \overline{x_2} \vee x_3) \wedge (\overline{x_1} \vee \overline{x_3}) \wedge (x_2 \vee x_3),$$

it is hard to compare two feasible individuals $p = (0,0,0)$ and $q = (1,0,0)$ (in both cases $F(p) = F(q) = 0$). De Jong and Spears [90] examined a few possibilities. For example, it is possible to define $eval_u$ to be a ratio of the number of conjuncts which evaluate to true; in that case

$$eval_f(p) = 0.666 \text{ and } eval_f(q) = 0.333.$$

It is also possible [305] to change the Boolean variables x_i into floating point numbers y_i and to assign:

$$eval_f(y) = |y_1 - 1||y_2 + 1||y_3 - 1| + |y_1 + 1||y_3 + 1| + |y_2 - 1||y_3 - 1|,$$

or

$$eval'_f(y) = (y_1 - 1)^2(y_2 + 1)^2(y_3 - 1)^2 + (y_1 + 1)^2(y_3 + 1)^2 + (y_2 - 1)^2(y_3 - 1)^2.$$

In the above cases the solution to the SAT problem corresponds to a set of global minimum points of the objective function: the *true* value of $F(x)$ is equivalent to the global minimum value 0 of $eval_u(y)$.

Let us also cite from [109], where the author rejected the idea of constructing a straightforward $eval_f$ for the bin packing problem (BBP):

> "Let's us define a suitable cost function for the BPP. The objective being to find the minimum number of bins required, the first cost function that comes to mind is simply the number of bins used to 'pack' all the objects. This is correct from a strictly mathematical point of view, but it is unusable in practice. Indeed, such a cost function leads to an extremely unfriendly landscape of the search space: a very small number of optimal points in the space are lost in an exponential number of points where this purported cost function is just one unit above the optimum. Worse, those slightly suboptimal points yield the same cost. The trouble is that such a cost function lacks any capacity of guiding an algorithm in the search, making the problem a 'needle in a haystack'.
>
> We thus settled for the following cost function for the BPP [...]: maximize
>
> $$f_{BPP} = \frac{\sum_{i=1}^{N}(F_i/C)^k}{N},$$
>
> with N being the number of bins actually used in the solution being evaluated, F_i the sum of sizes of the objects in (the fill of) the bin i, C is the bin capacity, k a constant, $k > 1$.
>
> The constant k expresses our concentration on the 'extremist' bins in comparison to the less filled ones. The larger k is, the more we prefer well-filled 'elite' groups as opposed to a collection of about

equally filled bins. In fact, the value of k gives us the possibility to vary the 'ruggedness' of the function to optimize, from the 'needle in a haystack' ($k = 1$, $f_{BPP} = 1/N$) up to the 'best-filled bin' ($k \to \infty$, $f_{BPP} \to \max_i[(F_i/C)^k]$)."

Clearly, the problem of selecting "perfect" $eval_f$ is far from trivial.

There is also another possibility: in some cases we do not need to define the evaluation function $eval_f$ at all! This function is necessary only if the evolutionary algorithm uses proportional selection (see Chapter 4). For other types of selection routines it is possible to establish only a linear ordering relation on individuals in the population. If a linear ordering relation ρ handles decisions of the type "is a feasible individual x better than a feasible individual y?",[6] then such a relation ρ is sufficient for tournament and ranking selections methods, which require either a selection of the best individual out of some number of individuals, or linear ordering of all individuals, respectively.

Of course, it might be necessary to use some heuristics to build such a linear ordering relation ρ. For example, for multi-objective optimization problems it is relatively easy to establish a partial ordering between individual solutions; additional heuristics might be necessary to order individuals which are not comparable by the partial relation.

In summary, it seems that tournament and ranking selections give some additional flexibility to the user: sometimes it is easier to compare two solutions than to provide their evaluation values as numbers. However, in these methods it is necessary to resolve additional problems of comparing two infeasible individuals (see part B) as well as comparing feasible and infeasible individuals (see part C).

B. Design of $eval_u$

This is a quite hard problem. We can avoid it altogether by rejecting infeasible individuals (see part D). Sometimes it is possible to extend the domain of function $eval_f$ to handle infeasible individuals, i.e., $eval_u(x) = eval_f(x) \pm Q(x)$, where $Q(x)$ represents either a penalty for infeasible individual x, or a cost for repairing such an individual (see part G). Another option is to design a separate evaluation function $eval_u$, independent of $eval_f$, but in a such case we have to establish some relationship between these two functions (see part C).

It is difficult to evaluate infeasible individuals. This is the case for the knapsack problem, where the amount of violation of capacity need not be a good measure of the individual's 'fitness' (see part G). This is also the case for many scheduling and timetable problems as well as the path planning problem: it is unclear whether path #1 or path #2 is better (Figure 15.4), since path #2 has more intersection points with obstacles and is longer than path #1; on the other hand most infeasible paths are "worse" using the above criteria than the straight line (path #1).

---

[6]The statement $\rho(x,y)$ is interpreted as x is better than y, for feasible x and y.

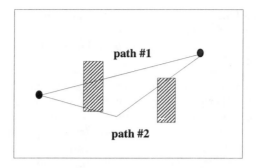

Fig. 15.4. Infeasible paths in an environment

As was the case with feasible solutions (part A), it is possible to develop an ordering relation for infeasible individuals (as opposed the construction of $eval_u$); in both cases it is necessary to establish a relationship between evaluations of feasible and infeasible individuals (part C).

C. Relationship between $eval_f$ and $eval_u$

Assume that we process both feasible and infeasible individuals in the population and that we evaluate them using two evaluation functions, $eval_f$ and $eval_u$, respectively. In other words, evaluations of a feasible individual x and infeasible individual y are $eval_f(x)$ and $eval_u(y)$, respectively. Now it is of great importance to establish a relationship between these two evaluation functions.

One possibility (as mentioned already in part B) is to design $eval_u$ by means of $eval_f$, i.e., $eval_u(y) = eval_f(y) \pm Q(y)$, where $Q(y)$ represents either a penalty for infeasible individual y, or a cost for repairing such an individual (we discuss this option in part G).

Another possibility is as follows. We can construct a global evaluation function $eval$ as

$$eval(p) = \begin{cases} q_1 \cdot eval_f(p) & \text{if } p \in \mathcal{F} \\ q_2 \cdot eval_u(p) & \text{if } p \in \mathcal{U} . \end{cases}$$

In other words, two weights, q_1 and q_2, are used to scale the relative importance of $eval_f$ and $eval_u$.

Both above methods allow infeasible individuals to be "better" than feasible individuals. In general, it is possible to have a feasible individual x and an infeasible one, y, such that $eval(y) \succ eval(x)$.[7] This may lead the algorithm to converge to an infeasible solution; it is why several researchers experimented with dynamic penalties Q (see part G) which increase pressure on infeasible

---

[7]The symbol $\succ$ is interpreted as 'is better than', i.e., 'greater than' for maximization and 'smaller than' for minimization problems.

individuals with respect to the current state of the search. An additional weakness of these methods lies in their problem dependence; often the problem of selecting $Q(x)$ (or weights q_1 and q_2) is almost as difficult as solving the original problem.

On the other hand, some researchers [312, 279] reported good results of their evolutionary algorithms, which worked under the assumption that any feasible individual was better than any infeasible one. Powell and Skolnick [312] applied this heuristic rule for numerical optimization problems (see Chapter 7): evaluations of feasible solutions were mapped into the interval $(-\infty, 1)$ and infeasible solutions into the interval $(1, \infty)$ (for minimization problems). Michalewicz and Xiao [279] experimented with the path planning problem (Chapter 11) and used two separate evaluation functions for feasible and infeasible individuals. The values of $eval_u$ were increased (i.e., made less attractive) by adding such a constant, so that the best infeasible individual was worse that the worst feasible one. However, it is not clear whether this should always be the case. In particular, it is doubtful whether the feasible individual 'b' (Figure 15.2) should have higher evaluation than infeasible individual 'm', which is "just next" to the optimal solution. A similar example can be drawn from the path planning problem: it is unclear whether a feasible path #2 (see Figure 15.5) deserves better evaluation than infeasible path #1!

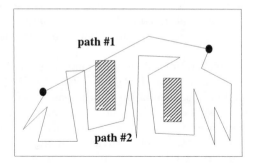

Fig. 15.5. Infeasible and feasible paths in an environment

The issue of establishing a relationship between evaluation functions for feasible and infeasible individuals is one of the most challenging problems to resolve while applying an evolutionary algorithm to a particular problem.

D. Rejection of infeasible individuals

This "death penalty" heuristic is a popular option in many evolutionary techniques (e.g., evolution strategies). Note that rejection of infeasible individuals offers a few simplifications of the algorithm: for example, there is no need to design $eval_u$ and to compare it with $eval_f$.

The method of eliminating infeasible solutions from a population may work reasonably well when the feasible search space is convex and it constitutes a reasonable part of the whole search space (e.g., evolution strategies do not allow equality constraints since with such constraints the ratio between the sizes of feasible and infeasible search spaces is zero). Otherwise such an approach has serious limitations. For example, for many search problems where the initial population consists of infeasible individuals only, it might be essential to improve them (as opposed to rejecting them). Moreover, quite often the system can reach the optimum solution more easily if it is possible to "cross" an infeasible region (especially in non-convex feasible search spaces).

E. Repair of infeasible individuals

Repair algorithms enjoy a particular popularity in the evolutionary computation community: for many combinatorial optimization problems (e.g., traveling salesman problem, knapsack problem, set covering problem, etc.) it is relatively easy to 'repair' an infeasible individual. Such a repaired version can be used either for evaluation only, i.e.,

$$eval_u(y) = eval_f(x),$$

where x is a repaired (i.e., feasible) version of y, or it can also replace (with some probability) the original individual in the population (see part F). Note, that the repaired version of solution 'm' (Figure 15.2) might be the optimum 'X'.

The process of repairing infeasible individuals is related to a combination of learning and evolution (so-called *Baldwin effect* [399]). Learning (as local search in general, and local search for the closest feasible solution, in particular) and evolution interact with each other: the fitness value of the improvement is transferred to the individual. In that way a local search is analogous to learning that occurs during one generation of a particular string.

The weakness of these methods is in their problem dependence. For each particular problem a specific repair algorithm should be designed. Moreover, there are no standard heuristics on design of such algorithms: usually it is possible to use a greedy repair, random repair, or any other heuristic which would guide the repair process. Also, for some problems the process of repairing infeasible individuals might be as complex as solving the original problem. This is the case for the nonlinear transportation problem, most scheduling and timetable problems, and many others. On the other hand, the GENOCOP III system (Chapter 7) for constrained numerical optimization (nonlinear constraints) is based on repair algorithms.

F. Replacement of individuals by their repaired versions

The question of replacing repaired individuals is related to so-called *Lamarckian evolution* [399], which assumes that an individual improves during its lifetime and that the resulting improvements are coded back into the chromosome. As stated in [399]:

> "Our analytical and empirical results indicate that Lamarckian strategies are often an extremely fast form of search. However, functions exist where both the simple genetic algorithm without learning and the Lamarckian strategy used [...] converge to local optima while the simple genetic algorithm exploiting the Baldwin effect converges to a global optimum."

This is why it is necessary to use the replacement strategy very carefully.

As discussed in section 4.5.2, Orvosh and Davis reported a so-called 5%-rule, which states that in many combinatorial optimization problems, an evolutionary computation technique with a repair algorithm provides the best results when 5% of repaired individuals replace their infeasible originals. In continuous domains, a new replacement rule is emerging. The GENOCOP III system (Chapter 7) for numerical optimization problems with nonlinear constraints is based on a repair approach. The first experiments (based on 10 test cases which have various numbers of variables, constraints, types of constraints, numbers of active constraints at the optimum, etc.) indicate that the 15% replacement rule is a clear winner: the results of the system are much better than with either lower or higher values of the replacement rate.

At present, it seems that the 'optimal' probability of replacement is problem-dependent and it may change over the evolution process as well. Further research is required for comparing different heuristics for setting this parameter, which is of great importance for all repair-based methods.

G. Penalizing infeasible individuals

This is the most common approach in the genetic algorithms community. The domain of function $eval_f$ is extended; the approach assumes that

$$eval_u(p) = eval_f(p) \pm Q(p),$$

where $Q(p)$ represents either a penalty for infeasible individual p, or a cost for repairing such an individual. The major question is, how should such a penalty function $Q(p)$ be designed? The intuition is simple: the penalty should be kept as low as possible, just above the limit below which infeasible solutions are optimal (so-called *minimal penalty rule* [239]. However, it is difficult to implement this rule effectively.

The relationship between infeasible individual 'p' and the feasible part $\mathcal{F}$ of the search space $\mathcal{S}$ plays a significant role in penalizing such individuals: an individual might be penalized just for being infeasible, the 'amount' of its infeasibility is measured to determine the penalty value, or the effort of 'repairing' the individual might be taken into account. For example, for the knapsack problem with capacity 99 we may have two infeasible solutions yielding the same profit, where the total weight of all items taken is 100 and 105, respectively. However, it is difficult to argue that the first individual with the total weight 100 is 'better' than the other one with the total weight 105, despite the fact that for this individual the violation of the capacity constraint is much smaller than for the other one. The reason is that the first solution may involve 5 items of the weight 20 each, and the second solution may contain (among other items) an item of a low profit and weight 6—removal of this item would yield a feasible solution, possibly much better than any repaired version of the first individual. However, in such cases a penalty function should consider the "easiness of repairing" an individual as well as the quality of its repaired version; designing such penalty functions is problem-dependent and, in general, quite hard.

In Chapter 7 we discussed many methods based on penalty functions; these include static methods [195], dynamic methods [210, 267], and adaptive methods [360, 27]. Also, in Chapter 7, we discussed segregated genetic algorithms [239], where low and high penalties are applied to two populations, which are run "in parallel".

It seems that the appropriate choice of the penalty method may depend on (1) the ratio between sizes of the feasible and the whole search space, (2) the topological properties of the feasible search space, (3) the type of the objective function, (4) the number of variables, (5) number of constraints, (6) types of constraints, and (7) number of active constraints at the optimum. Thus the use of penalty functions is not trivial and only some partial analysis of their properties is available. Also, a promising direction for applying penalty functions is the use of self-adaptive penalties: penalty factors can be incorporated in the chromosome structures in a similar way as some control parameters are represented in the structures of evolution strategies and evolutionary programming.

H. Maintaining feasible population by special representations and genetic operators

It seems that one of the most reasonable heuristics for dealing with the issue of feasibility is to use specialized representations and operators to maintain the feasibility of individuals in the population. This was the original idea behind this text, which was reflected in its title.

During the last decade several specialized systems were developed for particular optimization problems; these systems use a unique chromosomal representations and specialized 'genetic' operators which alter their composition. Some such systems were described in [78]; many other examples are described in this text. For example, GENOCOP system (Chapter 7) assumes linear constraints only and a feasible starting point (or feasible initial population). A closed set of operators maintains feasibility of solutions. For example, when a particular component x_i of a solution vector x is mutated, the system determines its current domain $dom(x_i)$ (which is a function of linear constraints and remaining values of the solution vector x) and the new value of x_i is taken from this domain (either with flat probability distribution for uniform mutation, or other probability distributions for non-uniform and boundary mutations). In any case the offspring solution vector is always feasible. Similarly, arithmetic crossover

$$ax + (1 - a)\overline{Y}$$

of two feasible solution vectors x and y yields always a feasible solution (for $0 \le a \le 1$) in convex search spaces (the system assumes linear constraints only which imply convexity of the feasible search space $\mathcal{F}$). Consequently, there is no need to define the function $eval_u$; the function $eval_f$ is (as usual) the objective function f.

Very often such systems are much more reliable than any other evolutionary techniques based on a penalty approach (e.g., Chapter 14), consequently, this is a quite popular trend. Many practitioners use problem-specific representations and specialized operators in building very successful evolutionary algorithms in many areas; these include numerical optimization, machine learning, optimal control, cognitive modeling, classic operations research problems (traveling salesman problem, knapsack problems, transportation problems, assignment problems, bin packing, scheduling, partitioning, etc.), engineering design, system integration, iterated games, robotics, signal processing, and many others.

Also, it is interesting to note that original evolutionary programming techniques [126] and genetic programming techniques [231] fall into this category of evolutionary algorithms: these techniques maintain feasibility of finite state machines or hierarchically structured programs by means of specialized representations and operators.

I. Use of decoders

Decoders offer an interesting option for all practitioners of evolutionary techniques. In these techniques a chromosome "gives instructions" on how to build a feasible solution. For example, a sequence of items for the knapsack problem can be interpreted as: "take an item if possible"—such interpretation would lead always to feasible solutions. Let us consider the following scenario: we try to solve the 0–1 knapsack problem with n items; the profit and weight of the i-th item are p_i and w_i, respectively. We can sort all items in decreasing order of p_i/w_i's and interpret the binary string

$$(1100110001001110101001010111010101...0010)$$

in the following way: take the first item from the list (i.e., item with the largest ratio profit per weight) if the item fits in the knapsack. Continue with second, fifth, sixth, tenth, etc. items from the sorted list, until the knapsack is full or there are no more items available. Note that the sequence of all 1's corresponds to a greedy solution. Any sequence of bits would translate into a feasible solution, every feasible solution may have many possible codes. We can apply classical binary operators (crossover and mutation): any offspring is clearly feasible.

However, it is important to point out that several factors should be taken into account while using decoders. Each decoder imposes a relationship T between a feasible solution and decoded solution (see Figure 15.6).

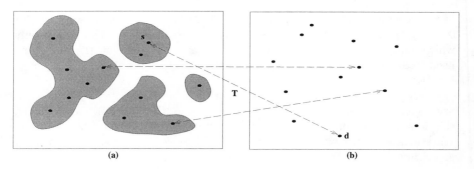

(a)	(b)

Fig. 15.6. Transformation T between solutions in original (a) and decoder's (b) space

It is important that several conditions are satisfied: (1) for each solution $s \in \mathcal{F}$ there is a decoded solution d, (2) each decoded solution d corresponds to a feasible solution s, and (3) all solutions in $\mathcal{F}$ should be represented by the same number of decodings d. Additionally, it is reasonable to request that (4) the transformation T is computationally fast and (5) it has locality feature in the sense that small changes in the decoded solution result in small changes in the solution itself. An interesting study on coding trees in genetic algorithms was reported by Palmer and Kershenbaum [304], where the above conditions were formulated.

J. Separation of individuals and constraints

This is a general and interesting heuristic. The first possibility would include utilization of multi-objective optimization methods, where the objective function f and constraint violation measures f_j (for m constraints) constitute a $(m + 1)$-dimensional vector $\boldsymbol{v}$:

$$\boldsymbol{v} = (f, f_1, \ldots, f_m).$$

Using some multi-objective optimization method, we can attempt to minimize its components: an ideal solution x would have $f_j(x) = 0$ for $1 \leq i \leq m$ and $f(x) \leq f(y)$ for all feasible y (minimization problems). As indicated in Chapter 7, a successful implementation of this approach was presented recently in [376].

Another heuristic is based on the idea of handling constraints in a particular order; Schoenauer and Xanthakis [346] called this method a "behavioral memory" approach. The initial steps of the method are devoted to sampling the feasible region; only in the final step is the objective function f optimized (see Chapter 7).

In Chapter 7 we also discussed briefly the possibility of incorporating the knowledge of the constraints into the belief space of cultural algorithms [325]; such algorithms provide a possibility of conducting an efficient search of the feasible search space [329].

K. Exploring boundaries between feasible and infeasible parts of the search space

One of the most recently developed approaches for constrained optimization is *strategic oscillation*. Strategic oscillation was originally proposed in conjunction with the evolutionary strategy of scatter search, and more recently has been applied to a variety of problem settings in combinatorial and nonlinear optimization (see, for example, the review of Glover [147]). The approach is based on identifying a *critical level*, which for our purposes represents a boundary between feasibility and infeasibility, but which also can include such elements as a stage of construction or a chosen interval of values for a functional. In the feasibility/infeasibility context, the basic strategy is to approach and cross the feasibility boundary, by a design that is implemented either by adaptive penalties and inducements (which are progressively relaxed or tightened according to whether the current direction of search is to move deeper into a particular region or to move back toward the boundary) or by simply employing modified gradients or subgradients to progress in the desired direction. Within the context of neighborhood search, the rules for selecting moves are typically amended to take account of the region traversed and the direction of traversal. During the process of repeatedly approaching and crossing the feasibility frontier from different directions, the possibility of retracing a prior trajectory is avoided by mechanisms of memory and probability.

The application of different rules (according to region and direction) is generally accompanied by crossing a boundary to different depths on different sides. An option is to approach and retreat from the boundary while remaining on a single side, without crossing. One-sided oscillations are especially relevant in a variety of scheduling and graph theory settings, where a useful structure can be maintained up to a certain point and then is lost (as by running out of jobs to assign or by going beyond the conditions that define a tree or tour, etc.). In these cases, a constructive process for building to the critical level is accompanied by a destructive process for dismantling the structure.

It is frequently important in strategic oscillation to spend additional search time in regions close to the boundary. This may be done by inducing a sequence of tight oscillations about the boundary as a prelude to each larger oscillation to a greater depth. If greater effort is allowed for executing each move, the method may use more elaborate moves (such as various forms of "exchanges") to stay at the boundary for longer periods. For example, such moves can be used to proceed to a local optimum each time a critical proximity to the boundary is reached. A strategy of applying such moves at additional levels is suggested by a *proximate optimality principle*, which states roughly that good constructions at one level are likely to be close to good constructions at another.

One of the useful forms of strategic oscillation operates by increasing and decreasing bounds for a function $g(x)$. Such an approach has been effective in a number of applications where $g(x)$ has represented such items as workforce assignments and function values (as well as feasibility/infeasibility levels), to guide the search to probe at various depths within the associated regions. In reference to degrees of feasibility and infeasibility, $g(x)$ may represent a vector-valued function associated with a set of problem constraints (which may be expressed, for example, as $g(x) \leq b$). In this instance, controlling the search by bounding $g(x)$ can be viewed as manipulating a parameterization of the selected constraint set. An often-used alternative is to make $g(x)$ a lagrangean or surrogate constraint penalty function, avoiding vector-valued functions and allowing tradeoffs between degrees of violation of different component constraints according to their importance. Surrogate constraint approaches are particularly useful for isolating such tradeoffs, accompanied by special memory to keep track of behavior that discloses the relative influence of constraints. Approaches that embody such ideas may be found, for example, in [133, 136, 221, 393, 410, 148].

L. Finding feasible solutions

There are problems for which any feasible solution would be of value. In such problems we are not really concern with the optimization issues (finding *the best* feasible solution) but rather we try to find *any* point in the feasible search space $\mathcal{F}$. Such problems are called *constraint satisfaction problems*. A classical example of a problem in this category is the well-known N-queens problem, where the task is to position N queens on a chess board with N rows and N columns in such a way that no two queens attack each other.

A few researchers have experimented with evolutionary systems to approach constraint satisfaction problems. Bowen and Dozier [48] designed a hybrid genetic algorithm with a complex chromosomal structure. Each chromosome, apart from values v_i of all variables, stores also (1) a so-called pivot (which tells the heuristic-based mutation which gene to mutate), (2) an individual's "family number", (3) a vector (for each gene there is one number) which represents numbers of violated constraints when it is assigned a value v_i, and (4) some heuristic value which is used to determine which gene should be the pivot. The system was successfully tested on many randomly generated constraint satisfaction problems [48] and compared with other constraint programming techniques.

Eiben et al. [99] approached the problem by designing specialized operators; the authors investigated several heuristic operators: mutations and crossovers (including multi-parent crossovers). Mutation operators select a number of positions in the parent chromosome and select new values for these positions, where the number of modified values, the criteria for identifying the position of the values to be modified, and the criteria for defining the new values are parameters of the operator. Multi-parent crossovers are based on scanning, which examine the positions of the parents consecutively and choose one of the values on the marked positions for the child. A system with such heuristic operators was implemented and tested on the N-queens problem and the graph coloring problem.

Paredis experimented with two different approaches to the constraint satisfaction problem. The first approach [306, 307] was based on clever representation of individuals; each gene was allowed to take, on top of values from the domain, an additional value '?', which represented choices left open. The initial population consisted of strings of all '?'s; a selection-assignment-propagation cycle replaced some '?' by values from the appropriate domain (this was done by selecting a variable whose domain had more than one element available), selecting a value from the domain and assigning it to the variable, and finally performing a propagation step, which ensured that the domains were consistent with the assignment made so far. The fitness of such partially-defined individuals was defined as the value of the objective function of the best complete solution found (when we start the search from a given partial individual). Operators were extended to incorporate a repair process (constraint checking step). The system was implemented and run on several N-queens problems [307] and some scheduling problems [306].

In the second approach [308] Paredis investigated a co-evolutionary model, where a population of potential solutions co-evolves with a population of constraints: fitter solutions satisfy more constraints, whereas fitter constraints are violated by more solutions. This means that individuals from the population of solutions are considered from the whole search space, and that there is no distinction between feasible and infeasible individuals. The evaluation of an individual is determined on the basis of constraint-violation measures f_j's; however, better f_j's (e.g., active constraints) would contribute more towards the evaluation of a solution. The approach was tested on the N-queens problem and compared with other single-population approaches [309, 308].

16. Conclusions

You ain't heard nothin' yet, folks.

Al Jolson, *The Jazz Singer*

The field of evolutionary computation has been growing rapidly over the last few years. Yet, there are still many gaps to be filled, many experiments to be done, many questions to be answered. In the final chapter of this text we examine a few important directions in which we can expect a lot of activity and significant results; we discuss them in turn.

Theoretical foundations

As indicated in this text, some evolution programs enjoy some theoretical foundations: for evolution strategies applied to regular problems (Chapter 8) a convergence property can be shown. Genetic algorithms, on the other hand, have a Schema Theorem (Chapter 3) which explains why they work. However, many techniques are modified when applied to particular real world problems. For example, to adapt a GA to the task of function optimization it was necessary to extend them by additional features (e.g., dynamic scaled fitness, rank-proportional selection, inclusion of elitist strategy, adaptation of various parameters of the search, various representations, new operators, etc.). Evolution strategies, applied to constrained numerical optimization problems, usually incorporate some heuristic method for constraint handling. Most of these modifications pushed simple algorithms away from their theoretical bases, but usually they enhanced the performance of the systems in a significant way. In the context of genetic algorithms, these modifications [89]:

> "... had pushed the application of simple GAs well beyond our initial theories and understanding, creating a need to revisit and extend them."

As indicated in the Preface to this edition, this might be one of the most challenging tasks for researchers in the field of evolutionary computation.

It is also important to continue research on factors affecting the ability of evolutionary systems to solve various (usually optimization) problems. What makes a problem hard or easy for an evolutionary method? This is a fundamental issue of evolutionary computation; some results related to deceptive problems, rugged fitness landscapes, epistasis, or royal road functions, are steps towards approximating an answer for this challenging question.

Function optimization

For many years, most evolutionary techniques were evaluated and compared with each other in the domain of function optimization. It seems also that this domain of function optimization would remain the primary test-bed for many new comparisons and new features of various algorithms. It is expected that new theories of evolutionary techniques for function optimization would emerge (e.g., breeder genetic algorithms [289]). Additionally, we should see progress in

- development of constraint-handling techniques. This is a very important area in general, and for function optimization in particular; most real problems of function optimization involve constraints. However, so far relatively few techniques have been proposed, analyzed, and compared with each other.

- development of systems for large-scale problems. Until now, most experiments have assumed a relatively small number of variables. It would be interesting to analyse how evolutionary techniques scale up with the problem size for problems with thousand variables.

- development of systems for mathematical programming problems. Very little work has been done in this area. There is a need to investigate evolutionary systems to handle integer/Boolean variables, and to experiment with mixed programming as well as integer programming problems.

Representation and operators

Traditionally, GAs work with binary strings, ESs with floating point vectors, and EPs with finite state machines (represented as matrices), whereas GP techniques use trees as a structure for the individuals. However, there is a need for systematic research on

- representation of complex, nonlinear objects of varying size, and, in particular, representation of 'blueprints' of complex objects, and

- development of evolutionary operators for such objects at the genotype level.

This direction can be perceived as a step towards building complex hybrid evolutionary systems which incorporate additional search techniques. For example, it seems worthwhile to experiment with Lamarckian operators, which would improve an individual during its lifetime—consequently, the improved, "learned" characteristics of such an individual would be passed to the next generation.

Additional issues to be resolved are connected with understanding the influence of various factors (representation and operators being the major two components) which affect the performance of evolutionary methods. In particular, we should come closer to answering the basic questions:

- which problems are easy for evolutionary techniques?

- which problems are hard?

- why?

- how to select the best representation and operators for a particular problem?

Research on deceptiveness [152], royal road functions [192, 212], properties of operators [315, 316, 317], or fitness distance correlation [214] are steps in this direction.

Non-random mating

Most current techniques which incorporate crossover operator use random mating, i.e, mating where individuals are paired randomly. It seems that with the trend of movement from simple to complex systems, the issue of non-random mating would be of growing importance. There are many possibilities to explore; these include introduction of sex or "family" relationships between individuals, or establishing some preferences (e.g., seduction [333]). Some simple schemes were already investigated by several researchers (e.g., Eshelman's incest prevention technique [105]), however, the ultimate goal seems to be to evolve rules for non-random mating. A few possibilities (see section 8.3.1) in the context of multimodal optimization were already explored; these include *sharing functions*, which permit the formation of stable subpopulations, and *tagging*, where individuals are assigned labels. However, very little has been done in this direction for complex chromosomal structures.

Self-adapting systems

Since evolutionary algorithms implement the idea of evolution, it is more than natural to expect some self-adapting characteristics of these techniques. Apart from evolutionary strategies, which incorporate some of their control parameters in the solution vectors, most other techniques use fixed representations, operators, and control parameters. One of the most promising research areas is based on inclusion of self-adapting mechanisms within the system for

- representation of individuals (as proposed by Shaefer [354]; the Dynamic Parameter Encoding technique [347] and messy genetic algorithms [155] also fall into this category).

- operators. It is clear that different operators play different roles at different stages of the evolutionary process. The operators should adopt (e.g., adaptive crossover [341, 367]). This is true especially for time-varying fitness landscapes.

- control parameters. There were already experiments aimed at these issues: adaptive population sizes [12] or adaptive probabilities of operators [77, 216, 368]. However, much more remains to be done.

It seems that this is one of the most promising directions of research; after all, the power of evolutionary algorithms lies in their adaptiveness.

Co-evolutionary systems

There is a growing interest in co-evolutionary systems, where more than one evolution process takes place: usually there are different populations there (e.g., additional populations of parasites or predators) which interact with each other. In such systems the evaluation function for one population may depend on the state of the evolution processes in the other population(s). This is an important topic for modeling artificial life, some business applications, etc.

Co-evolutionary systems might be important for approaching large-scale problems [311], where a (large) problem is decomposed into smaller subproblems; there is a separate evolutionary process for each of the subproblems, however, these evolutionary processes are connected with each other. Usually, evaluation of individuals in one population depends also on developments in other populations.

We have seen also (Chapter 7) a co-evolutionary algorithm (GENOCOP III) where two populations (of not necessarily feasible search points and fully feasible reference points) co-exist with each other. In this system, evaluation of search points depends on the current population of reference points. In the same chapter we presented also an approach proposed by Le Riche et al. [239], where two populations of individuals co-operate with each other and approach a feasible, optimum solution from two directions (from the feasible and infeasible parts of the search space). Also, Paredis [307] experimented with co-evolutionary systems in the context of constraint satisfaction problems (see sections 7.4 and 15.3 L).

Recently, a co-evolutionary system was used [295] to model the strategies of two competing companies (bus and rail companies) competing for passengers on the same routes. Clearly, the profits of one company depend on the current strategy (capacities and prices) of the other company; the study investigated the interrelationship between various strategies over time.

Diploid/polyploid versus haploid structures

Diploidy (or polyploidy) can be viewed as a way to incorporate memory into the individual's structure. Instead of a single chromosome (haploid structure) representing precise information about an individual, a diploid structure is made up of a pair of chromosomes: the choice between two values is made by some dominance function. The diploid (polyploid) structures are of particular significance in non-stationary environments (i.e., for time-varying objective functions) and for modeling complex systems (possibly using co-evolution models). However, there is no theory to support the incorporation of a dominance function into the system; there is also very little experimental data in this area.

Parallel models

Parallelism promises to put within our reach solutions to problems untractable before; clearly, it is one of the most important areas of computer science. Evolutionary algorithms are very suitable for parallel implementations; as Goldberg [154] observed:

> "In a world where serial algorithms are usually made parallel through countless tricks and contortions, it is no small irony that genetic algorithms (highly parallel algorithms) are made serial through equally unnatural tricks and turns."

However, there is no standard methodology for incorporating parallel ideas into GAs: existing parallel implementations can be classified into one of the following categories:

- *massively parallel GAs.* Such algorithms use a large number of processors (usually 2^{10} or more). Often a single processor is assigned to an individual in the population. In this model there are many possibilities for the selection method and mating (combining strings for crossover). Some experimental work in this area is reported by Mühlenbein [286].

- *parallel island models.* Such algorithms assume that several subpopulations evolve in parallel. The models include a concept of migration (movement of an individual string from one subpopulation to another) and crossovers between individuals from different subpopulations. There are many reported experiments in this parallel model; the reader is referred to Whitley's work [396] for full discussion.

- *parallel hybrid GAs.* This model is similar to the first (massively parallel GAs) in that there is one-to-one correspondence between processors and individuals. However, only a small number of processors is used. Additionally, such algorithms incorporate other (heuristic) algorithms (e.g., hillclimbing) to improve the performance of the system. Usually the experimental results are reported [286] [164], but an analysis of such systems is far from trivial.

Parallel models can also provide a natural embedding for other paradigms of evolutionary computation, like non-random mating, some aspects of self-adaptation, or co-evolutionary systems.

Other developments

There are also many other interesting developments in the field of evolutionary computation; one of them is an emergence of so-called Cultural Algorithms (see also section 7.4). As stated in [330]:

> "Cultural algorithms support two models of inheritance, one at the microevolutionary level in terms of traits, and the other at the macroevolutionary level in terms of beliefs. The two models interact via a communications channel that enables the behavior of individuals to alter the belief structure and allows the belief structure to constrain the ways in which individuals can behave. Thus, cultural

algorithms can be described in terms of three basic components; the belief structure, the population structure, and the communication channel."

The basic structure of a cultural algorithm is shown in Figure 16.1.

procedure cultural algorithm
begin
    $t \leftarrow 0$
    initialize population $P(t)$
    initialize belief network $B(t)$
    initialize communication channel
    evaluate $P(t)$
    while (not termination-condition) **do**
    begin
        communicate $(P(t), B(t))$
        adjust $B(t)$
        communicate $(B(t), P(t))$
        modulate fitness $P(t)$
        $t \leftarrow t + 1$
        select $P(t)$ from $P(t-1)$
        evolve $P(t)$
        evaluate $P(t)$
    end
end

Fig. 16.1. The structure of a cultural algorithm

Additionally, cultural algorithms seem to be appropriate for controlling the "evolution of evolution". The term "evolution of evolution" means that some elements in the evolution space (such as a man in sociological evolution, or a gene in biological evolution, or some units in the evolution of the matter) improve their learning abilities through their experience in the evolution process. This idea is a generalization of the idea of the analogy between the genetic evolution of biological species and the cultural evolution of human societies, for the abstract adaptive evolution process. These analogies have been brilliantly explored by Richard Dawkins in his book [81]. One of the arguments which supports the above ideas might be observed in Figure 16.2. The first column symbolizes the time from the beginning of the universe until the present. Every other column is an enlargement of the last 20 percent of the previous column. The star points represent the key events in the evolution process which leads to our current civilization. Note that the curve given by the star points grows much faster than exponential functions, say 5^x, since if the star points were on the same level, the growth would be precisely 5^x.

Let us quote from [371]:

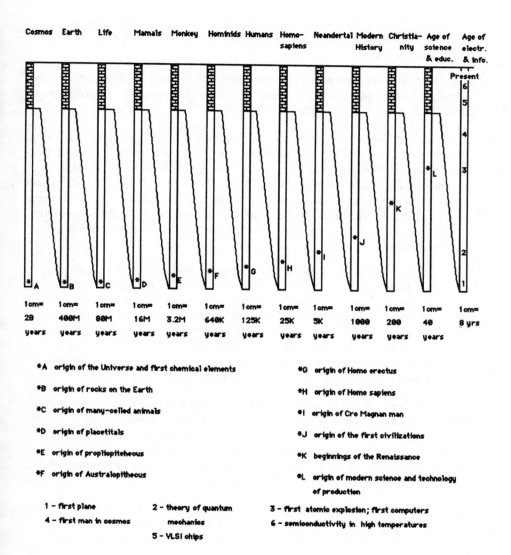

Fig. 16.2. The extra-exponential growth in the speed of evolution. The left column of this figure represents the entire time span from the origin of the universe to the present. The second column is an enlargement of the part of the first column, that is above the diagonal line connecting the two, and so on.

"Most physicists, chemists, and biological evolutionists agree that the evolution of organic molecules began about 4 billion years ago. The first living cell appeared about 3.5 billion years ago, and the first simple many-celled animal appeared roughly 6000 million years ago. The common ancestor of apes and humans appeared some 6 million years ago, and the beginning of recorded history was about 6000 years ago".

Clearly, this is an area which deserves some further investigations; several interesting results of applying cultural algorithms to a variety of problems are described in [60, 326, 327, 328, 330, 331, 377].

It is worthwhile to note that there are many other approaches to learning, optimization, and problem solving, which are based on other natural metaphors from nature — the best known examples include neural networks and simulated annealing. There is a growing interest in all these areas; the most fruitful and challenging direction seems to be a "recombination" of some ideas at present scattered in different fields. As Schwefel and Männer wrote in the introduction to the proceedings of the First Workshop on Parallel Problem Solving from Nature [351]:

"It is a matter of fact that in Europe evolution strategies and in the USA genetic algorithms have survived more than a decade of non-acceptance or neglect. It is also true, however, that so far both strata of ideas have evolved in geographical isolation and thus not led to recombined offspring. Now it is time for a new generation of algorithms which make use of the rich gene pool of ideas on both sides of Atlantic, and make use too of the favorable environment showing up in the form of massively parallel processor systems".

There were two main goals of this text. The first one was to convince the reader that evolution is a powerful and general concept which should find its place in many problem-solving techniques. In particular, the whole field of artificial intelligence should lean towards evolutionary techniques; as Lawrence Fogel stated [125] in his plenary talk during the World Congress on Computational Intelligence (Orlando, 27 June – 2 July 1994):

"If the aim is to generate artificial intelligence, that is, to solve new problems in new ways, then it is inappropriate to use any fixed set of rules. The rules required for solving each problem should simply evolve..."

The second goal was to emphasize similarities between various evolutionary techniques; these techniques were discussed in the book from the perspective of constructing evolutionary programs for particular classes of problems. In that way all differences between these techniques (e.g., genetic algorithms, evolution strategies, evolutionary programming, genetic programming, etc.) were hidden on a lower level; various techniques use just different chromosomal representations and appropriate sets of (more or less genetic) operators.

Appendix A

This is a very simple real-coded genetic algorithm built by Denis Cormier (North Carolina State University) and modified by Sita S. Raghavan (University of North Carolina at Charlotte). The code is minimal and virtually no error checking is performed; in many instances, efficiency has been sacrificed for clarity. To modify the code for a particular application, change the constants definitions and the user defined "evaluation function". Note that the code is designed for maximization problems where the objective function takes positive values only; there is no distinction between the objective value and the fitness of the individual. The system uses proportional selection, elitist model, one point crossover and uniform mutation (much better results can be obtained if uniform mutation is replaced by a Gaussian mutation; the reader is encouraged to incorporate such changes in the system — see exercise 6 from Appendix D).

The code does not make use of any graphics or even screen output, and should be highly portable between platforms; it is available from ftp.uncc.edu, directory coe/evol, file prog.c.

The required input file should be named as 'gadata.txt'; the system produces the output file 'galog.txt'. The input file consists of several lines: number of lines correspond to number of variables. Each line provides lower and upper bound for a variable in order (i.e., first line provides lower and upper bounds for the first variable, second line—for the second variable, etc.).

```
/**********************************************************************/
/* This is a simple genetic algorithm implementation where the      */
/* evaluation function takes positive values only and the           */
/* fitness of an individual is the same as the value of the         */
/* objective function                                               */
/**********************************************************************/

#include <stdio.h>
#include <stdlib.h>
#include <math.h>
```

```
/* Change any of these parameters to match your needs */

#define POPSIZE 50              /* population size */
#define MAXGENS 1000            /* max. number of generations */
#define NVARS 3                 /* no. of problem variables */
#define PXOVER 0.8              /* probability of crossover */
#define PMUTATION 0.15          /* probability of mutation */
#define TRUE 1
#define FALSE 0

int generation;                 /* current generation no. */
int cur_best;                   /* best individual */
FILE *galog;                    /* an output file */

struct genotype /* genotype (GT), a member of the population */
{
  double gene[NVARS];           /* a string of variables */
  double fitness;               /* GT's fitness */
  double upper[NVARS];          /* GT's variables upper bound */
  double lower[NVARS];          /* GT's variables lower bound */
  double rfitness;              /* relative fitness */
  double cfitness;              /* cumulative fitness */
};

struct genotype population[POPSIZE+1];      /* population */
struct genotype newpopulation[POPSIZE+1];   /* new population; */
                                            /* replaces the */
                                            /* old generation */

/* Declaration of procedures used by this genetic algorithm */

void initialize(void);
double randval(double, double);
void evaluate(void);
void keep_the_best(void);
void elitist(void);
void select(void);
void crossover(void);
void Xover(int,int);
void swap(double *, double *);
void mutate(void);
void report(void);
```

```
/*****************************************************************/
/* Initialization function: Initializes the values of genes     */
/* within the variables bounds. It also initializes (to zero)   */
/* all fitness values for each member of the population. It      */
/* reads upper and lower bounds of each variable from the       */
/* input file 'gadata.txt'. It randomly generates values        */
/* between these bounds for each gene of each genotype in the    */
/* population. The format of the input file 'gadata.txt' is      */
/* var1_lower_bound var1_upper bound                             */
/* var2_lower_bound var2_upper bound ...                         */
/*****************************************************************/

void initialize(void)
{
FILE *infile;
int i, j;
double lbound, ubound;

if ((infile = fopen("gadata.txt","r"))==NULL)
     {
     fprintf(galog,"\nCannot open input file!\n");
     exit(1);
     }

/* initialize variables within the bounds */

for (i = 0; i < NVARS; i++)
     {
   fscanf(infile, "%lf",&lbound);
   fscanf(infile, "%lf",&ubound);

     for (j = 0; j < POPSIZE; j++)
          {
          population[j].fitness = 0;
          population[j].rfitness = 0;
          population[j].cfitness = 0;
          population[j].lower[i] = lbound;
          population[j].upper[i]= ubound;
          population[j].gene[i] = randval(population[j].lower[i],
                              population[j].upper[i]);
          }
     }

fclose(infile);
}
```

```
/**************************************************************/
/* Random value generator: Generates a value within bounds */
/**************************************************************/

double randval(double low, double high)
{
double val;
val = ((double)(rand()%1000)/1000.0)*(high - low) + low;
return(val);
}

/***************************************************************/
/* Evaluation function: This takes a user defined function.  */
/* Each time this is changed, the code has to be recompiled. */
/* The current function is:   x[1]^2-x[1]*x[2]+x[3]          */
/***************************************************************/

void evaluate(void)
{
int mem;
int i;
double x[NVARS+1];

for (mem = 0; mem < POPSIZE; mem++)
      {
      for (i = 0; i < NVARS; i++)
            x[i+1] = population[mem].gene[i];

      population[mem].fitness = (x[1]*x[1]) - (x[1]*x[2]) + x[3];
      }
}

/***************************************************************/
/* Keep_the_best function: This function keeps track of the  */
/* best member of the population. Note that the last entry in */
/* the array Population holds a copy of the best individual   */
/***************************************************************/

void keep_the_best()
{
int mem;
int i;
cur_best = 0; /* stores the index of the best individual */
```

```
for (mem = 0; mem < POPSIZE; mem++)
      {
      if (population[mem].fitness > population[POPSIZE].fitness)
            {
            cur_best = mem;
            population[POPSIZE].fitness = population[mem].fitness;
            }
      }
/* once the best member in the population is found, copy the genes */
for (i = 0; i < NVARS; i++)
      population[POPSIZE].gene[i] = population[cur_best].gene[i];
}


/*****************************************************************/
/* Elitist function: The best member of the previous generation */
/* is stored as the last in the array. If the best member of    */
/* the current generation is worse then the best member of the  */
/* previous generation, the latter one would replace the worst  */
/* member of the current population                             */
/*****************************************************************/

void elitist()
{
int i;
double best, worst;                  /* best and worst fitness values */
int best_mem, worst_mem; /* indexes of the best and worst member */

best = population[0].fitness;
worst = population[0].fitness;
for (i = 0; i < POPSIZE - 1; ++i)
      {
      if(population[i].fitness > population[i+1].fitness)
            {
            if (population[i].fitness> = best)
                  {
                  best = population[i].fitness;
                  best_mem = i;
                  }
            if (population[i+1].fitness <= worst)
                  {
                  worst = population[i+1].fitness;
                  worst_mem = i + 1;
                  }
            }
```

```
        else
            {
            if (population[i].fitness < = worst)
                {
                worst = population[i].fitness;
                worst_mem = i;
                }
            if (population[i+1].fitness >= best)
                {
                best = population[i+1].fitness;
                best_mem = i + 1;
                }
            }
        }
/* if best individual from the new population is better than */
/* the best individual from the previous population, then    */
/* copy the best from the new population; else replace the    */
/* worst individual from the current population with the      */
/* best one from the previous generation                      */

if (best >= population[POPSIZE].fitness)
    {
    for (i = 0; i < NVARS; i++)
        population[POPSIZE].gene[i] = population[best_mem].gene[i];
    population[POPSIZE].fitness = population[best_mem].fitness;
    }
else
    {
    for (i = 0; i < NVARS; i++)
        population[worst_mem].gene[i] = population[POPSIZE].gene[i];
    population[worst_mem].fitness = population[POPSIZE].fitness;
    }
}
/***************************************************************/
/* Selection function: Standard proportional selection for    */
/* maximization problems incorporating elitist model - makes  */
/* sure that the best member survives                         */
/***************************************************************/

void select(void)
{
int mem, i, j, k;
double sum = 0;
double p;
```

```
/* find total fitness of the population */
for (mem = 0; mem < POPSIZE; mem++)
      {
      sum += population[mem].fitness;
      }

/* calculate relative fitness */
for (mem = 0; mem < POPSIZE; mem++)
      {
      population[mem].rfitness =  population[mem].fitness/sum;
      }
population[0].cfitness = population[0].rfitness;

/* calculate cumulative fitness */
for (mem = 1; mem < POPSIZE; mem++)
      {
      population[mem].cfitness =  population[mem-1].cfitness +
                        population[mem].rfitness;
      }

/* finally select survivors using cumulative fitness. */

for (i = 0; i < POPSIZE; i++)
      {
      p = rand()%1000/1000.0;
      if (p < population[0].cfitness)
            newpopulation[i] = population[0];
      else
            {
            for (j = 0; j < POPSIZE;j++)
                  if (p >= population[j].cfitness &&
                            p<population[j+1].cfitness)
                        newpopulation[i] = population[j+1];
            }
      }
/* once a new population is created, copy it back */

for (i = 0; i < POPSIZE; i++)
      population[i] = newpopulation[i];
}
```

```
/****************************************************************/
/* Crossover selection: selects two parents that take part in  */
/* the crossover. Implements a single point crossover          */
/****************************************************************/

void crossover(void)
{
int i, mem, one;
int first  =  0; /* count of the number of members chosen */
double x;

for (mem = 0; mem < POPSIZE; ++mem)
      {
      x = rand()%1000/1000.0;
      if (x < PXOVER)
            {
            ++first;
            if (first % 2 == 0)
                  Xover(one, mem);
            else
                  one = mem;
            }
      }
}
/****************************************************************/
/* Crossover: performs crossover of the two selected parents. */
/****************************************************************/

void Xover(int one, int two)
{
int i;
int point; /* crossover point */

/* select crossover point */
if(NVARS > 1)
   {
   if(NVARS == 2)
         point = 1;
   else
         point = (rand() % (NVARS - 1)) + 1;

   for (i = 0; i < point; i++)
         swap(&population[one].gene[i], &population[two].gene[i]);
   }
}
```

```
/**************************************************************/
/* Swap: A swap procedure that helps in swapping 2 variables */
/**************************************************************/

void swap(double *x, double *y)
{
double temp;

temp = *x;
*x = *y;
*y = temp;


}


/***************************************************************/
/* Mutation: Random uniform mutation. A variable selected for */
/* mutation is replaced by a random value between lower and   */
/* upper bounds of this variable                              */
/***************************************************************/

void mutate(void)
{
int i, j;
double lbound, hbound;
double x;

for (i = 0; i < POPSIZE; i++)
     for (j = 0; j < NVARS; j++)
             {
             x = rand()%1000/1000.0;
             if (x < PMUTATION)
                     {
                     /* find the bounds on the variable to be mutated */
                     lbound = population[i].lower[j];
                     hbound = population[i].upper[j];
                     population[i].gene[j] = randval(lbound, hbound);
                     }
             }
}
```

```
/****************************************************************/
/* Report function: Reports progress of the simulation. Data   */
/* dumped into the  output file are separated by commas        */
/****************************************************************/

void report(void)
{
int i;
double best_val;        /* best population fitness */
double avg;             /* avg population fitness */
double stddev;          /* std. deviation of population fitness */
double sum_square;      /* sum of square for std. calc */
double square_sum;      /* square of sum for std. calc */
double sum;             /* total population fitness */

sum = 0.0;
sum_square = 0.0;

for (i = 0; i < POPSIZE; i++)
     {
     sum += population[i].fitness;
     sum_square += population[i].fitness * population[i].fitness;
     }

avg = sum/(double)POPSIZE;
square_sum = avg * avg * (double)POPSIZE;
stddev = sqrt((sum_square - square_sum)/(POPSIZE - 1));
best_val = population[POPSIZE].fitness;

fprintf(galog, "\n%5d,        %6.3f, %6.3f, %6.3f \n\n", generation,
                                    best_val, avg, stddev);
}
```

```
/***************************************************************/
/* Main function: Each generation involves selecting the best */
/* members, performing crossover & mutation and then          */
/* evaluating the resulting population, until the terminating */
/* condition is satisfied                                     */
/***************************************************************/

void main(void)
{
int i;

if ((galog = fopen("galog.txt","w"))==NULL)
    {
    exit(1);
    }
generation = 0;

fprintf(galog, "\n generation  best  average  standard \n");
fprintf(galog, " number       value fitness  deviation \n");

initialize();
evaluate();
keep_the_best();
while(generation<MAXGENS)
    {
    generation++;
    select();
    crossover();
    mutate();
    report();
    evaluate();
    elitist();
    }
fprintf(galog,"\n\n Simulation completed\n");
fprintf(galog,"\n Best member: \n");

for (i = 0; i < NVARS; i++)
   {
   fprintf (galog,"\n var(%d) = %3.3f",i,population[POPSIZE].gene[i]);
   }
fprintf(galog,"\n\n Best fitness = %3.3f",population[POPSIZE].fitness);
fclose(galog);
printf("Success\n");
}

/***************************************************************/
```

Appendix B

There are several test functions which can be used for various experiments (for additional problems, see also Appendix A in [350], [186], and [113]):[1]

1. De Jong function F1:

   $\sum_{i=1}^{3} x_i^2$, where $-5.12 \leq x_i \leq 5.12$.

   The function has a global minimum value of 0 at $(x_1, x_2, x_3) = (0, 0, 0)$.

2. De Jong function F2:

   $100(x_1^2 - x_2)^2 + (1 - x_1)^2$, where $-2.048 \leq x_i \leq 2.048$.

   The function has a global minimum value of 0 at $(x_1, x_2) = (1, 1)$.

3. De Jong function F3:

   $\sum_{i=1}^{5} \text{integer}(x_i)$, where $-5.12 \leq x_i \leq 5.12$.

   The function has a global minimum value of -30 for all $-5.12 \leq x_i < -5.0$.

4. De Jong function F4:

   $\sum_{i=1}^{30} ix_i^4 + \text{Gauss}(0,1)$, where $-1.28 \leq x_i \leq 1.28$

   The function (without Gaussian noise) has a global minimum value of 0 at $(x_1, x_2, \ldots, x_{30}) = (0, 0, \ldots, 0)$.

5. De Jong function F5:

   $\frac{1}{1/K + \sum_{j=1}^{25} f_j^{-1}(x_1, x_2)}$, where $f_j(x_1, x_2) = c_j + \sum_{i=1}^{2}(x_i - a_{ij})^6$,

   where $-65.536 \leq x_i \leq 65.536$, $K = 500$, $c_j = j$, and

   $$[a_{ij}] = \begin{bmatrix} -32 & -16 & 0 & 16 & 32 & -32 & -16 & \ldots & 0 & 16 & 32 \\ -32 & -32 & -32 & -32 & -32 & -16 & -16 & \ldots & 32 & 32 & 32 \end{bmatrix}$$

   The function has a global minimum value of 0.998 at $(x_1, x_2) = (-32, -32)$.

---

[1] See also [401] for a discussion on building test functions.

6. Schaffer function F6:

$$0.5 + \frac{\sin^2 \sqrt{x_1^2 + x_2^2} - 0.5}{[1.0 + 0.001(x_1^2 + x_2^2)]^2}, \text{ where } -100 \leq x_i \leq 100.$$

The function has a global minimum value of 0 at $(x_1, x_2) = (0, 0)$.

7. Schaffer function F7:

$$(x_1^2 + x_2^2)^{0.25} [\sin^2(50(x_1^2 + x_2^2)^{0.1}) + 1.0], \text{ where } -100 \leq x_i \leq 100.$$

The function has a global minimum value of 0 at $(x_1, x_2) = (0, 0)$.

8. Goldstein-Price function:

$$[1 + (x_1 + x_2 + 1)^2(19 - 14x_1 + 3x_1^2 - 14x_2 + 6x_1x_2 + 3x_2^2)]$$
$$\cdot [30 + (2x_1 - 3x_2)^2(18 - 32x_1 + 12x_1^2 + 48x_2 - 36x_1x_2 + 27x_2^2)],$$

where $-2 \leq x_i \leq 2$.

The function has a global minimum value of 3 at $(x_1, x_2) = (0, -1)$.

9. Branin RCOS function:

$$a(x_2 - bx_1^2 + cx_1 - d)^2 + e(1 - f)\cos(x_1) + e,$$

where $-5 \leq x_1 \leq 10$, $0 \leq x_2 \leq 15$, and $a = 1$, $b = 5.1/(4\pi^2)$, $c = 5/\pi$, $d = 6$, $e = 10$, $f = 1/(8\pi)$.

The function has a global minimum value of 0.397887 at three different points: $(x_1, x_2) = (-\pi, 12.275), (\pi, 2.275)$, and $(9.42478, 2.475)$.

10. The Shekel SQRN5, SQRN7, SQRN10 family of 4-dimensional functions:

$$s3(x_1, x_2, x_3, x_4) = -\sum_{j=1}^{5} \frac{1}{\sum_{i=1}^{4}(x_i - a_{ij})^2 + c_j}$$
$$s4(x_1, x_2, x_3, x_4) = -\sum_{j=1}^{7} \frac{1}{\sum_{i=1}^{4}(x_i - a_{ij})^2 + c_j}$$
$$s5(x_1, x_2, x_3, x_4) = -\sum_{j=1}^{10} \frac{1}{\sum_{i=1}^{4}(x_i - a_{ij})^2 + c_j},$$

where $0 \leq x_i \leq 10$ for $i = 1, 2, 3, 4$, and a_{ij}'s and c_j's are listed in Table B.1.

These functions have a global minimum value at the point $(x_1, x_2, x_3, x_4) = (4, 4, 4, 4)$, with the following values: $s3_{min} = -10.15320$, $s4_{min} = -10.402820$, and $s5_{min} = -10.53628$.

11. The six-hump camel back function:

$$(4 - 2.1x_1^2 + \frac{x_1^4}{3})x_1^2 + x_1x_2 + (-4 + 4x_2^2)x_2^2,$$

where $-3 \leq x_1 \leq 3$ and $-2 \leq x_2 \leq 2$.

The function has a global minimum value of -1.0316 at two different points: $(x_1, x_2) = (-0.0898, 0.7126)$ and $(0.0898, -0.7126)$.

j	a_{1j}	a_{2j}	a_{3j}	a_{4j}	c_j
1	4.0	4.0	4.0	4.0	0.1
2	1.0	1.0	1.0	1.0	0.2
3	8.0	8.0	8.0	8.0	0.2
4	6.0	6.0	6.0	6.0	0.4
5	3.0	7.0	3.0	7.0	0.6
6	2.0	9.0	2.0	9.0	0.6
7	5.0	5.0	3.0	3.0	0.3
8	8.0	1.0	8.0	1.0	0.7
9	6.0	2.0	6.0	2.0	0.5
10	7.0	3.6	7.0	3.6	0.5

Table B.1. Data for functions $s3$, $s4$, and $s5$

12. Shubert function:

$$\sum_{i=1}^{5} i \cos[(i+1)x_1 + i] \cdot \sum_{i=1}^{5} i \cos[(i+1)x_2 + i],$$

where $-10 \le x_i \le 10$ for $i = 1, 2$.

The function has 760 local minima, 18 of which are global minima with -186.73.

13. The Stuckman function:

$$f8(x_1, x_2) = \begin{cases} \lfloor(\lfloor m_1\rfloor + \frac{1}{2})\sin(a_1)/a_1\rfloor & \text{if } 0 \le x_1 \le b \\ \lfloor(\lfloor m_2\rfloor + \frac{1}{2})\sin(a_2)/a_2\rfloor & \text{if } b < x_1 \le 10 \end{cases}$$

where $0 \le x_i \le 10$ for $i = 1, 2$, and m_i is a random variable between 0 and 100 $(i = 1, 2)$, b is a random variable between 0 and 10, and

$$a_i = \lfloor|x_1 - r_{1i}|\rfloor + \lfloor|x_2 - r_{2i}|\rfloor,$$

where r_{11} is a random variable between 0 and b, r_{12} is a random variable between b and 10, r_{21} is a random variable between 0 and 10, and r_{22} is a random variable between 0 and 10 (all random variables are uniform).

The global maximum is located at

$$(x_1, x_2) = \begin{cases} (r_{11}, r_{21}) & \text{if } m_1 > m_2 \\ (r_{12}, r_{22}) & \text{otherwise} \end{cases}$$

14. The Easom function:

$$-\cos(x_1)\cos(x_2)e^{-(x_1 - \pi)^2 - (x_2 - \pi)^2},$$

where $-100 \le x_i \le 100$ for $i = 1, 2$.

The function has a global minimum at $(x_1, x_2) = (\pi, \pi)$ with -1.

15. the Bohachevsky function #1:

$$x_1^2 + 2x_2^2 - 0.3\cos(3\pi x_1) - 0.4\cos(4\pi x_2) + 0.7$$

where $-50 \le x_i \le 50$.

The function has a global minimum value of 0 at $(x_1, x_2) = (0, 0)$.

16. the Bohachevsky function #2:

$$x_1^2 + 2x_2^2 - 0.3(\cos(3\pi x_1)\cos(4\pi x_2)) + 0.3$$

where $-50 \le x_i \le 50$.

The function has a global minimum value of 0 at $(x_1, x_2) = (0, 0)$.

17. the Bohachevsky function #3:

$$x_1^2 + 2x_2^2 - 0.3\cos(3\pi x_1) + \cos(4\pi x_2) + 0.3$$

where $-50 \le x_i \le 50$.

The function has a global minimum value of 0 at $(x_1, x_2) = (0, 0)$.

18. the Colville function:

$$100(x_2 - x_1^2)^2 + (1 - x_1)^2 + 90(x_4 - x_3^2)^2 + (1 - x_3)^2 + \\ +10.1((x_2 - 1)^2 + (x_4 - 1)^2) + 19.8(x_2 - 1)(x_4 - 1),$$

where $-10 \le x_i \le 10$.

The function has a global minimum value of 0 at
$(x_1, x_2, x_3, x_4) = (1, 1, 1, 1)$.

Appendix C

There are several test functions which can be used for various experiments for constrained optimization (for additional problems, see also Appendix A in [350], [186], and [113]):

1. The problem [114] is

$$\text{minimize } G1(\boldsymbol{x}, \boldsymbol{y}) = 5x_1 + 5x_2 + 5x_3 + 5x_4 - 5\sum_{i=1}^{4} x_i^2 - \sum_{i=1}^{9} y_i,$$

   subject to:

$$
\begin{array}{ll}
2x_1 + 2x_2 + y_6 + y_7 \leq 10, & 2x_1 + 2x_3 + y_6 + y_8 \leq 10, \\
2x_2 + 2x_3 + y_7 + y_8 \leq 10, & -8x_1 + y_6 \leq 0, \\
-8x_2 + y_7 \leq 0, & -8x_3 + y_8 \leq 0, \\
-2x_4 - y_1 + y_6 \leq 0, & -2y_2 - y_3 + y_7 \leq 0, \\
-2y_4 - y_5 + y_8 \leq 0, & 0 \leq x_i \leq 1,\ i = 1, 2, 3, 4, \\
0 \leq y_i \leq 1,\ i = 1, 2, 3, 4, 5, 9, & 0 \leq y_i,\ i = 6, 7, 8.
\end{array}
$$

   The global solution is $(\boldsymbol{x}^*, \boldsymbol{y}^*) = (1, 1, 1, 1, 1, 1, 1, 1, 1, 3, 3, 3, 1)$, and $G1(\boldsymbol{x}^*, \boldsymbol{y}^*) = -15$.

2. The problem [186] is to minimize a function:

$$G2(\overline{X}) = x_1 + x_2 + x_3,$$

   where

$$
\begin{array}{ll}
1 - 0.0025(x_4 + x_6) \geq 0, & x_1 x_6 - 833.33252 x_4 - 100 x_1 + 83333.333 \geq 0, \\
1 - 0.0025(x_5 + x_7 - x_4) \geq 0, & x_2 x_7 - 1250 x_5 - x_2 x_4 + 1250 x_4 \geq 0, \\
1 - 0.01(x_8 - x_5) \geq 0, & x_3 x_8 - 1250000 - x_3 x_5 + 2500 x_5 \geq 0, \\
100 \leq x_1 \leq 10000, & 1000 \leq x_i \leq 10000,\quad i = 2, 3, \\
10 \leq x_i \leq 1000,\quad i = 4, \dots, 8. &
\end{array}
$$

   The problem has 3 linear and 3 nonlinear constraints; the function $G2$ is linear and has its global minimum at

$$
\begin{aligned}
\overline{X}^* = (&579.3167, 1359.943, 5110.071, 182.0174, \\
&295.5985, 217.9799, 286.4162, 395.5979),
\end{aligned}
$$

where $G2(\overline{X}^*) = 7049.330923$. All six constraints are active at the global optimum.

3. The problem [186] is to minimize a function:

$$G3(\overline{X}) = (x_1 - 10)^2 + 5(x_2 - 12)^2 + x_3^4 + 3(x_4 - 11)^2 + 10x_5^6 + 7x_6^2 + x_7^4 - 4x_6x_7 - 10x_6 - 8x_7,$$

where

$127 - 2x_1^2 - 3x_2^4 - x_3 - 4x_4^2 - 5x_5 \geq 0,$
$282 - 7x_1 - 3x_2 - 10x_3^2 - x_4 + x_5 \geq 0,$
$196 - 23x_1 - x_2^2 - 6x_6^2 + 8x_7 \geq 0,$
$-4x_1^2 - x_2^2 + 3x_1x_2 - 2x_3^2 - 5x_6 + 11x_7 \geq 0$
$-10.0 \leq x_i \leq 10.0, \ i = 1, \ldots, 7.$

The problem has 4 nonlinear constraints; the function $G3$ is nonlinear and has its global minimum at

$$\overline{X}^* = (2.330499, 1.951372, -0.4775414, 4.365726, -0.6244870, 1.038131, 1.594227),$$

where $G3(\overline{X}^*) = 680.6300573$. Two (out of four) constraints are active at the global optimum (the first and the last one).

4. The problem [186] is to minimize a function:

$$G4(\overline{X}) = e^{x_1 x_2 x_3 x_4 x_5},$$

subject to

$x_1^2 + x_2^2 + x_3^2 + x_4^2 + x_5^2 = 10, \quad x_2x_3 - 5x_4x_5 = 0, \quad x_1^3 + x_2^3 = -1,$
$-2.3 \leq x_i \leq 2.3, \qquad\qquad i = 1, 2, \ -3.2 \leq x_i \leq 3.2, \quad i = 3, 4, 5.$

The problem has 3 nonlinear equations; nonlinear function $G4$ has its global minimum at

$$\overline{X}^* = (-1.717143, 1.595709, 1.827247, -0.7636413, -0.7636450),$$

where $G4(\overline{X}^*) = 0.0539498478$.

5. The problem [186] is to minimize a function:

$$G5(\overline{X}) = x_1^2 + x_2^2 + x_1x_2 - 14x_1 - 16x_2 + (x_3 - 10)^2 + 4(x_4 - 5)^2 + (x_5 - 3)^2 + 2(x_6 - 1)^2 + 5x_7^2 + 7(x_8 - 11)^2 + 2(x_9 - 10)^2 + (x_{10} - 7)^2 + 45,$$

where

$105 - 4x_1 - 5x_2 + 3x_7 - 9x_8 \geq 0,$
$-3(x_1 - 2)^2 - 4(x_2 - 3)^2 - 2x_3^2 + 7x_4 + 120 \geq 0,$
$-10x_1 + 8x_2 + 17x_7 - 2x_8 \geq 0,$

$-x_1^2 - 2(x_2 - 2)^2 + 2x_1x_2 - 14x_5 + 6x_6 \geq 0,$
$8x_1 - 2x_2 - 5x_9 + 2x_{10} + 12 \geq 0,$
$-5x_1^2 - 8x_2 - (x_3 - 6)^2 + 2x_4 + 40 \geq 0,$
$3x_1 - 6x_2 - 12(x_9 - 8)^2 + 7x_{10} \geq 0,$
$-0.5(x_1 - 8)^2 - 2(x_2 - 4)^2 - 3x_5^2 + x_6 + 30 \geq 0,$
$-10.0 \leq x_i \leq 10.0, \quad i = 1, \ldots, 10.$

The problem has 3 linear and 5 nonlinear constraints; the function $G5$ is quadratic and has its global minimum at

$$\overline{X}^* = (2.171996, 2.363683, 8.773926, 5.095984, 0.9906548,$$
$$1.430574, 1.321644, 9.828726, 8.280092, 8.375927),$$

where $G5(\overline{X}^*) = 24.3062091$. Six (out of eight) constraints are active at the global optimum (all except the last two).

6. The problem [220] is to maximize a function:

$$G6(\boldsymbol{x}) = |\frac{\sum_{i=1}^n cos^4(x_i) - 2\prod_{i=1}^n cos^2(x_i)}{\sqrt{\sum_{i=1}^n ix_i^2}}|,$$

where

$$\prod_{i=1}^n x_i > 0.75, \ \sum_{i=1}^n x_i < 7.5n, \text{ and } 0 < x_i < 10 \text{ for } 1 \leq i \leq n.$$

The problem has 2 nonlinear constraints; the function $G6$ is nonlinear and its global maximum is unknown. Some good solutions (found by Genocop III, see Chapter 7) are the following. For $n = 20$:

$$\boldsymbol{x} = (3.16311359, 3.13150430, 3.09515858, 3.06016588, 3.03103566,$$
$$2.99158549, 2.95802593, 2.92285895, 0.48684388, 0.47732279,$$
$$0.48044473, 0.48790911, 0.48450437, 0.44807032, 0.46877760,$$
$$0.45648506, 0.44762608, 0.44913986, 0.44390863, 0.45149332),$$

where $G6(\boldsymbol{x}) = 0.80351067$. For $n = 50$:

$$\boldsymbol{x} = (6.28006029, 3.16155291, 3.15453815, 3.14085174, 3.12882447,$$
$$3.11211085, 3.10170507, 3.08703685, 3.07571769, 3.06122732,$$
$$3.05010581, 3.03667951, 3.02333045, 3.00721049, 2.99492717,$$
$$2.97988462, 2.96637058, 2.95589066, 2.94427204, 2.92796040,$$
$$0.40970641, 2.90670991, 0.46131119, 0.48193336, 0.46776962,$$
$$0.43887550, 0.45181099, 0.44652876, 0.43348753, 0.44577143,$$
$$0.42379948, 0.45858049, 0.42931050, 0.42928645, 0.42943302,$$
$$0.43294361, 0.42663351, 0.43437257, 0.42542559, 0.41594154,$$
$$0.43248957, 0.39134723, 0.42628688, 0.42774364, 0.41886297,$$
$$0.42107263, 0.41215360, 0.41809589, 0.41626775, 0.42316407),$$

where $G6(\boldsymbol{x}) = 0.83319378$.

7. The problem [114] is to minimize

$$G7(\boldsymbol{x}, y) = -10.5x_1 - 7.5x_2 - 3.5x_3 - 2.5x_4 - 1.5x_5 - 10y - 0.5\sum_{i=1}^{5} x_i^2,$$

subject to:

$$6x_1 + 3x_2 + 3x_3 + 2x_4 + x_5 \le 6.5, \quad 10x_1 + 10x_3 + y \le 20,$$
$$0 \le x_i \le 1, \qquad\qquad\qquad\qquad 0 \le y.$$

The global solution is $(\boldsymbol{x}^*, y^*) = (0, 1, 0, 1, 1, 20)$, and $G7(\boldsymbol{x}^*, y^*) = -213$.

8. The problem [186] is to minimize

$$G8(\boldsymbol{x}) = \sum_{j=1}^{10} x_j(c_j + \ln \frac{x_j}{x_1 + \ldots + x_{10}}),$$

subject to:

$$x_1 + 2x_2 + 2x_3 + x_6 + x_{10} = 2, \quad x_4 + 2x_5 + x_6 + x_7 = 1,$$
$$x_3 + x_7 + x_8 + 2x_9 + x_{10} = 1, \quad x_i \ge 0.000001, \ (i = 1, \ldots, 10),$$

where

$$c_1 = -6.089;\ c_2 = -17.164;\ c_3 = -34.054;\ c_4 = -5.914;\ c_5 = -24.721;$$
$$c_6 = -14.986;\ c_7 = -24.100;\ c_8 = -10.708;\ c_9 = -26.662;\ c_{10} = -22.179;$$

The best solution found by Genocop (Chapter 7) is

$$\boldsymbol{x}^* = (.04034785, .15386976, .77497089, .00167479, .48468539,$$
$$.00068965, .02826479, .01849179, .03849563, .10128126),$$

for which the value of the objective function is equal to -47.760765.

9. The problem [113] is to maximize

$$G9(\boldsymbol{x}) = \frac{3x_1 + x_2 - 2x_3 + 0.8}{2x_1 - x_2 + x_3} + \frac{4x_1 - 2x_2 + x_3}{7x_1 + 3x_2 - x_3},$$

subject to:

$$x_1 + x_2 - x_3 \le 1, \qquad\qquad -x_1 + x_2 - x_3 \le -1,$$
$$12x_1 + 5x_2 + 12x_3 \le 34.8, \quad 12x_1 + 12x_2 + 7x_3 \le 29.1,$$
$$-6x_1 + x_2 + x_3 \le -4.1, \qquad 0 \le x_i, \ i = 1, 2, 3.$$

The global solution is $\boldsymbol{x}^* = (1, 0, 0)$, and $G9(\boldsymbol{x}^*) = 2.471428$.

10. The problem[114] is to minimize

$$G10(\boldsymbol{x}) = x_1^{0.6} + x_2^{0.6} - 6x_1 - 4x_3 + 3x_4,$$

subject to:

$$-3x_1 + x_2 - 3x_3 = 0, \quad x_1 + 2x_3 \le 4,$$
$$x_2 + 2x_4 \le 4, \qquad\qquad x_1 \le 3,$$
$$x_4 \le 1, \qquad\qquad\quad\; 0 \le x_i, \; i = 1, 2, 3, 4.$$

The best known global solution is $\boldsymbol{x}^* = (\frac{4}{3}, 4, 0, 0)$, and $G10(\boldsymbol{x}^*) = -4.5142$.

11. The problem [114] is to minimize

$$G11(x, \boldsymbol{y}) = 6.5x - 0.5x^2 - y_1 - 2y_2 - 3y_3 - 2y_4 - y_5,$$

subject to:

$$x + 2y_1 + 8y_2 + y_3 + 3y_4 + 5y_5 \le 16,$$
$$-8x - 4y_1 - 2y_2 + 2y_3 + 4y_4 - y_5 \le -1,$$
$$2x + 0.5y_1 + 0.2y_2 - 3y_3 - y_4 - 4y_5 \le 24,$$
$$0.2x + 2y_1 + 0.1y_2 - 4y_3 + 2y_4 + 2y_5 \le 12,$$
$$-0.1x - 0.5y_1 + 2y_2 + 5y_3 - 5y_4 + 3y_5 \le 3,$$
$$y_3 \le 1, \; y_4 \le 1, \text{ and } y_5 \le 2,$$
$$x \ge 0, \; y_i \ge 0, \text{ for } 1 \le i \le 5.$$

The global solution is $(x, \boldsymbol{y}^*) = (0, 6, 0, 1, 1, 0)$, and $G11(x, \boldsymbol{y}^*) = -11$.

12. The problem was constructed from three separate problems [186] in the following way: minimize

$$G12(\boldsymbol{x}) = \begin{cases} f_1 = x_2 + 10^{-5}(x_2 - x_1)^2 - 1.0 & if \; 0 \le x_1 < 2 \\ f_2 = \frac{1}{27\sqrt{3}}((x_1 - 3)^2 - 9)x_2^3 & if \; 2 \le x_1 < 4 \\ f_3 = \frac{1}{3}(x_1 - 2)^3 + x_2 - \frac{11}{3} & if \; 4 \le x_1 \le 6 \end{cases}$$

subject to:

$$x_1/\sqrt{3} - x_2 \ge 0,$$
$$-x_1 - \sqrt{3}x_2 + 6 \ge 0,$$
$$0 \le x_1 \le 6, \text{ and } x_2 \ge 0.$$

The function $G12$ has three global solutions:

$$\boldsymbol{x}_1^* = (0, 0), \; \boldsymbol{x}_2^* = (3, \sqrt{3}), \text{ and } \boldsymbol{x}_3^* = (4, 0),$$

in all cases $G12(\boldsymbol{x}_i^*) = -1 \; (i = 1, 2, 3)$.

Appendix D

Probably the best way to run a course on evolutionary computation techniques is to organize it as a project-oriented course. After some preliminary project (to implement, say, a simple genetic algorithm), which would allow the students to grasp the basic concepts of this evolutionary technique (this can be done during the first few weeks of the semester), the class is ready for something more challenging.

There are several possible experimental projects one can try; after all, this text should provide more questions than answers! The last chapter, Conclusions, gives a list of the current research areas; several projects can emerge from there. Of course, projects can vary in complexity and time needed for their completion; some of them are quite simple, others may require a team of a few students working together. Anyway, it is important to remember that the list of projects contains just an arbitrary collection of possible problems, which may (and should) trigger some other ideas.

So in this appendix we provide a list of various projects together with a few remarks on reporting computational experiments — usually it happens that some results are quite interesting and are worth publishing!

A few possible projects

1. Compare several algorithms: hill-climbing, stochastic hill-climbing, simulated annealing, genetic algorithms, evolutionary strategies on several test functions.

2. Compare several selection methods for GAs (proportional, ranking, tournament) on a few test cases.

3. Compare 3 versions of genetic algorithms with different chromosomal coding (binary, Gray, and floating point numbers) on a few test cases.

4. For a typical constrained problem (e.g., knapsack problem) experiment with different constraint-handling methods (decoders, repair algorithms, penalty functions, etc.).

5. Compare different operators for floating point representation (e.g., Gaussian mutation versus non-uniform mutation, experiments with heuristic crossover, etc.).

6. Take a simple code from Appendix A and make it useful for numerical optimization problems. This can be done in many ways:

   (a) provide a better input file, where a user can define the number of variables as well as other parameters of the method,

   (b) modify the system to handle minimization problems,

   (c) modify the system to handle all values of the objective function (not only positive),

   (d) replace uniform mutation by Gaussian mutation (to approximate a random number Q which follows the normal distribution with expected value of μ and the variance σ^2, you may generate 12 random numbers R_i , $i = 1, 2, \ldots, 12$, from the range $[0..1]$ (uniform distribution); then $Q = \mu + \sigma(\sum_{i=1}^{12} R_i - 6))$,

   (e) replace one-point crossover by arithmetical crossover,

   (f) introduce additional operators, including multiple-parent crossovers (e.g., calculate a 'center' of a few parents as an average of their coordinates and move from the weakest individual towards this center),

   (g) introduce various types for variables (Boolean, integer),

   (h) etc.

7. Compare evolutionary strategies against genetic algorithm with floating point representation and appropriate operators.

8. Design and experiment with adaptive mechanism for parameters of some evolutionary method (probabilities of operators, population size, step of mutation, type of crossover, etc.).

9. Take the current public software (like GENOCOP) and adopt it to handle integer and Boolean variables. Experiment with integer programming problems.

10. Introduce and experiment with some non-standard features of GAs (sex, family relationships, cooperation between individuals, couples producing multiple children). Develop a justification for such additional heuristics and experiment with them on some test problems. Be careful to take into account the tradeoff between possible improvements and increased complexity of the system.

11. Consider a problem where the objective function varies over time. Experiment with haploid versus diploid chromosomal structures.

12. Develop a graphical interface to some evolutionary system to display the current statistics of the search.

13. Take any nontrivial optimization problem from your area of expertise (databases, operations research, image processing, engineering design, fuzzy controllers, artificial neural networks, games, robotics, etc.). Develop an evolutionary application for this class of problems. Compare it with the other known methods for this problem (see the following section).

A few remarks on reporting computational experiments with heuristic methods

Many evolutionary techniques are evaluated by experimenting with several test cases (such as those listed in Appendices B and C); very often it is quite difficult to generalize these experimental results to make some "global" claim about a particular technique. It is possible, however, to demonstrate the usefulness of a new method on several, carefully selected cases, by comparing the method against other well-established techniques. Following [26], contributions of a new heuristic method may include the following:

- it produces high-quality solutions more quickly than other approaches,

- it identifies higher-quality solutions than other approaches,

- it is less sensitive to differences in problem characteristics, data quality, or tuning parameters than other approaches,

- it is easy to implement,

- it has applications to a broad range of problems.

Additionally [26]:

> "research reports about heuristics are valuable if they are *revealing* — offering insight into general heuristic design or the problem structure by establishing the reasons for an algorithm's performance and explaining its behavior, and *theoretical* — providing theoretical insights, such as bounds on solution quality."

The paper by Barr et al. [26] gives excellent overview material on how to design and report on computational experiments with heuristic methods. For example, in preparing and reporting your experiments, it might be desirable if you follow five steps (listed in [26]):

- define the goals of the experiment,

- choose measures of performance and factors to explore,

- design and execute the experiment,

- analyse the data and draw conclusions,

- report the experiment's results.

It is important to address all these issues. For example, the goal of experiments may vary; as stated in [26]:

> "Computational experiments with algorithms are usually undertaken (a) to compare the performance of different algorithms for the same class of problems, or (b) to characterize or describe an algorithm's performance in isolation. While these goals are somewhat interrelated, the investigator should identify what, specifically, is to be accomplished by the testing (e.g., what questions are to be answered, what hypotheses are to be tested)."

Also, you may select as a measure of performance the quality of the best solution found, or the time to get there, or the algorithm's time to reach an "acceptable" solution, or the robustness of the method, to list a few possibilities. In most cases it is essential to compare the new method with the established techniques for a given class of problems. It is important to remember to analyze the key factors (like the influence of the problem size on the quality of the solution and the computational effort). The final report should contain also all information to allow the reader to reproduce the results.

There are many libraries of standard test problems available on the *World Wide Web*; these should be frequently used by any experimental research (e.g., for a collection of operations research test problems, see the OR-Library (o.rlibrary@ic.ac.uk); it has ftp access available at mscmga.ms.ic.ac.uk and WWW access available at http://mscmga.ms.ic.ac.uk).

References

1. Aarts, E.H.L. and Korst, J., *Simulated Annealing and Boltzmann Machines*, John Wiley, Chichester, UK, 1989.

2. Aarts, E.H.L. and Lenstra, J.K., (Editors), *Local Search in Combinatorial Optimization*, John Wiley, Chichester, UK, 1995.

3. Abuali, F.N., Wainwright, R.L., Schoenefeld, D.A., *Determinant Factorization: A New Encoding Scheme for Spanning Trees Applied to the Probabilistic Minimum Spanning Tree Problem*, in [103], pp.470–477.

4. Ackley, D.H., *An Empirical Study of Bit Vector Function Optimization*, in [73], pp.170–204.

5. Adler, D., *Genetic Algorithms and Simulated Annealing: A Marriage Proposal*, Proceedings of the IEEE International Conference on Neural Networks, March 28–April 1, 1993, pp.1104–1109.

6. Akl, S.G., *The Design and Analysis of Parallel Algorithms*, Prentice-Hall, Englewood Cliffs, NJ, 1989.

7. Alander, J.T., *Proceedings of the First Finnish Workshop on Genetic Algorithms and their Applications*, Helsinki University of Technology, Finland, November 4–5, 1992.

8. Angeline, P.J. and Kinnear, K.E. (Editors), *Advances in Genetic Programming II*, MIT Press, Cambridge, MA, 1996.

9. Antonisse, H.J., *A New Interpretation of Schema Notation that Overturns the Binary Encoding Constraint*, in [344], pp.86–91.

10. Antonisse, H.J. and Keller, K.S., *Genetic Operators for High Level Knowledge Representation*, in [171], pp.69–76.

11. Arabas, J., Mulawka, J., and Pokraśniewicz, J., *A New Class of the Crossover Operators for the Numerical Optimization*, in [103], pp.42–48.

12. Arabas, J., Michalewicz, Z., and Mulawka, J., *GAVaPS — a Genetic Algorithm with Varying Population Size*, in [275], pp.73–78.

13. Attia, N.F., *New Methods of Constrained Optimization Using Penalty Functions*, Ph.D. Thesis, Essex University, England, 1985.

14. Axelrod, R., *Genetic Algorithm for the Prisoner Dilemma Problem*, in [73], pp.32–41.

15. Bäck, T., *Evolutionary Algorithms in Theory and Practice*, Oxford University Press, New York, 1995.

16. Bäck, T., and Hoffmeister, F., *Extended Selection Mechanisms in Genetic Algorithms*, in [32], pp.92–99.

17. Bäck, T., Fogel, D.B., and Michalewicz, Z., *Handbook of Evolutionary Computation*, University Oxford Press, New York, 1996.

18. Bäck, T., Hoffmeister, F., and Schwefel, H.-P., *A Survey of Evolution Strategies*, in [32], pp.2–9.

19. Bäck, T., Rudolph, G., and Schwefel, H.-P., *Evolutionary Programming and Evolution Strategies: Similarities and Differences*, in [124], pp.11–22.

20. Bäck, T. and Schwefel, H.-P., *An Overview of Evolutionary Algorithms for Parameter Optimization*, Evolutionary Computation, Vol.1, No.1, pp.1–23, 1993.

21. Bagchi, S., Uckun, S., Miyabe, Y., and Kawamura, K., *Exploring Problem-Specific Recombination Operators for Job Shop Scheduling*, in [32], pp.10–17.

22. Baker, J.E., *Adaptive Selection Methods for Genetic Algorithms*, in [167], pp.101–111.

23. Baker, J.E., *Reducing Bias and Inefficiency in the Selection Algorithm*, in [171], pp.14–21.

24. Bala, J., De Jong, K.A., Pachowicz, P., *Using Genetic Algorithms to Improve the Performance of Classification Rules Produced by Symbolic Inductive Method*, Proceedings of the 6th International Symposium on Methodologies of Intelligent Systems, Lecture Notes in Artificial Intelligence, Vol.542, pp.286–295, Springer-Verlag, 1991.

25. Banach, S., *Sur les opérations dans les ensembles abstraits et leur applications aux équations intégrales*, Fundamenta Mathematica, Vol.3, pp.133–181, 1922.

26. Barr, R.S., Golden, B.L., Kelly, J.P., Resende, M.G.C., Stewart, W.R., *Designing and Reporting on Computational Experiments with Heuristic Methods*, Proceedings of the International Conference on Metaheuristics for Optimization, Kluwer Publishing, pp.1–17, 1995.

27. Bean, J.C. and Hadj-Alouane, A.B., *A Dual Genetic Algorithm for Bounded Integer Programs*, Department of Industrial and Operations Engineering, The University of Michigan, TR 92-53, 1992.

28. Beasley, D., Bull, D.R., and Martin, R.R., *An Overview of Genetic Algorithms: Part 1, Foundations*, University Computing, Vol.15, No.2, pp.58–69, 1993.

29. Beasley, D., Bull, D.R., and Martin, R.R., *An Overview of Genetic Algorithms: Part 2, Research Topics*, University Computing, Vol.15, No.4, pp.170–181, 1993.

30. Beasley, D., Bull, D.R., and Martin, R.R., *A Sequential Niche Technique for Multimodal Function Optimization*, Evolutionary Computation, Vol.1, No.2, pp.101–125, 1993.

31. Beasley, J.E. and Chu, P.C., *A Genetic Algorithm for the Set Covering Problem*, Technical Report, The Management School, Imperial College, July 1994.

32. Belew, R. and Booker, L. (Editors), Proceedings of the Fourth International Conference on Genetic Algorithms, Morgan Kaufmann Publishers, San Mateo, CA, 1991.

33. Bellman, R., *Dynamic Programming*, Princeton University Press, Princeton, NJ, 1957.

34. Bennett, K., Ferris, M.C., and Ioannidis, Y.E., *A Genetic Algorithm for Database Query Optimization*, in [32], pp.400–407.

35. Bersini, H. and Varela, F.J., *The Immune Recruitment Mechanism: A Selective Evolutionary Strategy*, in [32], pp.520–526.

36. Bertoni, A. and Dorigo, M., *Implicit Parallelism in Genetic Algorithms*, Artificial Intelligence, Vol.61, no.2, pp.307–314, 1993.

37. Bertsekas, D.P., *Dynamic Programming. Deterministic and Stochastic Models*, Prentice-Hall, Englewood Cliffs, NJ, 1987.

38. Bethke, A.D., *Genetic Algorithms as Function Optimizers*, Doctoral Dissertation, University of Michigan, 1980.

39. Betts, J.T., *An Accelarated Multiplier Method for Nonlinear Programming*, Journal of Optimization Theory and Applications, Vol.21, No.2, pp.137–174, 1977.

40. Biggs, M.C., *Constrained Minimization Using Recursive Quadratic Programming: Some Alternative Subproblem Formulations*, Towards Global Optimization (Eds. L.C.W. Dixon and G.P. Szego), North-Holland, 1975.

41. Biggs, M.C., *A Numerical Comparison Between Two Approaches to the Nonlinear Programming Problem*, Numerical Optimization Centre Technical Report No. 77, The Hatfield Polytechnic, 1976.

42. Bilchev, G. and Parmee, I.C., *Ant Colony Search vs. Genetic Algorithms*, Technical Report, Plymouth Engineering Design Centre, University of Plymouth, 1995.

43. Bilchev, G., *Private communication*, April 1995.

44. Bland, R.G. and Shallcross, D.F., *Large Traveling Salesman Problems Arising From Experiments in X-Ray Crystallography: A Preliminary Report on Computation*, Operations Research Letters, Vol.8, pp.125–128, 1989.

45. Booker, L.B., *Intelligent Behavior as an Adaptation to the Task Environment*, Doctoral Dissertation, University of Michigan, 1982.

46. Booker, L.B., *Improving Search in Genetic Algorithms*, in [73], pp.61–73.

47. Bosworth, J., Foo, N., and Zeigler, B.P., *Comparison of Genetic Algorithms with Conjugate Gradient Methods*, Washington, DC, NASA (CR–2093), 1972.

48. Bowen, J. and Dozier, G., *Solving Constraint Satisfaction Problems Using a Genetic/Systematic Search Hybrid that Realizes when to Quit*, in [103], pp.122–129.

49. Brindle, A., *Genetic Algorithms for Function Optimization*, Doctoral Dissertation, University of Alberta, Edmonton, 1981.

50. Brooke, A., Kendrick, D., and Meeraus, A., *GAMS: A User's Guide*, The Scientific Press, 1988.

51. Broyden, C.G. and Attia, N.F., *A Smooth Sequential Penalty Function Method for Solving Nonlinear Programming Problem*, System Modelling and Optimization, Proceedings of the 11th IFIP Conference, July 1983. Lecture Notes in Control and Information Sciences, Vol.59, Springer-Verlag, 1983.

52. Broyden, C.G. and Attia, N.F., *Penalty Functions, Newton's Method, and Quadratic Programming*, Journal of Optimization Theory and Applications, Vol.58, No.3., 1988.

53. Bruns, R., *Direct Chromosome Representation and Advanced Genetic Operators for Production Scheduling*, in [129], pp.352–359.

54. Bruns, R., *Scheduling*, in [17], section F1.5.

55. Bui, T.N. and Eppley, P.H., *A Hybrid Genetic Algorithm for the Maximum Clique Problem*, in [103], pp.478–483.

56. Bui, T.N. and Moon, B.R., *On Multi-Dimensional Encoding/Crossover*, in [103], pp.49–56.

57. Burke, E.K., Elliman, D.G., and Weare, R.F., *A Hybrid Genetic Algorithm for Highly Constrained Timetabling Problems*, in [103], pp.605–610.

58. Carey, M.R. and Johnson, D.S., *Computers and Intractability – A Guide to the Theory of NP-Completeness*, W.H. Freeman, San Francisco, 1979.

59. Cartwright, H.M., and Mott, G.F., *Looking Around: Using Clues from the Data Space to Guide Genetic Algorithm Searches*, in [32], pp.108–114.

60. Cavaretta, M.J., *Using Cultural Algorithm to Control Genetic Operators*, in [378], pp.158–166.

61. Četverikov, S.S., *On Some Aspects of the Evolutionary Process From the Viewpoint of Modern Genetics* (in Russian), Journal Exper. Biol., Vol.2, No.1, pp.3–54, 1926.

62. Cleveland, G.A. and Smith, S.F., *Using Genetic Algorithms to Schedule Flow Shop Releases*, in [344], pp.160–169.

63. Colorni, A., Dorigo, M., and Maniezzo, V., *Genetic Algorithms and Highly Constrained Problems: The Time-Table Case*, in [351], pp.55–59.

64. Colville, A.R., *A Comparative Study on Nonlinear Programming Codes*, IBM Scientific Center Report 320-2949, New York, 1968.

65. Coombs, S., and Davis, L., *Genetic Algorithms and Communication Link Speed Design: Theoretical Considerations*, in [171], pp.252–256.

66. Cooper, L., and Steinberg, D., *Introduction to Methods of Optimization*, W.B. Saunders, London, 1970.

67. Cormier, D.R., *Pluto: A General Purpose Evolution Programming Shell*, Technical Report, August 17, 1993, Department of Industrial Engineering, North Carolina State University.

68. Craighurst, R. and Martin, W., *Enhancing GA Preformance Through Crossover Prohibitions Based on Ancestry*, in [103], pp.130–135.

69. Davidor, Y., *Genetic Algorithms and Robotics*, World Scientific, 1991.

70. Davidor, Y., Schwefel, H.-P., and Männer, R. (Editors), Proceedings of the Third International Conference on Parallel Problem Solving from Nature (PPSN), Lecture Notes in Computer Science, Vol.866, Springer-Verlag, 1994.

71. Davis, L., *Applying Adaptive Algorithms to Epistatic Domains*, Proceedings of the International Joint Conference on Artificial Intelligence, pp.162–164, 1985.

72. Davis, L., *Job Shop Scheduling with Genetic Algorithms*, in [167], pp.136–140.

73. Davis, L., (Editor), *Genetic Algorithms and Simulated Annealing*, Morgan Kaufmann Publishers, San Mateo, CA, 1987.

74. Davis, L. and Steenstrup, M., *Genetic Algorithms and Simulated Annealing: An Overview*, in [73], pp.1–11.

75. Davis, L. and Ritter, F., *Schedule Optimization with Probabilistic Search*, Proceedings of the Third Conference on Artificial Intelligence Applications, Computer Society Press, pp.231–236, 1987.

76. Davis, L., and Coombs, S., *Genetic Algorithms and Communication Link Speed Design: Constraints and Operators*, in [171], pp.257–260.

77. Davis, L., *Adapting Operator Probabilities in Genetic Algorithms*, in [344], pp.61–69.

78. Davis, L., (Editor), *Handbook of Genetic Algorithms*, Van Nostrand Reinhold, New York, 1991.

79. Davis, L., *Bit-Climbing, Representational Bias, and Test Suite Design*, in [32], pp.18–23.

80. Davis, T.E. and Principe, J.C., *A Simulated Annealing Like Convergence Theory for the Simple Genetic Algorithm*, in [32], pp.174–181.

81. Dawkins, R., *The Selfish Gene*, Oxford University Press, New York, 1976.

82. De Jong, K.A., "An Analysis of the Behavior of a Class of Genetic Adaptive Systems", (Doctoral dissertation, University of Michigan), *Dissertation Abstract International*, 36(10), 5140B. (University Microfilms No 76-9381).

83. De Jong, K.A., *Adaptive System Design: A Genetic Approach*, IEEE Transactions on Systems, Man, and Cybernetics, Vol.10, No.3, pp.556–574, 1980.

84. De Jong, K.A., *Genetic Algorithms: A 10 Year Perspective*, in [167], pp.169–177.

85. De Jong, K.A., *On Using Genetic Algorithms to Search Program Spaces*, in [171], pp.210–216.

86. De Jong, K.A., *Learning with Genetic Algorithm: An Overview*, Machine Learning, Vol.3, pp.121–138, 1988.

87. De Jong, K.A., (Editor), *Evolutionary Computation*, MIT Press, 1993.

88. De Jong K.A., *Genetic-Algorithm-Based Learning*, in [226], pp.611–638.

89. De Jong, K., *Genetic Algorithms: A 25 Year Perspective*, in [415], pp.125–134.

90. De Jong K.A., Spears, W.M., *Using Genetic Algorithms to Solve NP-Complete Problems*, in [344], pp.124–132.

91. De Jong K.A., Spears, W.M., *Using Genetic Algorithms for Supervised Concept Learning*, Proceedings of the Second International Conference on Tools for AI, pp.335–341, 1990.

92. De La Maza, M. and Yuret, D., *Dynamic Hill Climbing*, AI Expert, March 1994, pp.26–31.

93. Dhar, V. and Ranganathan, N., *Integer Programming vs. Expert Systems: An Experimental Comparison*, Communications of the ACM, Vol.33, No.3, pp.323–336, 1990.

94. Dixmier, J., *General Topology*, Springer-Verlag, 1984.

95. Dozier, G., Bahler, D., and Bowen, J., *Solving Small and Large Scale Constraint Satisfaction Problems Using a Heuristic-Based Microgenetic Algorithm*, in [275], pp.306–311.

96. Eades, P. and Lin, X., *How to Draw a Directed Graph*, Technical Report, Department of Computer Science, University of Queensland, Australia, 1989.

97. Eades, P. and Tamassia, R., *Algorithms for Drawing Graphs: An Annotated Bibliography*, Technical Report No. CS-89-09, Brown University, Department of Computer Science, October, 1989.

98. Eiben, A.E., Aarts, E.H.L., Van Hee, K.M., *Global Convergence of Genetic Algorithms: On Infinite Markov Chain Analysis*, in [351], pp.4–12.

99. Eiben, A.E., Raue, P.-E., and Ruttkay, Zs., *Solving Constraint Satisfaction Problems Using Genetic Algorithms*, in [276], pp.542–547.

100. Eiben, A.E., Raue, P.-E., and Ruttkay, Zs., *Genetic Algorithms with Multi-parent Recombination*, in [70], pp.78–87.

101. Esbensen, H., *Finding (Near-)Optimal Steiner Trees in Large Graphs*, in [103], pp.485–491.

102. Eshelman, L.J., *The CHC Adaptive Search Algorithm: How to Have Safe Search When Engaging in Nontraditional Genetic Recombination*, in [318], pp.265–283.

103. Eshelman, L.J., (Editor), Proceedings of the Sixth International Conference on Genetic Algorithms, Morgan Kaufmann, San Mateo, CA, 1995.

104. Eshelman, L.J., Caruana, R.A., and Schaffer, J.D., *Biases in the Crossover Landscape*, in [344], pp.10–19.

105. Eshelman, L.J., and Schaffer, J.D., *Preventing Premature Convergence in Genetic Algorithms by Preventing Incest*, in [32], pp.115–122.

368   References

106. Even, S., Itai, A., and Shamir, A., *On the Complexity of Timetable and Multicommodity Flow Problems*, SIAM Journal on Computing, Vol.5, No.4, 1976, pp.691–703.

107. Evolutionary Computation, Vol.2, No.1, Spring 1994; special issue on classifier systems.

108. Falkenauer, E., *The Grouping Genetic Algorithms - Widening the Scope of the GAs*, Belgian Journal of Operations Research, Statistics and Computer Science, Vol.33, 1993, pp.79–102.

109. Falkenauer, E., *A New Representation and Operators for GAs Applied to Grouping Problems*, Evolutionary Computation, Vol.2, No.2, pp.123–144.

110. Falkenauer, E., *Solving Equal Piles with a Grouping Genetic Algorithm*, in [103], pp.492–497.

111. Fiacco, A.V., and McCormick, G.P., *Nonlinear Programming*, John Wiley, Chichester, UK, 1968.

112. Fletcher, R., *Practical Methods of Optimization*, Vol.2, of *Constrained Optimization*, John Wiley, Chichester, UK, 1981.

113. Floudas, C.A. and Pardalos, P.M., *A Collection of Test Problems for Constrained Global Optimization Algorithms*, Lecture Notes in Computer Science, Vol.455, Springer-Verlag, 1987.

114. Floudas, C.A. and Pardalos, P.M., *Recent Advances in Global Optimization*, Princeton Series in Computer Science, Princeton University Press, Princeton, NJ, 1992.

115. Fogarty, T.C., *Varying the Probability of Mutation in the Genetic Algorithm*, in [344], pp.104–109.

116. Fogel, D.B., *An Evolutionary Approach to the Traveling Salesman Problem*, Biol. Cybern., Vol.60, pp.139–144, 1988.

117. Fogel, D.B., *Evolving Artificial Intelligence*, PhD Thesis, University of California, San Diego, 1992.

118. Fogel, D.B. (Editor), IEEE Transactions on Neural Networks, special issue on Evolutionary Computation, Vol.5, No.1, 1994.

119. Fogel, D.B., *An Introduction to Simulated Evolutionary Optimization*, IEEE Transactions on Neural Networks, special issue on EP, Vol.5, No.1, 1994.

120. Fogel, D.B., *Evolving Behaviours in the Iterated Prisoner's Dilemma*, Evolutionary Computation, Vol.1, No.1, pp.77–97, 1993.

121. Fogel, D.B., *Evolutionary Computation: Toward a New Philosophy of Machine Intelligence*, IEEE Press, Piscataway, NJ, 1995.

122. Fogel, D.B. and Atmar, J.W., *Comparing Genetic Operators with Gaussian Mutation in Simulated Evolutionary Process Using Linear Systems*, Biol. Cybern., Vol.63, pp.111–114, 1990.

123. Fogel, D.B. and Atmar, W., *Proceedings of the First Annual Conference on Evolutionary Programming*, La Jolla, CA, 1992, Evolutionary Programming Society.

124. Fogel, D.B. and Atmar, W., *Proceedings of the Second Annual Conference on Evolutionary Programming*, La Jolla, CA, 1993, Evolutionary Programming Society.

125. Fogel, L.J., *Evolutionary Programming in Perspective: The Top-Down View*, in [415], pp.135–146.

126. Fogel, L.J., Owens, A.J., and Walsh, M.J., *Artificial Intelligence Through Simulated Evolution*, John Wiley, Chichester, UK, 1966.

127. Fonseca, C.M. and Fleming, P.J., *An Overview of Evolutionary Algorithms in Multiobjective Optimization*, Evolutionary Computation, Vol.3, No.1, 1995, pp.165–180.

128. Forrest, S., *Implementing Semantic Networks Structures Using the Classifier System*, in [167], pp.24–44.

129. Forrest, S. (Editor), Proceedings of the Fifth International Conference on Genetic Algorithms, Morgan Kaufmann, San Mateo, CA, 1993.

130. Foux, G., Heymann, M., Bruckstein, A., *Two-Dimensional Robot Navigation Among Unknown Stationary Polygonal Obstacles*, IEEE Transactions on Robotics and Automation, Vol.9, pp.96–102, 1993.

131. Fox, B.R., and McMahon, M.B., *Genetic Operators for Sequencing Problems*, in [318], pp.284–300.

132. Fox, M.S., *Constraint–Directed Search: A Case Study of Job–Shop Scheduling*, Morgan Kaufmann Publishers, San Mateo, CA, 1987.

133. Freville, A., and Plateau, G., *Heuristics and Reduction Methods for Multiple Constraint 0-1 Linear Programming Problems*, European Journal of Operational Research, Vol. 24, pp.206–215.

134. Garey, M. and Johnson, D., *Computers and Intractability*, W.H. Freeman, San Francisco, 1979.

135. Gen, M., Wasserman, G.S., and Smith, A.E., (Guest Editors), Computers and Industrial Engineering Journal, special issue on genetic algorithms and industrial engineering, Vol.30, No.2, 1996.

136. Gendreau, M., Hertz, A., and Laporte, G., *A Tabu Search Heuristic for Vehicle Routing*, CRT-7777, Centre de Recherche sur les transports, Universite de Montreal, 1991.

137. Gill, P.E., Murray, W., and Wright, M.H., *Practical Optimization*, Academic Press, London, 1978.

138. Gillies, A.M., *Machine Learning Procedures for Generating Image Domain Feature Detectors*, Doctoral Dissertation, University of Michigan, 1985.

139. Giordana, A. and Saitta, L., *REGAL: An Integrated System for Learning Relations Using Genetic Algorithms*, Proceedings of the International Workshop on Multistrategy Learning, MSL-93, Harpers Ferry, VA, pp.234–249, 1993.

140. Giordana, A., Saitta, L., and Zini, F., *Learning Disjunctive Concepts with Distributed Genetic Algorithms*, in [275], pp.115–119.

141. Glover, D.E., *Solving a Complex Keyboard Configuration Problem Through Generalized Adaptive Search*, in [73], pp.12–27.

142. Glover, F., *Heuristics for Integer Programming Using Surrogate Constraints*, Decision Sciences, Vol.8, No.1, pp.156–166, 1977.

143. Glover, F., *Tabu Search — Part I*, ORSA Journal on Computing, Vol.1, No.3, pp.190–206, 1989.

144. Glover, F., *Tabu Search — Part II*, ORSA Journal on Computing, Vol.2, No.1, pp.4–32, 1990.

145. Glover, F., *Tabu Search for Nonlinear and Parametric Optimization*, Technical Report, Graduate School of Business, University of Colorado at Boulder; preliminary version presented at the EPFL Seminar on OR and AI Search Methods for Optimization Problems, Lausanne, Switzerland, November 1990.

146. Glover, F., *Genetic Algorithms and Scatter Search: Unsuspected Potentials*, Statistics and Computing, Vol.4, pp.131–140.

147. Glover, F., *Tabu Search Fundamentals and Uses*, Graduate School of Business, University of Colorado, 1995.

148. Glover, F. and Kochenberger, G., *Critical Event Tabu Search for Multidimensional Knapsack Problems*, Proceedings of the International Conference on Metaheuristics for Optimization, Kluwer Publishing, pp.113–133, 1995.

149. Goldberg, D.E., *Genetic Algorithm and Rule Learning in Dynamic Control System*, in [167], pp.8–15.

150. Goldberg, D.E., *Dynamic System Control Using Rule Learning and Genetic Algorithms*, in Proceedings of the International Joint Conference on Artificial Intelligence, 9, pp.588–592, 1985.

151. Goldberg, D.E., *Optimal Initial Population Size for Binary-Coded Genetic Algorithms*, TCGA Report No.85001, Tuscaloosa, University of Alabama, 1985.

152. Goldberg, D.E., *Simple Genetic Algorithms and the Minimal, Deceptive Problem*, in [73], pp.74–88.

153. Goldberg, D.E., *Sizing Populations for Serial and Parallel Genetic Algorithms*, in [344], pp.70–79.

154. Goldberg, D.E., *Genetic Algorithms in Search, Optimization and Machine Learning*, Addison-Wesley, Reading, MA, 1989.

155. Goldberg, D.E., *Messy Genetic Algorithms: Motivation, Analysis, and First Results*, Complex Systems, Vol.3, pp.493–530, 1989.

156. Goldberg, D.E., *Zen and the Art of Genetic Algorithms*, in [344], pp.80–85.

157. Goldberg, D.E., *Real-Coded Genetic Algorithms, Virtual Alphabets, and Blocking*, University of Illinois at Urbana-Champaign, Technical Report No. 90001, September 1990.

158. Goldberg, D.E., *Messy Genetic Algorithms Revisited: Studies in Mixed Size and Scale*, Complex Systems, Vol.4, pp.415–444, 1990.

159. Goldberg, D.E., Deb, K., and Korb, B., *Do not Worry, Be Messy*, in [32], pp.24–30.

160. Goldberg, D.E. and Lingle, R., *Alleles, Loci, and the TSP*, in [167], pp.154–159.

161. Goldberg, D.E., Milman, K., and Tidd, C., *Genetic Algorithms: A Bibliography*, IlliGAL Technical Report 92008, 1992.

162. Goldberg, D.E. and Richardson, J., *Genetic Algorithms with Sharing for Multimodal Function Optimization*, in [171], pp.41–49.

163. Goldberg, D.E. and Segrest, P., *Finite Markov Chain Analysis of Genetic Algorithms*, in [171], pp.1–8.

164. Gorges-Schleuter, M., *ASPARAGOS An Asynchronous Parallel Genetic Optimization Strategy*, in [344], pp.422–427.

165. Greene, F., *A Method for Utilizing Diploid/Dominance in Genetic Search*, in [275], pp.439–444.

166. Grefenstette, J.J., *GENESIS: A System for Using Genetic Search Procedures*, Proceedings of the 1984 Conference on Intelligent Systems and Machines, pp.161–165.

167. Grefenstette, J.J., (Editor), Proceedings of the First International Conference on Genetic Algorithms, Lawrence Erlbaum Associates, Hillsdale, NJ, 1985.

168. Grefenstette, J.J., Gopal, R., Rosmaita, B., and Van Gucht, D., *Genetic Algorithm for the TSP*, in [167], pp.160–168.

169. Grefenstette, J.J., *Optimization of Control Parameters for Genetic Algorithms*, IEEE Transactions on Systems, Man, and Cybernetics, Vol. 16, No.1, pp.122–128, 1986.

170. Grefenstette, J.J., *Incorporating Problem Specific Knowledge into Genetic Algorithms*, in [73], pp.42–60.

171. Grefenstette, J.J., (Editor), Proceedings of the Second International Conference on Genetic Algorithms, Lawrence Erlbaum Associates, Hillsdale, NJ, 1987.

172. Gregory, J.W., *Nonlinear Programming FAQ*, 1995, Usenet sci.answers; available via anonymous ftp from rtfm.mit.edu in /pub/usenet/sci.answers/nonlinear-programming-faq.

173. Groves, L., Michalewicz, Z., Elia, P., Janikow, C., *Genetic Algorithms for Drawing Directed Graphs*, Proceedings of the Fifth International Symposium on Methodologies of Intelligent Systems, North-Holland, Amsterdam, pp.268–276, 1990.

174. Hadj-Alouane, A.B. and Bean, J.C., *A Genetic Algorithm for the Multiple-Choice Integer Program*, Department of Industrial and Operations Engineering, The University of Michigan, TR 92-50, 1992.

175. Han, S.P., *Superlinearly Convvergent Variable Metric Algorithm for General Nonlinear Programming Problems*, Mathematical Programming, Vol.11, pp.263–282, 1976.

176. Handley, S., *The Genetic Planner: The Automatic Generation of Plans for a Mobile Robot via Genetic Programming*, Proceedings of 8th IEEE International Symposium on Intelligent Control, August 25-27, 1993.

177. Heitkötter, J., (Editor), *The Hitch-Hiker's Guide to Evolutionary Computation*, FAQ in comp.ai.genetic, issue 1.10, 20 December 1993.

178. Herdy, M., *Application of the Evolution Strategy to Discrete Optimization Problems*, in [351], pp.188–192.

179. Held, M. and Karp, R.M., *The Traveling Salesman Problem and Minimum Spanning Trees*, Operations Research, Vol.18, pp.1138–1162.

180. Held, M. and Karp, R.M., *The Traveling Salesman Problem and Minimum Spanning Trees: Part II*, Mathematical Programming, Vol.1, pp.6–25.

181. Hesser, J., Männer, R., and Stucky, O., *Optimization of Steiner Trees Using Genetic Algorithms*, in [344], pp.231–236.

182. Hillier, F.S. and Lieberman, G.J., *Introduction to Operations Research*, Holden–Day, San Francisco, CA, 1967.

183. Hinterding, R., *Mapping, Order-independent Genes and the Knapsack Problem*, in [275], pp.13–17.

184. Hinterding, R., Gielewski, H., and Peachey, T.C., *The Nature of Mutation in Genetic Algorithms*, in [103], pp.65–72.

185. Hobbs, M.F., *Genetic Algorithms, Annealing, and Dimension Alleles*, MSc. Thesis, Victoria University of Wellington, 1991.

186. Hock, W. and Schittkowski K., *Test Examples for Nonlinear Programming Codes*, Lecture Notes in Economics and Mathematical Systems, Vol.187, Springer-Verlag, 1981.

187. Hoffmeister, F. and Bäck, T., *Genetic Algorithms and Evolution Strategies*, in [351], pp.455–470.

188. Holland, J.H., *Adaptation in Natural and Artificial Systems*, University of Michigan Press, Ann Arbor, 1975.

189. Holland, J.H. and Reitman, J.S., *Cognitive Systems Based on Adaptive Algorithms*, in D.A. Waterman and F. Hayes-Roth (Editors), *Pattern-Directed Inference Systems*, Academic Press, New York, 1978.

190. Holland, J.H., *Properties of the Bucket Brigade*, in [167], pp.1–7.

191. Holland, J.H., *Escaping Brittleness*, in R.S. Michalski, J.G. Carbonell, T.M. Mitchell (Editors), *Machine Learning II*, Morgan Kaufmann Publishers, San Mateo, CA, 1986.

192. Holland, J.H., *Royal Road Functions*, Genetic Algorithm Digest, Vol.7, No.22, 12 August 1993.

193. Holland, J.H., Holyoak, K.J., Nisbett, R.E., and Thagard, P.R., *Induction*, MIT Press, Cambridge, MA, 1986.

194. Homaifar, A. and Guan, S., *A New Approach on the Traveling Salesman Problem by Genetic Algorithm*, Technical Report, North Carolina A & T State University, 1991.

195. Homaifar, A., Lai, S. H.-Y., and Qi, X., *Constrained Optimization via Genetic Algorithms*, Simulation, Vol.62, 1994, pp.242–254.

196. Horn, J. and Nafpliotis, N., *Multiobjective Optimization Using the Niched Pareto Genetic Algorithm*, Department of Computer Science, University of Illinois at Urbana-Champaign, IlliGAL Report 93005, 1993.

197. Horowitz, E. and Sahni, S., *Fundamentals of Computer Algorithms*, Computer Science Press, Potomac, MD, 1978.

198. Husbands, P., Mill, F., and Warrington, S., *Genetic Algorithms, Production Plan Optimization, and Scheduling*, in [351], pp.80–84.

199. Ingber, L. and Rosen, B., *Genetic Algorithms and Very Fast Simulated Reannealing: A Comparison*, Mathematical and Computer Modelling, Vol.16, No.11, pp.87–100, 1992.

200. Janikow, C., *Inductive Learning of Decision Rules in Attribute-Based Examples: a Knowledge-Intensive Genetic Algorithm Approach*, PhD Dissertation, University of North Carolina at Chapel Hill, July 1991.

201. Janikow, C., and Michalewicz, Z., *Specialized Genetic Algorithms for Numerical Optimization Problems*, Proceedings of the International Conference on Tools for AI, pp.798–804, 1990.

202. Janikow, C., and Michalewicz, Z., *On the Convergence Problem in Genetic Algorithms*, UNCC Technical Report, 1990.

203. Janikow, C. and Michalewicz, Z., *An Experimental Comparison of Binary and Floating Point Representations in Genetic Algorithms*, in [32], pp.31–36.

204. Jankowski, A., Michalewicz, Z., Ras, Z., and Shoff, D., *Issues on Evolution Programming*, in Computing and Information, (R. Janicki and W.W. Koczkodaj, Editors), North-Holland, Amsterdam, 1989, pp.459–463.

205. Jarvis, R.A., and Byrne, J.C., *Robot Navigation: Touching, Seeing, and Knowing*, Proceedings of the 1st Australian Conference on AI, 1986.

206. Jog, P., Suh, J.Y., Gucht, D.V., *The Effects of Population Size, Heuristic Crossover, and Local Improvement on a Genetic Algorithm for the Traveling Salesman Problem*, in [344], pp.110–115.

207. Johnson, D.S., *Local Optimization and the Traveling Salesman Problem*, in M.S. Paterson (Editor), Proceedings of the 17th Colloquium on Automata, Languages, and Programming, Lecture Notes in Computer Science, Vol.443, pp.446–461, Springer-Verlag, 1990.

208. Johnson, D.S., *The Traveling Salesman Problem: A Case Study in Local Search*, presented during the Metaheuristics International Conference, Breckenridge, Colorado, July 22–26, 1995.

209. Johnson, D.S., *Private communication*, October 1995.

210. Joines, J.A. and Houck, C.R., *On the Use of Non-Stationary Penalty Functions to Solve Nonlinear Constrained Optimization Problems With GAs*, in [276], pp.579–584.

211. Jones, D.R. and Beltramo, M.A., *Solving Partitioning Problems with Genetic Algorithms*, in [32], pp.442–449.

212. Jones, T., *A Description of Holland's Royal Road Function*, Evolutionary Computation, Vol.2, No.4, 1994, pp.409–415.

213. Jones, T., *Crossover, Macromutation, and Population-based Search*, in [103], pp.73–80.

214. Jones, T. and Forrest, S., *Fitness Distance Correlation as a Measure of Problem Difficulty for Genetic Algorithms*, in [103], pp.184–192.

215. Juliff, K., *A Multi-chromosome Genetic Algorithm for Pallet Loading*, in [129], pp.467–473.

216. Julstrom, B.A., *What Have You Done for Me Lately? Adapting Operator Probabilities in a Steady-State Genetic Algorithm*, in [103], pp.81–87.

217. Kaufman, K.A., Michalski, R.S., and Scultz, A.C., *EMERALD 1: An Integrated System for Machine Learning and Discovery Programs for Education and Research*, Center for AI, George Mason University, User's Guide, No. MLI–89–12, 1989.

218. Kambhampati, S.K., and Davis, L.S., *Multi-resolution Path Planning for Mobile Robots*, IEEE Journal of Robotics and Automation, RA-2, pp.135–145, 1986.

219. Karp, R.M., *Probabilistic Analysis of Partitioning Algorithm for the Traveling Salesman Problem in the Plane*, Mathematics of Operations Research, Vol.2, No.3, 1977, pp.209–224.

220. Keane, A., Genetic Algorithms Digest, Thursday, May 19, 1994, Volume 8, Issue 16.

221. Kelly, J.P., Golden, B.L., and Assad, A.A., *Large Scale Controlled Rounding Using Tabu Search with Strategic Oscillation*, Annals of Operations Research, Vol.41, pp.69–84, 1993.

222. Khatib O., *Real-Time Obstacles Avoidance for Manipulators and Mobile Robots*, International Journal of Robotics Research, Vol.5, pp.90–98, 1986.

223. Khuri, S., Bäck, T., and Heitkötter, J., *The Zero/One Multiple Knapsack Problem and Genetic Algorithms*, Proceedings of the ACM Symposium of Applied Computation (SAC '94).

224. Kingdon, J., *Genetic Algorithms: Deception, Convergence and Starting Conditions*, Technical Report, Dept. of Computer Science, University College, London, 1992.

225. Kinnear, K.E. (Editor), *Advances in Genetic Programming*, MIT Press, Cambridge, MA, 1994.

226. Kodratoff, Y. and Michalski, R., *Machine Learning: An Artificial Intelligence Approach*, Vol.3, Morgan Kaufmann Publishers, San Mateo, CA, 1990.

227. Korte, B., *Applications of Combinatorial Optimization*, talk at the 13th International Mathematical Programming Symposium, Tokyo, 1988.

228. Koza, J.R., *Genetic Programming: A Paradigm for Genetically Breeding Populations of Computer Programs to Solve Problems*, Report No. STAN–CS–90–1314, Stanford University, 1990.

229. Koza, J.R., *A Hierarchical Approach to Learning the Boolean Multiplexer Function*, in [318], pp.171–192.

230. Koza, J.R., *Evolving a Computer Program to Generate Random Numbers Using the Genetic Programming Paradigm*, in [32], pp.37–44.

231. Koza, J.R., *Genetic Programming*, MIT Press, Cambridge, MA, 1992.

232. Koza, J.R., *Genetic Programming – 2*, MIT Press, Cambridge, MA, 1994.

233. Laarhoven, P.J.M. van, and Aarts, E.H.L., *Simulated Annealing: Theory and Applications*, D. Reidel, Dordrecht, Holland, 1987.

234. Langley, P., *On Machine Learning*, Machine Learning, Vol.1, No.1, pp.5–10, 1986.

235. von Laszewski, G., *Intelligent Structural Operators for the k-Way Graph Partitioning Problem*, in [32], pp.45–52.

236. Lawer, E.L., Lenstra, J.K., Rinnooy Kan, A.H.G., and Shmoys, D.B., *The Traveling Salesman Problem*, John Wiley, Chichester, UK, 1985.

237. Lee, F.N., *The Application of Commitment Utilization Factor (CUF) to Thermal Unit Commitment*, IEEE Transactions on Power Systems, Vol.6, No.2, pp.691–698, May 1991.

238. Le Riche, R., and Haftka, R.T., *Improved Genetic Algorithm for Minimum Thickness Composite Laminate Design*, Composites Engineering, Vol.3, No.1, pp.121-139, 1995.

239. Le Riche, R., Knopf-Lenoir, C., and Haftka, R.T., *A Segregated Genetic Algorithm for Constrained Structural Optimization*, in [103], pp.558–565.

240. Leler, W., *Constraint Programming Languages: Their Specification and Generation*, Addison-Wesley, Reading, MA, 1988.

241. Lidd, M.L., *Traveling Salesman Problem Domain Application of a Fundamentally New Approach to Utilizing Genetic Algorithms*, Technical Report, MITRE Corporation, 1991.

242. Liepins, G.E., and Vose, M.D., *Representational Issues in Genetic Optimization*, Journal of Experimental and Theoretical Artificial Intelligence, Vol.2, No.2, pp.4–30, 1990.

243. Lin, H.-S., Xiao, J., and Michalewicz, Z., *Evolutionary Algorithm for Path Planning in Mobile Robot Environment*, in [275], pp.211–216.

244. Lin, H.-S., Xiao, J., and Michalewicz, Z., *Evolutionary Navigator for a Mobile Robot*, Proceedings of the IEEE Conference on Robotics and Automation, IEEE Computer Society Press, New York, 1994, pp.2199–2204.

245. Lin, S. and Kernighan, B.W., *An Effective Heuristic Algorithm for the Traveling Salesman Problem*, Operations Research, pp.498–516, 1972.

246. Litke, J.D., *An Improved Solution to the Traveling Salesman Problem with Thousands of Nodes*, Communications of the ACM, Vol.27, No.12, pp.1227–1236, 1984.

247. Logan, T.D., *GENOCOP: AN Evolution Program for Optimization Problems with Linear Constraints*, Master Thesis, Department of Computer Science, University of North Carolina at Charlotte, 1993.

248. Lozano-Perez, T., and Wesley, M.A., *An Algorithm for Planning Collision Free Paths among Polyhedral Obstacles*, Communications of the ACM, Vol.22, pp.560–570, 1979.

249. Mahfoud, S.W., *A Comparison of Parallel and Sequential Niching Methods*, in [103], pp.136–143.

250. Maniezzo, V., *Granularity Evolution*, Dipartimento di Elettronica e Informazione, Politecnico di Milano, Technical Report, 1993.

251. Männer, R. and Manderick, B. (Editors), Proceedings of the Second International Conference on Parallel Problem Solving from Nature (PPSN), North-Holland, Elsevier Science Publishers, Amsterdam, 1992.

252. Martello, S. and Toth, P., *Knapsack Problems*, John Wiley, Chichester, UK, 1990.

253. McCormick, G.P., *Computability of Global Solutions to Factorable Nonconvex Programs*, Part I: Convex Underestimating Problems, Mathematical Programming, Vol.10, No.2 (1976), pp.147–175.

254. McDonnell, J.R., Reynolds, R.G., and Fogel, D.B. (Editors), Proceedings of the Fourth Annual Conference on Evolutionary Programming, The MIT Press, 1995.

255. Messa, K. and Lybanon, M., *A New Technique for Curve Fitting*, Naval Oceanographic and Atmospheric Research Report JA 321:021:91, 1991.

256. Michalewicz, Z. (Editor), Proceedings of the 5th International Conference on Statistical and Scientific Databases, Lecture Notes in Computer Science, Vol.420, Springer-Verlag, 1990.

257. Michalewicz, Z., *A Genetic Algorithm for Statistical Database Security*, IEEE Bulletin on Database Engineering, Vol.13, No.3, September 1990, pp.19–26.

258. Michalewicz, Z., *EVA Programming Environment*, UNCC Technical Report, 1990.

259. Michalewicz, Z., *Optimization of Communication Networks*, Proceedings of the SPIE International Symposium on Optical Engineering and Photonics in Aerospace Sensing, SPIE, Bellingham, WA, 1991, pp.112–122.

260. Michalewicz, Z. (Editor), *Statistical and Scientific Databases*, Ellis Horwood, London, 1991.

261. Michalewicz, Z., *A Hierarchy of Evolution Programs: An Experimental Study*, Evolutionary Computation, Vol.1, No.1, 1993, pp.51–76.

262. Michalewicz, Z., *Evolutionary Computation Techniques for Nonlinear Programming Problems*, International Transactions in Operational Research, 1994.

263. Michalewicz, Z. (Ed.), Statistics & Computing, special issue on evolutionary computation, 1994.

264. Michalewicz, Z., *A Hierarchy of Evolution Programs: An Experimental Study*, Evolutionary Computation, Vol.1, No.1, 1993, pp.51–76.

265. Michalewicz, Z. (Editor), Statistics & Computing, special issue on evolutionary computation, 1994.

266. Michalewicz, Z., *Genetic Algorithms, Numerical Optimization and Constraints*, in [103], pp.151–158.

267. Michalewicz, Z., and Attia, N., *Evolutionary Optimization of Constrained Problems*, in [378], pp.98–108.

268. Michalewicz, Z. and Janikow, C., *Genetic Algorithms for Numerical Optimization*, Statistics and Computing, Vol.1, No.1, 1991.

269. Michalewicz, Z. and Janikow, C., *Handling Constraints in Genetic Algorithms*, in [32], pp.151–157.

270. Michalewicz, Z. and Janikow, C., *GENOCOP: A Genetic Algorithm for Numerical Optimization Problems with Linear Constraints*, accepted for publication, Communications of the ACM, 1992.

271. Michalewicz, Z., Janikow, C., and Krawczyk, J., *A Modified Genetic Algorithm for Optimal Control Problems*, Computers & Mathematics with Applications, Vol.23, No.12, pp.83–94, 1992.

272. Michalewicz, Z., Jankowski, A., Vignaux, G.A., *The Constraints Problem in Genetic Algorithms*, in *Methodologies of Intelligent Systems: Selected Papers*, M.L. Emrich, M.S. Phifer, B. Huber, M. Zemankova, Z. Ras (Editors), ICAIT, Knoxville, TN, pp.142–157, 1990.

273. Michalewicz, Z., Krawczyk, J., Kazemi, M., Janikow, C., *Genetic Algorithms and Optimal Control Problems*, Proceedings of the 29th IEEE Conference on Decision and Control, Honolulu, pp.1664–1666, 1990.

274. Michalewicz, Z., Logan, T.D., and Swaminathan, S., *Evolutionary Operators for Continuous Convex Parameter Spaces*, in [378], pp.84–97.

275. Michalewicz, Z., Schaffer, D., Schwefel, H.-P., Fogel, D., Kitano, H. (Editors), *Proceedings of the First IEEE International Conference on Evolutionary Computation*, IEEE Service Center, Piscataway, NJ, Volume 1, Orlando, 27–29 June, 1994.

276. Michalewicz, Z., Schaffer, D., Schwefel, H.-P., Fogel, D., Kitano, H. (Editors), *Proceedings of the First IEEE International Conference on Evolutionary Computation*, IEEE Service Center, Piscataway, NJ, Volume 2, Orlando, 27–29 June, 1994.

277. Michalewicz, Z., Vignaux, G.A., Groves, L., *Genetic Algorithms for Optimization Problems*, Proceedings of the 11th NZ Computer Conference, Wellington, New Zealand, pp.211–223, 1989.

278. Michalewicz, Z., Vignaux, G.A., Hobbs, M., *A Non-Standard Genetic Algorithm for the Nonlinear Transportation Problem*, ORSA Journal on Computing, Vol.3, No.4, pp.307–316, 1991.

279. Michalewicz, Z. and Xiao, J., *Evaluation of Paths in Evolutionary Planner/Navigator*, Proceedings of the 1995 International Workshop on Biologically Inspired Evolutionary Systems, Tokyo, Japan, May 30–31, 1995, pp.45–52.

280. Michalski, R., *A Theory and Methodology of Inductive Learning*, in R. Michalski, J. Carbonell, T. Mitchell (Editors), *Machine Learning: An Artificial Intelligence Approach*, Vol.1, Tioga Publishing Co., Palo Alto, CA, 1983, and Springer-Verlag, 1994, pp.83–134.

281. Michalski, R., Mozetic, I., Hong, J., Lavrac, N., *The AQ15 Inductive Learning System: An Overview and Experiments*, Technical Report, Department of Computer Science, University of Illinois at Urbana–Champaign, 1986.

282. Michalski, R. and Watanabe, L., *Constructive Closed–Loop Learning: Fundamental Ideas and Examples*, MLI–88, George Mason University, 1988.

283. Michalski, R., *A Methodological Framework for Multistrategy Task–Adaptive Learning*, Methodologies for Intelligent Systems, Vol.5, in Z. Ras, M. Zemankowa, M. Emrich (Editors), North-Holland, Amsterdam, 1990.

284. Michalski, R. and Kodratoff, Y., *Research in Machine Learning: Recent Progress, Classification of Methods, and Future Directions*, in [226], pp.3–30.

285. Montana, D.J., and Davis, L., *Training Feedforward Neural Networks Using Genetic Algorithms*, Proceedings of the 1989 International Joint Conference on Artificial Intelligence, Morgan Kaufmann Publishers, San Mateo, CA, 1989.

286. Mühlenbein, H., *Parallel Genetic Algorithms, Population Genetics and Combinatorial Optimization*, in [344], pp.416-421.

287. Mühlenbein, H., *Parallel Genetic Algorithms in Combinatorial Optimization*, in O. Balci, R. Sharda, and S.A. Zenios (Editors), *Computer Science and Operations Research—New Developments in Their Interfaces*, Pergamon Press, New York, 1992, pp.441–453.

288. Mühlenbein, H., Gorges-Schleuter, M., Krämer, O., *Evolution Algorithms in Combinatorial Optimization*, Parallel Computing, Vol.7, pp.65–85, 1988.

289. Mühlenbein, H. and Schlierkamp-Vosen, D., *Predictive Models for the Breeder Genetic Algorithm*, Evolutionary Computation, Vol.1, No.1, pp.25–49, 1993.

290. Mühlenbein, H. and Voigt, H.-M., *Gene Poool Recombination for the Breeder Genetic Algorithm*, Proceedings of the Metaheuristics International Conference, Breckenridge, Colorado, July 22–26, 1995, pp.19–25.

291. Murray, W., *An Algorithm for Constrained Minimization*, Optimization, (Ed: R. Fletcher), pp.247–258, Academic Press, London, 1969.

292. Murtagh, B.A. and Saunders, M.A., *MINOS 5.1 User's Guide*, Report SOL 83-20R, December 1983, revised January 1987, Stanford University.

293. Muselli, M. and Ridella, S., *Global Optimization of Functions with the Interval genetic Algorithm*, Complex Systems, Vol.6, pp.193–212, 1992.

294. Myung, H., Kim, J.-H., and Fogel, D.B., *Preliminary Investigations into a Two-Stage Method of Evolutionary Optimization on Constrained Problems*, in [254], pp.449–463.

295. Nadhamuni, P.V.R., *Application of Co-evolutionary Genetic Algorithm to a Game*, Master Thesis, Department of Computer Science, University of North Carolina, Charlotte, 1995.

296. von Neumann, J., *Theory of Self-Reproducing Automata*, edited by Burks, University of Illinois Press, 1966.

297. Nissen, V., *Evolutionary Algorithms in Management Science: An Overview and List of References*, European Study Group for Evolutionary Economics, 1993.

298. Ng, K.P, and Wong, K.C., *A New Diploid Scheme and Dominance Change Mechanism for Non-Stationary Function Optimization*, in [103], pp.159–166.

299. Oliver, I.M., Smith, D.J., and Holland, J.R.C., *A Study of Permutation Crossover Operators on the Traveling Salesman Problem*, in [171], pp.224–230.

300. Olsen, A., *Penalty Functions and the Knapsack Problem*, in [276], pp.554–558.

301. Orvosh, D. and Davis, L., *Shall We Repair? Genetic Algorithms, Combinatorial Optimization, and Feasibility Constraints*, in [129], p.650.

302. Padberg, M., and Rinaldi, G., *Optimization of a 532-City Symmetric Travelling Salesman Problem*, Technical Report IASI–CNR, Italy, 1986.

303. Paechter, B., Luchian, H., and Petruic, M., *Two Solutions to the General Timetable Problem Using Evolutionary Methods*, in [275], pp.300–305.

304. Palmer, C.C. and Kershenbaum, A., *Representing Trees in Genetic Algorithms*, in [275], pp.379–384.

305. Pardalos, P., *On the Passage from Local to Global in Optimization*, in "Mathematical Programming", J.R. Birge and K.G. Murty (Editors), The University of Michigan, 1994.

306. Paredis, J., *Exploiting Constraints as Background Knowledge for Genetic Algorithms: a Case-study for Scheduling*, in [251], pp.229–238.

307. Paredis, J., *Genetic State-Space Search for Constrained Optimization Problems*, Proceedings of the Thirteen International Joint Conference on Artificial Intelligence, Morgan Kaufmann, San Mateo, CA, 1993.

308. Paredis, J., *Co-evolutionary Constraint Satisfaction*, in [70], pp.46–55.

309. Paredis, J., *The Symbiotic Evolution of Solutions and their Representations*, in [103], pp.359–365.

310. Pavlidis, T., *Algorithms for Graphics and Image Processing*, Computer Science Press, 1982.

311. Potter, M. and De Jong, K., *A Cooperative Coevolutionary Approach to Function Optimization*, in [70], pp.249–257.

312. Powell, D. and Skolnick, M.M., *Using Genetic Algorithms in Engineering Design Optimization with Non-linear Constraints*, in [129], pp.424–430.

313. Powell, M.J.D., *Variable Metric Methods for Constrained Optimization*, Mathematical Programming: The State of the Art, (Eds: A.Bachem, M. Grötschel and B. Korte), pp.288–311, Springer-Verlag, 1983.

314. Quinlann, J.R., *Induction of Decision Trees*, Machine Learning, Vol.1, No.1, 1986.

315. Radcliffe, N.J., *Forma Analysis and Random Respectful Recombination*, in [32], pp.222–229.

316. Radcliffe, N.J., *Genetic Set Recombination*, in [398], pp.203–219.

317. Radcliffe, N.J., and George, F.A.W., *A Study in Set Recombination*, in [129], pp.23–30.

318. Rawlins, G., *Foundations of Genetic Algorithms*, First Workshop on the Foundations of Genetic Algorithms and Classifier Systems, Morgan Kaufmann Publishers, San Mateo, CA, 1991.

319. Rechenberg, I., *Evolutionsstrategie: Optimierung technischer Systeme nach Prinzipien der biologischen Evolution*, Frommann–Holzboog Verlag, Stuttgart, 1973.

320. Reeves, C.R., *Modern Heuristic Techniques for Combinatorial Problems*, Blackwell Scientific Publications, London, 1993.

321. Reinelt, G., *TSPLIB – A Traveling Salesman Problem Library*, ORSA Journal on Computing, Vol.3, No.4, pp.376–384, 1991.

322. Reinke, R.E. and Michalski, R., *Incremental Learning of Concept Descriptions*, in D. Michie (Editor), *Machine Intelligence XI*, 1985.

323. Rendell, Larry A., *Genetic Plans and the Probabilistic Learning System: Synthesis and Results*, in [167], pp.60–73.

324. Renders, J.-M., and Bersini, H., *Hybridizing Genetic Algorithms with Hill-climbing Methods for Global Optimization: Two Possible Ways*, in [275], pp.312–317.

325. Reynolds, R.G., *An Introduction to Cultural Algorithms*, in [378], pp.131–139.

326. Reynolds, R.G., Brown, W., and Abinoja, E.O., *Guiding Parallel Bi-Directional Search Using Cultural Algorithms*, in [378], pp.167–174.

327. Reynolds, R.G. and Maletic, J.I., *The Use of Version Space Controlled Genetic Algorithms to Solve the Boole Problem*, International Journal on Artificial Intelligence Tools, Vol.2, No.2, pp.219–234, 1993.

328. Reynolds, R.G. and Maletic, J.I., *Learning to Cooperate Using Cultural Algorithms*, in [378], pp.140–149.

329. Reynolds, R.G., Michalewicz, Z., and Cavaretta, M., *Using Cultural Algorithms for Constraint Handling in Genocop*, in [254], pp.289–305.

330. Reynolds, R.G. and Sverdlik, W., *Solving Problems in Hierarchically Structured Systems Using Cultural Algorithms*, in [124], pp.144–153.

331. Reynolds, R.G., Zannoni, E., and Posner, R.M., *Learning to Understand Software Using Cultural Algorithms*, in [378], pp.150–157.

332. Richardson, J.T., Palmer, M.R., Liepins, G., and Hilliard, M., *Some Guidelines for Genetic Algorithms with Penalty Functions*, in [344], pp.191–197.

333. Ronald, E., *When Selection Meets Seduction*, in [103], pp.167–173.

334. Rudolph, G., *Convergence Analysis of Canonical Genetic Algorithms*, IEEE Transactions on Neural Networks, special issue on evolutionary computation, Vol.5, No.1, 1994.

335. Rumelhart, D. and McClelland, J., *Parallel Distributed Processing: Exploration in the Microstructure of Cognition*, Vol.1, MIT Press, Cambridge, MA, 1986.

336. Saravanan, N. and Fogel, D.B., *A Bibliography of Evolutionary Computation & Applications*, Department of Mechanical Engineering, Florida Atlantic University, Technical Report No. FAU-ME-93-100, 1993.

337. Sarle, W., *Kangaroos*, article posted on *comp.ai.neural-nets* on 1 September 1993.

338. Sasieni, M., Yaspan, A., and Friedman, L., *Operations Research Methods and Problems*, John Wiley, Chichester, UK, 1959.

339. Schaffer, J.D., *Some Experiments in Machine Learning Using Vector Evaluated Genetic Algorithms*, PhD Dissertation, Vanderbilt University, Nashville, 1984.

340. Schaffer, J.D., *Learning Multiclass Pattern Discrimination*, in [167], pp.74–79.

341. Schaffer, J.D. and Morishima, A., *An Adaptive Crossover Distribution Mechanism for Genetic Algorithms*, in [171], pp.36–40.

342. Schaffer, J.D., *Some Effects of Selection Procedures on Hyperplane Sampling by Genetic Algorithms*, in [73], pp.89–103.

343. Schaffer, J., Caruana, R., Eshelman, L., and Das, R., *A Study of Control Parameters Affecting Online Performance of Genetic Algorithms for Function Optimization*, in [344], pp.51–60.

344. Schaffer, J., (Editor), Proceedings of the Third International Conference on Genetic Algorithms, Morgan Kaufmann Publishers, San Mateo, CA, 1989.

345. Schaffer, J.D., Whitley, D., and Eshelman, L.J., *Combinations of Genetic Algorithms and Neural Networks: A Survey of the State of the Art*, Proceedings of the International Workshop on Combinations of Genetic Algorithms and Neural Networks, Baltimore, MD, June 6, 1992, pp.1–37.

346. Schoenauer, M., and Xanthakis, S., *Constrained GA Optimization*, in [129], pp.573–580.

347. Schraudolph, N. and Belew, R., *Dynamic Parameter Encoding for Genetic Algorithms*, Machine Learning, Vol.9, No.1, pp.9–21, 1992.

348. Schwefel, H.-P., *Numerical Optimization for Computer Models*, John Wiley, Chichester, UK, 1981.

349. Schwefel, H.-P., *Evolution Strategies: A Family of Non-Linear Optimization Techniques Based on Imitating Some Principles of Organic Evolution*, Annals of Operations Research, Vol.1, pp.165–167, 1984.

350. Schwefel, H.-P., *Evolution and Optimum Seeking*, John Wiley, Chichester, UK, 1995.

351. Schwefel, H.-P. and Männer, R. (Editors), Proceedings of the First International Conference on Parallel Problem Solving from Nature (PPSN), Lecture Notes in Computer Science, Vol.496, Springer-Verlag, 1991.

352. Schwefel, H.-P., *Private communication*, July 1991.

353. Seniw, D., *A Genetic Algorithm for the Traveling Salesman Problem*, MSc Thesis, University of North Carolina at Charlotte, 1991.

354. Shaefer, C.G., *The ARGOT Strategy: Adaptive Representation Genetic Optimizer Technique*, in [171], pp.50–55.

355. Shibata, T., and Fukuda, T., *Robot Motion Planning by Genetic Algorithm with Fuzzy Critic*, Proceedings of the 8th IEEE International Symposium on Intelligent Control, August 25-27, 1993.

356. Shing, M.T., and Parker, G.B., *Genetic Algorithms for the Development of Real-Time Multi-Heuristic Search Strategies*, in [129], pp.565–570, 1993.

357. Shonkwiler, R. and Van Vleck, E., *Parallel Speed-up of Monte Carlo Methods for Global Optimization*, unpublished manuscript, 1991.

358. Siedlecki, W. and Sklanski, J., *Constrained Genetic Optimization via Dynamic Reward–Penalty Balancing and Its Use in Pattern Recognition*, in [344], pp.141–150.

359. Sirag, D.J. and Weisser, P.T., *Toward a Unified Thermodynamic Genetic Operator*, in [171], pp.116–122.

360. Smith, A. and Tate, D., *Genetic Optimization Using a Penalty Function*, in [129], pp.499–503.

361. Smith, D., *Bin Packing with Adaptive Search*, in [167], pp.202–207.

362. Smith, R.E., *Adaptively Resizing Populations: An Algorithm and Analysis*, in [129], pp.653.

363. Smith, S.F., *A Learning System Based on Genetic Algorithms*, PhD Dissertation, University of Pittsburgh, 1980.

364. Smith, S.F., *Flexible Learning of Problem Solving Heuristics through Adaptive Search*, Proceedings of the Eighth International Conference on Artificial Intelligence, Morgan Kaufmann Publishers, San Mateo, CA, 1983.

365. Spears, W.M., *Simple Subpopulation Schemes*, in [378], pp.296–307.

366. Spears, W.M. and De Jong, K.A., *On the Virtues of Parametrized Uniform Crossover*, in [32], pp.230–236.

367. Spears, W.M., *Adapting Crossover in Evolutionary Algorithms*, in [254], pp.367–384.

368. Srinivas, M. and Patnaik, L.M., *Adaptive Probabilities of Crossover and Mutation in Genetic Algorithms*, IEEE Transactions on Systems, Man, and Cybernetics, Vol.24, No.4, 1994, pp.17–26.

369. Srinivas, N. and Deb, K., *Multiobjective Optimization Using Nondominated Sorting in Genetic Algorithms*, Evolutionary Computation, Vol.2, No.3, 1994, pp.221–248.

370. Starkweather, T., McDaniel, S., Mathias, K., Whitley, C., and Whitley, D., *A Comparison of Genetic Sequencing Operators*, in [32], pp.69–76.

371. Stebbins, G.L., *Darwin to DNA, Molecules to Humanity*, W.H. Freeman, New York, 1982.

372. Stein, D., *Scheduling Dial a Ride Transportation Systems: An Asymptotic Approach*, PhD Dissertation, Harvard University, 1977.

373. Stevens, J., *A Genetic Algorithm for the Minimum Spanning Tree Problem*, MSc Thesis, University of North Carolina at Charlotte, 1991.

374. Stuckman, B.E. and Easom, E.E., *A Comparison of Bayesian/Sampling Global Optimization Techniques*, IEEE Transactions on Systems, Man, and Cybernetics, Vol.22, No.5, pp.1024–1032, 1992.

375. Suh, J.-Y. and Gucht, Van D., *Incorporating Heuristic Information into Genetic Search*, in [171], pp.100–107.

376. Surry, P.D., Radcliffe, N.J., and Boyd, I.D., *A Multi-objective Approach to Constrained Optimization of Gas Supply Networks*, presented at the AISB-95 Workshop on Evolutionary Computing, Sheffield, UK, April 3–4, 1995.

377. Sverdlik, W. and Reynolds, R.G., *Incorporating Domain Specific Knowledge into Version Space Search*, Proceedings of the 1993 IEEE International Conference on Tools with AI, Boston, MA, November 1993, pp.216–223.

378. Sebald, A.V. and Fogel, L.J., *Proceedings of the Third Annual Conference on Evolutionary Programming*, San Diego, CA, 1994, World Scientific.

379. Steele, J.M., *Probabilistic Algorithm for the Directed Traveling Salesman Problem*, Mathematics of Operations Research, Vol.11, No.2, 1986, pp.343–350.

380. Svirezhev, Yu.M., and Passekov, V.P., *Fundamentals of Mathematical Evolutionary Genetics*, Kluwer Academic Publishers, Mathematics and Its Applications (Soviet Series), Vol.22, London, 1989.

381. Sysło, M.M., Deo, N., Kowalik, J.S., *Discrete Optimization Algorithms*, Prentice-Hall, Englewood Cliffs, NJ, 1983.

382. Syswerda, G., *Uniform Crossover in Genetic Algorithms*, in [344], pp.2–9.

383. Syswerda, G., *Schedule Optimization Using Genetic Algorithms*, in [78], pp.332–349.

384. Syswerda, G. and Palmucci, J., *The Application of Genetic Algorithms to Resource Scheduling*, in [32], pp.502–508.

385. Suh, J.-Y. and Lee C.-D., *Operator–Oriented Genetic Algorithm and Its Application to Sliding Block Puzzle Problem*, in [351], pp.98–103.

386. Szałas, A., and Michalewicz, Z., *Contractive Mapping Genetic Algorithms and Their Convergence*, Department of Computer Science, University of North Carolina at Charlotte, Technical Report 006-1993.

387. Taha, H.A., *Operations Research: An Introduction*, 4th ed., Collier Macmillan, London, 1987.

388. Tamassia, R., Di Battista, G. and Batini, C., *Automatic Graph Drawing and Readability of Diagrams*, IEEE Transactions Systems, Man, and Cybernetics, Vol.18, No.1, pp.61–79, 1988.

389. Ulder, N.L.J., Aarts, E.H.L., Bandelt, H.-J., van Laarhoven, P.J.M., Pesch, E., *Genetic Local Search Algorithms for the Traveling Salesman Problem*, in [351], pp.109–116.

390. Valenzuela, C.L. and Jones, A.J., *Evolutionary Divide and Conquer (I): A Novel Genetic Approach to the TSP*, Evolutionary Computation, Vol.1, No.4, 1994, pp.313–333.

391. Vignaux, G.A. and Michalewicz, Z., *Genetic Algorithms for the Transportation Problem*, Proceedings of the 4th International Symposium on Methodologies for Intelligent Systems, North-Holland, Amsterdam, pp.252–259, 1989.

392. Vignaux, G.A. and Michalewicz, Z., *A Genetic Algorithm for the Linear Transportation Problem*, IEEE Transactions on Systems, Man, and Cybernetics, Vol.21, No.2, pp.445–452, 1991.

393. Voss, S., *Tabu Search: Applications and Prospects*, Technical Report, Technische Hochshule Darmstadt, 1993.

394. Whitley, D., *GENITOR: A Different Genetic Algorithm*, Proceedings of the Rocky Mountain Conference on Artificial Intelligence, Denver, 1988.

395. Whitley, D., *The GENITOR Algorithm and Selection Pressure: Why Rank–Based Allocation of Reproductive Trials is Best*, in [344], pp.116–121.

396. Whitley, D., *GENITOR II: A Distributed Genetic Algorithm*, Journal of Experimental and Theoretical Artificial Intelligence, Vol.2, pp.189–214.

397. Whitley, D., *Genetic Algorithms: A Tutorial*, in [263].

398. Whitley, D. (Editor), *Foundations of Genetic Algorithms–2*, Second Workshop on the Foundations of Genetic Algorithms and Classifier Systems, Morgan Kaufmann Publishers, San Mateo, CA, 1993.

399. Whitley, D., Gordon, V.S., and Mathias, K., *Lamarckian Evolution, the Baldwin Effect and Function Optimization*, in [70], pp.6–15.

400. Whitley, D., Mathias, K., Fitzhorn, P., *Delta Coding: An Iterative Search Strategy for Genetic Algorithms*, in [32], pp.77–84.

401. Whitley, D., Mathias, K., Rana, S., and Dzubera, J., *Building Better Test Functions*, in [103], pp.239–246.

402. Whitley, D., Starkweather, T., and Fuquay, D'A., *Scheduling Problems and Traveling Salesman: The Genetic Edge Recombination Operator*, in [344], pp.133–140.

403. Whitley, D., Starkweather, T., and Shaner, D., *Traveling Salesman and Sequence Scheduling: Quality Solutions Using Genetic Edge Recombination*, in [78], pp.350–372.

404. Wilson, R.B., *Some Theory and Methods of Mathematical Programming*, Ph.D. Dissertation, Harvard University Graduate School of Business Administration, 1963.

405. Winston, W.L., *Operations Research: Applications and Algorithms*, Duxbury, Boston, 1987.

406. Wirth, N., *Algorithms + Data Structures = Programs*, Prentice-Hall, Englewood Cliffs, NJ, 1976.

407. Wnęk, J., Sarma, J., Wahab, A.A., and Michalski, R.S., *Comparing Learning Paradigms via Diagrammatic Visualization*, Proceedings of the 5th International Symposium on Methodologies for Intelligent Systems, North-Holland, Amsterdam, pp.428–437, 1990.

408. Wright, A.H., *Genetic Algorithms for Real Parameter Optimization*, in [318], pp. 205–218.

409. Xiao, J., *Evolutionary Planner/Navigator in a Mobile Robot Environment*, in [17], section G3.11.

410. Xu, J. and Kelly, J.P., *A Robust Network Flow-Based Tabu Search Approach for the Vehicle Routing Problem*, Graduate School of Business, University of Colorado, Boulder, 1995.

411. Yagiura, Y. and Ibaraki, T., *GA and Local Search Algorithms as Robust and Simple Optimization Tools*, Proceedings of the Metaheuristics International Conference, Breckenridge, Colorado, July 22–26, 1995, pp.129–134.

412. Yao, X., *A Review of Evolutionary Artificial Neural Networks*, Technical Report, CSIRO, Highett, Victoria, Australia, 1993.

413. Yao, X. and Darwen, P., *An Experimental Study of N-person Prisoner's Dilemma Games*, Proceedings of the Workshop on Evolutionary Computation, University of New England, November 21–22, 1994, pp.94–113.

414. Zhou, H.H., and Grefenstette, J.J., *Learning by Analogy in Genetic Classifier Systems*, in [344], pp.291–297, 1989.

415. Zurada, J., Marks, R., and Robinson, C. (Editors), *Computational Intelligence: Imitating Life*, IEEE Press, 1994.

Index